Julius Weisbach, Gustav Herrmann, Karl P Dahlstrom

The Mechanics of Pumping Machinery

A Textbook for Technical Schools and a Guide for Practical Engineers

Julius Weisbach, Gustav Herrmann, Karl P Dahlstrom

The Mechanics of Pumping Machinery
A Textbook for Technical Schools and a Guide for Practical Engineers

ISBN/EAN: 9783743686922

Printed in Europe, USA, Canada, Australia, Japan

Cover: Foto ©berggeist007 / pixelio.de

THE

MECHANICS OF PUMPING MACHINERY

THE MECHANICS

OF

PUMPING MACHINERY

A TEXT-BOOK FOR TECHNICAL SCHOOLS AND A GUIDE
FOR PRACTICAL ENGINEERS

BY

DR. JULIUS WEISBACH

AND

PROFESSOR GUSTAV HERRMANN

AUTHORISED TRANSLATION FROM THE SECOND GERMAN EDITION

BY

KARL P. DAHLSTROM, M.E.,

LATE INSTRUCTOR OF MECHANICAL ENGINEERING AT THE LEHIGH UNIVERSITY

WITH 197 ILLUSTRATIONS

London
MACMILLAN AND CO., LIMITED
NEW YORK: THE MACMILLAN COMPANY
1897

TRANSLATOR'S PREFACE

The volume herewith presented might be introduced in the same words as its recent predecessor on *The Mechanics of Hoisting Machinery*. It is a translation of a portion of Professor Herrmann's revised edition of Weisbach's work on Engineering Mechanics, which has not before appeared in English. Moreover, it is destined mainly to serve as a text-book for the advanced courses in the Mechanics of Machinery, though it may also prove of value to the practical engineer.

Responding to the suggestion made by some critics of the volume on *Hoisting Machinery*, the translator has undertaken to supplement the text by a few remarks and references respecting some developments in pumping practice since the German original was published, and a few new illustrations of modern types of pumps have been inserted.

References in the text to previous volumes of Weisbach's Mechanics allude to the English translations unless otherwise specified.

In editing the work, the translator has been greatly aided by having had recourse to notes previously prepared by Professor J. F. Klein of the Lehigh University.

August 1896.

CONTENTS

INTRODUCTION

§ 1. The simplest, or at least the oldest, method of raising liquids consisted in *bailing*, the liquid being elevated, together with the vessels containing it, either by hand or by special machines driven by prime movers. Various machines have been devised for raising water in this manner, most of them being known to the ancients, and almost exclusively employed by them.

For very small lifts another method has been long in use, namely, to *throw* up the water either by shovels operated by hand or by the use of so-called *flash-wheels*. Both of the above methods of raising water have found their principal application in draining excavations or low lands, as well as for irrigation and procuring water for agricultural or building purposes.

As they permit of moderate lifts only, another plan was introduced at a later date, when it became necessary to overcome greater elevations. This consisted in subjecting the water in a vessel to a pressure sufficient to balance a column of water of a height exceeding the required lift. As a consequence the water is forced from the vessel upward through a delivery pipe and discharged at the orifice of the pipe. This principle is embodied in all water-raising machines known as *pumps*, their action differing only in the manner of exerting the pressure on the water. In ordinary *piston-pumps*, which are by far the most widely used machines for elevating water, the pressure is produced by a piston fitting closely in a cylindrical tube. This piston is acted upon in the direction of its axis by a force sufficient to move it, thereby effecting the lifting operation.

With these machines should also be classed the *rotary pumps*, in which the vessel has the form of one or more solids of revolution, and the piston or pistons consist of vanes or wheels attached to a revolving shaft, since in all constructions of this nature the space occupied by the water may be alternately increased and diminished by the motion of the pistons, the vessel being thus alternately filled and emptied.

It is only in special cases that the direct pressure of compressed air or steam on the surface of the liquid has been employed for raising the latter. Compressed air, for instance, was used in *Hoell's* water-raising machine, which is now of only historical interest, and, as is well known, the direct pressure of steam was tried by *Papin*, and has been largely introduced in sugar refineries for forcing the sugar juice upward. More recently this method has been applied in a water-raising contrivance known as the *pulsometer*. On the whole it may be said, however, that the employment of direct pressure of steam or compressed air for raising water is very limited.

On the other hand, extensive use has of late been made of the living force of the water itself as a means of overcoming the opposing resistance, the water being given the velocity required for this purpose. The contrivances based on this principle differ essentially in the manner in which the living force is communicated to the water. This is accomplished either by rapidly rotating vanes, as in centrifugal pumps; or by means of a steam jet, as in *Gifford's* injector; or by a jet of water, as in *Thompson's* pump and other related constructions.

The pressure which forces the water upward is, of course, to be understood to be the *excess* of pressure acting on the water in the pump chamber above that of the atmosphere, as the latter acts at the orifice of the efflux, and its resisting influence must be overcome before the water can be raised. It is also possible to produce the difference of pressure necessary to move the water by removing the atmospheric pressure originally existing in the pump chamber by any of the above-mentioned methods. The external atmospheric pressure will then act on the free surface of the water to be lifted, and cause the latter to rise to a height corresponding to the vacuum created, the maximum height being equal to that of the water barometer (10·336 m. [33·9 ft.]). This action is known as *suction*, and is usually found to exist in pumps, but never occurs in the scoop and flash apparatus, and therefore the existence of suction, or the possibility of it, is occasionally regarded as the characteristic feature of a pump, although there are pumps, of course, in the operation of which no suction is exerted.

CHAPTER I

EARLY FORMS OF WATER ELEVATORS

§ 2. **Bailing.**—The simplest means of bailing water consists in the employment of pails or buckets having a cubic contents of about 10 litres [0·35 cub. ft.] In this manner one man may lift the water to a height of 1 to 1·2 m. [3·28 to 3·94 ft.], and when it is necessary to lift it to a greater elevation, two or more men must be located one above the other, and the buckets be allowed to pass from man to man. It is estimated that one man can bail out fifteen buckets full per minute, and each time lift it to a height of 1 m. [3·28 ft.] This is equivalent to an effect of 150 metre kilograms per minute [1085 ft. lbs.], and for an actual working time of 6 hours per day, the day's work per man will therefore be only $6 \times 60 \times 150 =$ 54,000 m. kg. [390,590 ft. lbs.], that is, equal to the work done by one horse-power in 12 minutes.

If it is desired to raise the water higher by one man alone, the bucket may be provided with a handle about 2 m. [6·56 ft.] in length, which can serve as a lever. In this case the buckets must evidently be smaller or may be only partly filled. The daily performance in bailing water with a bucket provided with a handle does not materially differ from the results obtained with an ordinary bucket. It may be employed to advantage when the labourer is located above, and not below, the surface of the water, as he would otherwise be obliged to bend over a great deal and have to raise a portion of his body as well.

When it is required to raise water to greater heights, say 4 or 6 m. [13 to 20 ft.], from wells, for instance, the bucket is suspended from a *sweep* or lever ACG with a counter-weight, as in Fig. 1. If the bucket, when half full, is balanced by the counter-weight G, it will require the same effort to hoist the full bucket as to lower it empty. The chord AB of the arc described by the point of suspension A of the bucket E should be at least equal to the depth of the well, in order that the bucket may be sufficiently immersed into the water W. It

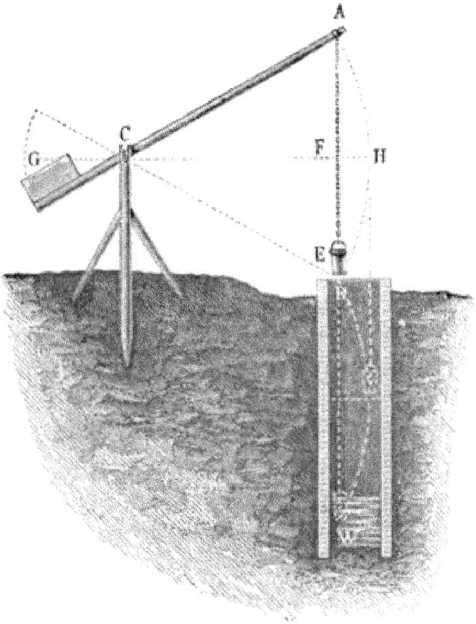

Fig. 1.

is also necessary to make the width of the well at least equal to that of the bucket plus the height FH of the arc, in order to have the bucket clear the walls of the well.

Water may be drawn from still greater depth by the use of buckets or tubs if a simple *guide pulley* or a drum be employed; for instance, an ordinary *winch* or a *whin* with a bucket or tub secured to each end of the hoisting rope. The two receptacles will then balance each other, and therefore the driving effort will equal the weight of the water in the ascending bucket.

At mines water is sometimes raised in barrels by means of a vertical drum or a whin. The barrels are then filled through a valve in their bottom opening inward when the water is reached.

At the top of the shaft the discharging is done either by overturning the barrel or by pulling open the valve.

§ 3. **Raising Water by Throwing.**—Like other heavy bodies water may be thrown or flung to a moderate height. The simplest implement used for this purpose is a *shovel* or *scoop*, which may be employed either as an ordinary hand-shovel or may be swung around a fixed point. In the former case it consists of a large wooden ladle with a handle about 1 or 1·5 m. [3 to 5 ft.] in length, and is made use of when it is desired to remove the water entirely from a space, as from a boat, for instance. The *swinging* or *Dutch scoop* (AB, Fig. 2) is made of boards or sheet-iron, and is provided with a handle BC, 3 to 4 m. [10 to 13 ft.] in length; it is suspended by a rope D from a horse or a tripod. The scoop proper has a length of 0·5 to 0·6 m. [1½ to 2 ft.], a width of 0·3 m. [1 ft.], and a depth of 0·2 m. [8 ins.] The operator, in pushing the scoop away from him by the handle, dips out about 15 litres [½ cub. ft.] of water, and sends it about 1 m. [3¼ ft.] high and 2 m. [6½ ft.] horizontally. Usually two helpers are placed opposite the first workman, and assist in swinging the scoop by pulling in ropes. One man cannot accomplish materially better results by the use of the scoop than by bailing with a bucket. When three men are engaged in the work they can give the scoop 28 oscillations per minute, each time throwing 25 litres [0·9 cub. ft.] of water 1·1 m. [3·6 ft.] high. The work done per minute will then be $25 \times 28 \times 1 \cdot 1 = 770$ m. kg. [5580 ft. lbs.], and consequently the daily performance, with an actual working day of 6 hours,

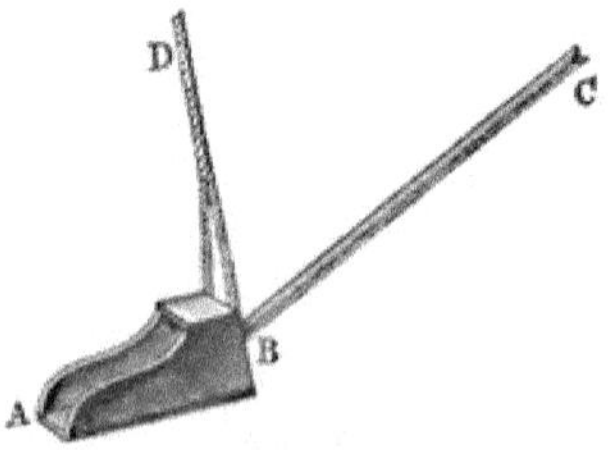

Fig. 2.

$$770 \times 60 \times 6 = 277{,}200 \text{ m. kg. } [2{,}006{,}200 \text{ ft. lbs.}],$$

so that the work done by each workman can be placed at 92,400 m. kg. [661,120 ft. lbs.]

The throwing operation may be performed by the aid of machines, such as the *water-balance*, shown in Fig. 3, and the *flash-wheels*.

The former consists of a swinging scoop CD attached to a balance beam ACB and moving in a channel or *chase* EF of circular profile. Four or six labourers give the beam an oscillating motion by pulling in ropes L, as in an ordinary ringing pile-driver, 10 or 12 oscillations being made per

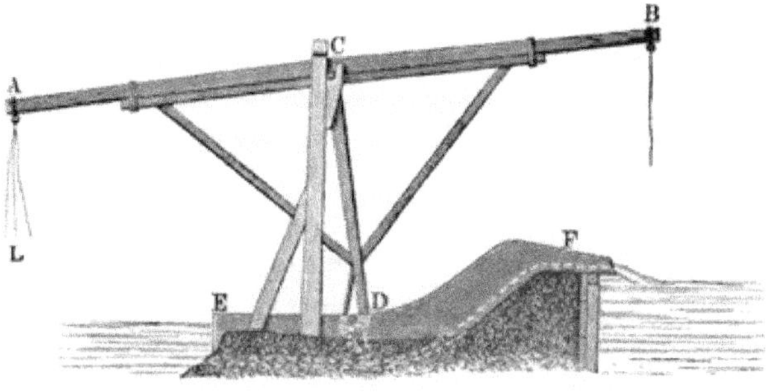

Fig. 3.

minute and 2·2 to 2·6 cub. m. [78 to 92 cub. ft.] of water being in this time raised to a height of 1·2 m. [4 ft.] In order to facilitate the return of the scoop D through the water, flap-valves K_1, K_2, K_3 (Fig. 4) are introduced, having a length of 0·5 m. [1·64 ft.] and a width of 0·2 m. [·66 ft.], and being surrounded by an iron frame secured to the end of the arm CD; in this frame the bearings for the gudgeons of the valves are located.

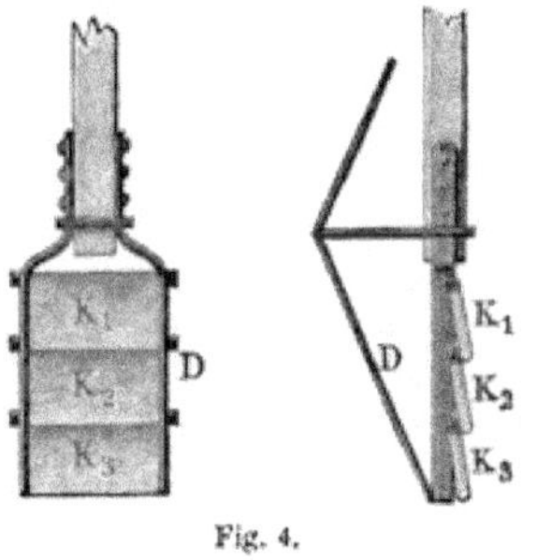

Fig. 4.

§ 4. **Flash-Wheels.**—These are of the form of ordinary water-wheels running in a *chase*, only turning in the opposite direction, so that the floats or paddles ascend in the curved waterway and carry the water along to the desired level. In order that as little water as possible may flow back in the clearance space between the paddles and the chase, it is necessary to rotate the wheel at a great velocity so as to fling the water rather than lift it. When the prime mover is a water-wheel, the flash-wheel is therefore not attached directly to the water-wheel shaft but driven by means of a large gear on the latter engaging a smaller one on the flash-wheel shaft.

In Holland flash-wheels are commonly used for draining low lands, the rotary motion being obtained from ordinary windmills. A side elevation of a wheel of this kind is shown in Fig 5. This wheel is 5 m. [16·4 ft.] in diameter, and has 28 floats or paddles 0·3 to 0·5 m. [1 to 1·6 ft.] in length and 0·15 to 0·24 m. [6 to 9 ins.] in width. The clearance between the paddles and the chase is 25 mm. [1 in.], both at the sides and at the bottom, and the water is raised from 1 to 1·2 m. [3¼ to 4 ft.] The four spokes A, B, C, D

Fig. 5.

embrace the square shaft E, and their outer ends form four of the paddles; the spokes are not only tenoned into each other but also tied together by two sets of braces and two rims, which also serve as means of securing the other paddles. On the shaft of the flash-wheel is a large bevel-gear which meshes with a smaller bevel on the upright shaft of the windmill, the latter making twice the number of revolutions of the flash-wheel. A gate G is placed at the beginning of the channel F, which serves as a conduit for the raised water, and is kept open by the action of the water flung against it; when the wind fails to turn the wheel or only revolves it slowly, the gate closes automatically.

Another flash-wheel, which raises water from the river Seine into the basin that forms a harbour at St. Ouen near Paris, is shown in Fig. 6, in $\frac{1}{125}$ of its natural size. It is operated by a steam-engine, and lifts about 1 cub. m. [35 cub. ft.] of water per second in a chase built of stone and having a height of 4 m. [13 ft.] The driving is accomplished by means of a small pinion C on the engine-shaft

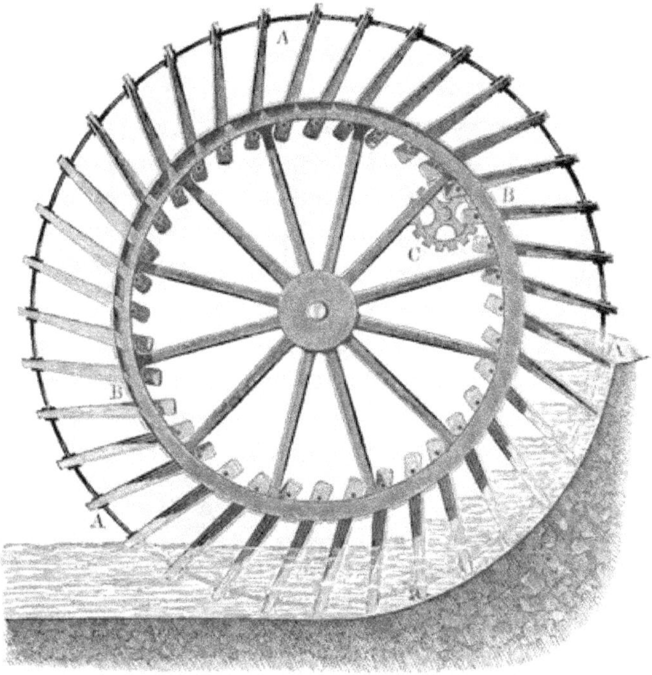

Fig. 6.

engaging a large internal gear attached to the rim BB of the flash-wheel A, which makes three revolutions per minute while the engine makes eighteen. A circular sluice-board surrounds the wheel on the side of the lower water-level, and when raised or lowered by means of a gear segment, serves to regulate the quantity of water lifted.

§ 5. **Scoop-Wheels** differ from flash-wheels in that they raise the water in receptacles or *buckets*, while the flash-wheels simply make use of paddles. We can distinguish between—

(1) Scoop-wheels with movable receptacles,
(2) Scoop-wheels with fixed receptacles,
(3) Scoop-wheels with buckets, and
(4) Scoop-wheels with spiral ducts.

A simple scoop-wheel with movable receptacles, as described by *Belidor*, is shown in Fig. 7. The vessels E, E_1 are suspended from iron rods A, secured to the side of the rim or between two rims of the wheel. As a rule, the water-wheel which gives motion to the apparatus is attached directly to the shaft C of the scoop-wheel. When the latter revolves there are always a few of the receptacles immersed in the water HR, and, after being filled, these ascend on one side, discharging their contents at the top into a trough T, after striking against the edge of the trough by means of a lifter cam or projection B.

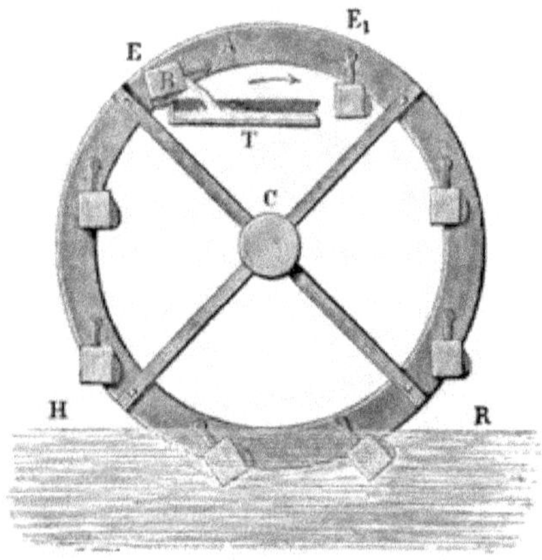

Fig. 7.

The scoop-wheels with fixed receptacles are of various forms.

In Fig. 8 the so-called *Chinese scoop-wheel* is illustrated. The water is here lifted in pails E, E (in China made of bamboo) secured in inclined positions in the circumference of the wheel. Motion is imparted to the wheel by the current of the water HR acting on the floats A, which are firmly secured to the rim. Also here the water is discharged into a trough located at the upper end, and evidently a portion of the total lift is thereby lost. Wheels of this kind are still in use in Europe. A peculiarity of the arrangement as found in Tyrol consists in mounting the wheel in a framework provided with a counter-weight, thus enabling the wheel to be adjusted at different heights, according to the variations of the water-level.

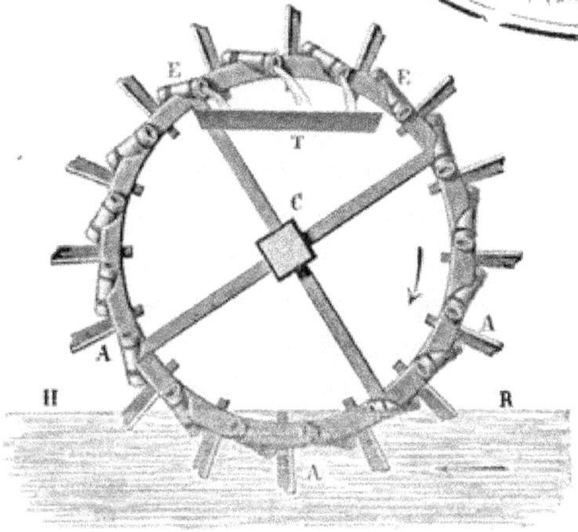

Fig. 8.

The so-called *Frankish scoop-wheel*, as employed on the river Rednitz at Erlangen, is shown in Fig 9. It consists of

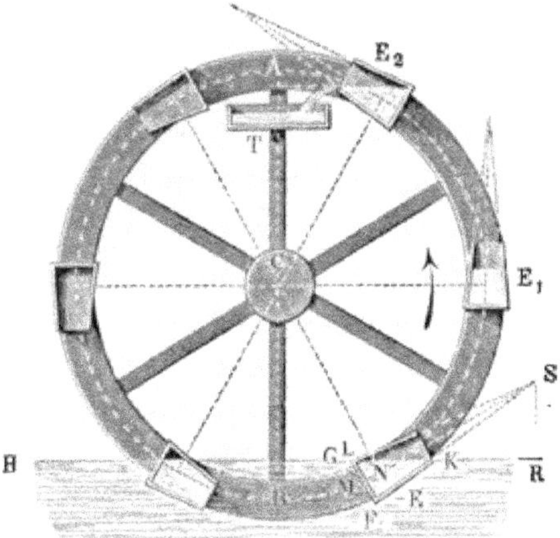

Fig. 9.

a wheel ACB hanging freely in the river and provided with tapering vessels E, E_1 . . . secured to the shroudings in such a manner that their axes are parallel to the chords of the segments covered. During the motion of the wheel the vessels are filled to the greater portion with water at HR and discharge the latter into the conduit T closely beneath the crown A of the wheel. In order to obtain the maximum effect, the wheel is placed in the water at such a depth that the inclination to the horizon of the axis of the vessel E, when the latter emerges from the water HR (Fig. 9), is equal to the inclination of the axis of the vessel E_2 at the point where discharge commences.

In other constructions of scoop-wheels the receptacles are made of box shape, being put together of boards and located either in inclined position as in Fig. 8, or tangentially as in Fig. 9. As a rule the boxes are then enclosed on all sides, having only an opening at one side for taking in and pouring out the water. This is made clear from Fig. 10, where, as in Fig. 9, E is being filled, E_1 is ascending, and E_2 is discharging the water contained.

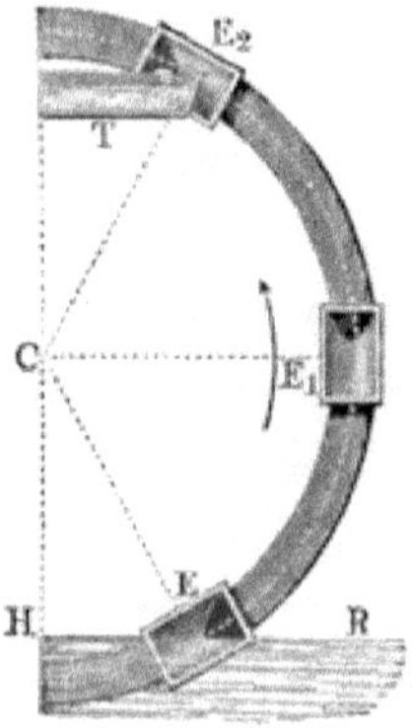

Fig. 10.

The well-known *bucket-wheels* may also serve as scoop-wheels when revolved in the opposite direction to what is the case when they are used as prime movers. A scoop-wheel of this type, which was constructed by *Laurenz* and *Thomas* for the irrigation of meadows at Soissons,[1] is illustrated in $\frac{1}{120}$ of its natural size in Figs. 11 and 12. As the wheel revolves, the buckets immersed are filled with water through the outside circumference, and, after lifting it to a certain height, they discharge it through openings in the sole of the wheel into the conduits BD and CE which embrace the latter. The three sets of arms F, together with the rims which are inserted between them and serve to connect the buckets rigidly with the shaft, are confined to the central portion of the wheel in order to leave a clear space for the heads BB and CC of the water conduits (Fig. 12). A breast-wheel located somewhat lower down gives motion to the bucket-wheel from the shaft G through the gears HK.

[1] See *Bulletin de la Société d'encouragement*, 1848; also *Der Ingenieur*, vol. ii.

Fig. 13 represents a vertical section of a scoop-wheel which may be seen at Ranstedt in the district of Nidda, the scale being

Fig. 11.

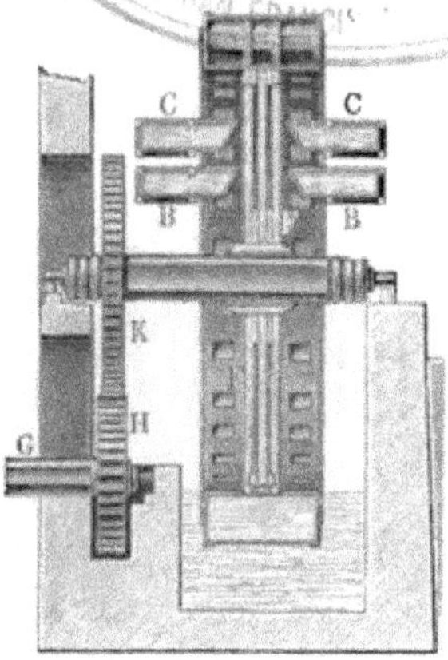

Fig. 12.

$\frac{1}{40}$. The left half AAA of the cut shows an ordinary breast-wheel, and the right half BBB the scoop-wheel which is attached

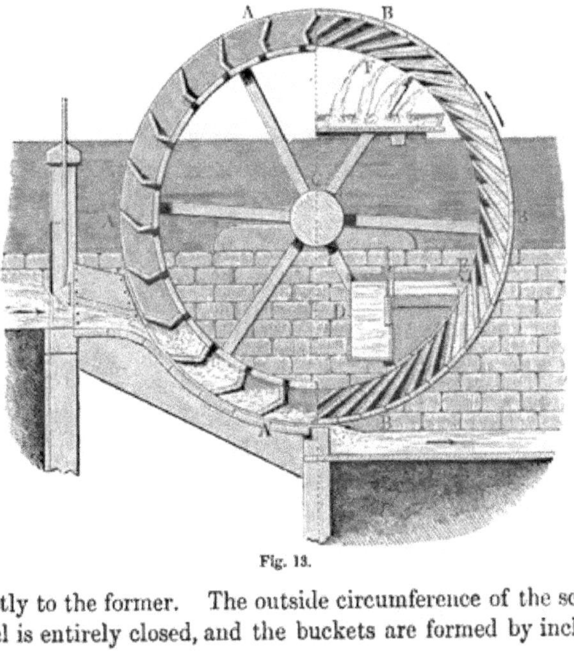

Fig. 13.

directly to the former. The outside circumference of the scoop-wheel is entirely closed, and the buckets are formed by inclined partitions. Water is admitted to the buckets at E from a side-channel, and the discharge takes place at F.

The *drum-wheel* or *tympanum*, according to *Vitruvius*, consisted of a hollow drum AB, Fig. 14, with a hollow shaft C, and divided by radial partitions into sector-shaped compartments. Each compartment had an opening *a* at the circumference for the reception of the water, and was also placed in connection with the interior of the hollow shaft through an orifice *b* at the side of the latter. When now the wheel, which was partly submerged, was set in motion, each section carried up a small quantity of water to the level of the axis of the wheel and discharged it into the hollow shaft, from where it was conducted through other side-outlets into reservoirs.

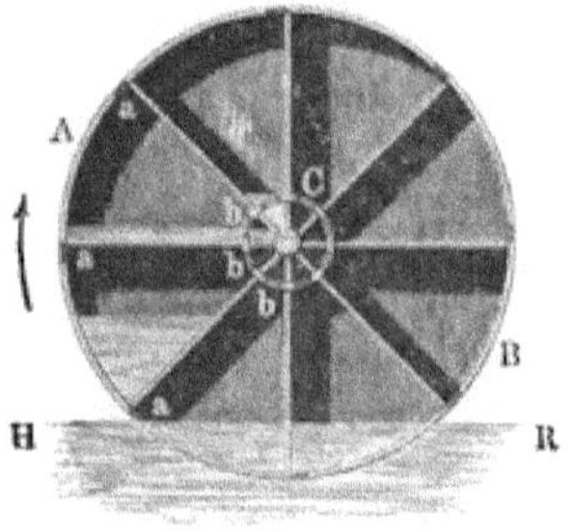

Fig. 14.

The *spiral-wheels* do not differ in principle from the drum-wheels, as they also admit the water through openings in the periphery and deliver it through the hollow shaft, only the radial partitions are replaced by spiral ducts. In *De Lafaye's* wheel, pipes wound in the shape of spirals are made use of for conducting the water from the circumference to the interior of the shaft, whereas *Perronet's* wheel is provided with spiral-shaped partitions, the sides being plane surfaces.

In Figs. 15 and 16 a spiral tympanum ACB, designed by *Cavé*, is illustrated in $\frac{1}{100}$ of its natural size. It is made entirely of iron, the shaft CC and the two sets of arms D_1D being of cast-iron, and the spiral partitions, which are given the shape of an involute of a circle, as well as the sides, are made of iron plate 3 mm. [$\frac{1}{8}$ in.] in thickness. Motion is transmitted to the wheel from the shaft *w* through the gears K and L, and during the slow rotation the elevated water flows away at both sides through the passages contained between the arms and the funnel-shaped rim which surrounds them.

Such spirally wound partitions are often arranged in the interior of the steam-heated drying-cylinder of finishing machines for the purpose of raising the condensed water from the bottom of the cylinder to the level of its axis, from where it is afterwards removed through the hollow gudgeons.

§ 6. **Mechanical Effect of Scoop-Wheels.**—The effect of scoop-wheels may be calculated simply as follows. Let V denote the quantity of water contained in each bucket or receptacle, n the number of receptacles in the circumference of the wheel, u the number of revolutions of the latter per minute, and h the vertical distance through which the water

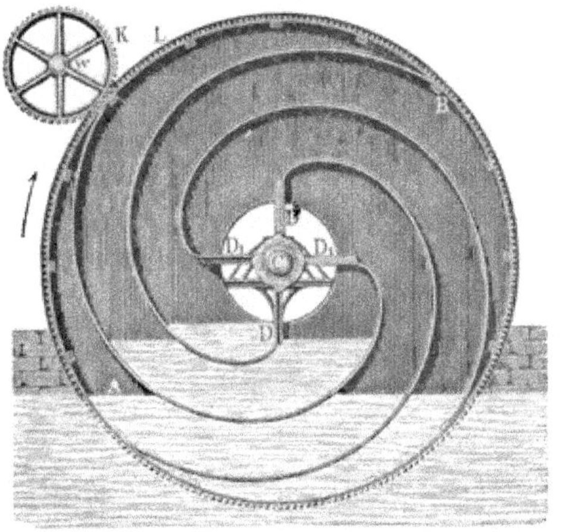

Fig. 15.

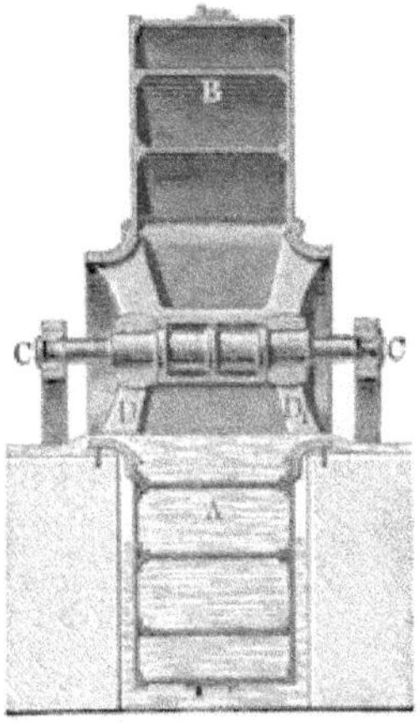

Fig. 16.

is to be lifted. We then obtain as the quantity of water raised per second

$$Q = \frac{nu\,V}{60} \quad . \qquad (1),$$

and consequently the work required per second will be

$$L = Qh\gamma = \frac{nu}{60} Vh\gamma \quad . \qquad (2),$$

where γ is the weight of water per unit of volume.

Evidently both formulæ are also applicable to the flash-wheels, if n denotes the number of paddles and V the water raised by each paddle.

Owing to the wasteful resistances, however, it is necessary

to increase the value of L, both for flash- and scoop-wheels by a considerable amount.

The velocity of the elevated water at the point of discharge being equal to the velocity v of the scoop-wheel, it is evident that an additional amount of work equal to $\frac{v^2}{2g}Q\gamma$ has been absorbed in overcoming the inertia of the water, and besides, in order that the latter may be discharged conveniently, it is necessary to lift it to a certain height h_1 above the upper water level, which requires the further expenditure of an amount of work $= Qh_1\gamma$. Thus, if we take these two sources of loss into account, the total amount of work required for turning the scoop-wheel will be

$$L = \left(h + h_1 + \frac{v^2}{2g}\right)Q\gamma = \left(h + h_1 + \frac{v^2}{2g}\right)\frac{nu}{60}V\gamma \qquad (3).$$

In order to compute, by the aid of this formula, the power necessary for driving a scoop-wheel, we must determine in the first place the volume of water contained in each bucket or vessel. This is done for each particular case from the drawing or by calculation.

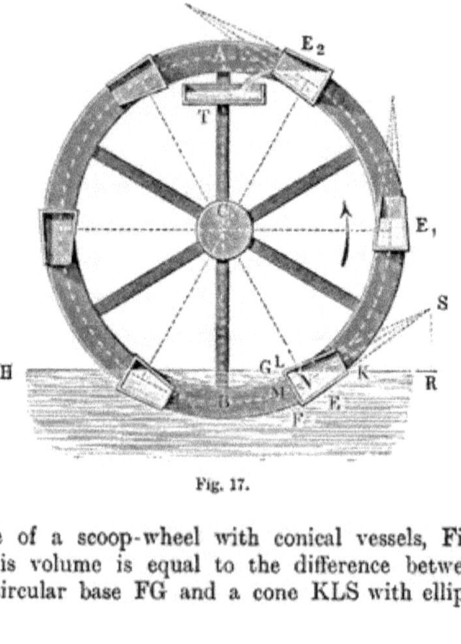

Fig. 17.

In the case of a scoop-wheel with conical vessels, Fig. 17, for instance, this volume is equal to the difference between a cone FGS with circular base FG and a cone KLS with elliptical base.

Denoting the height SM by s, and half the angle at the apex FSM $=$ GSM by α, then the volume of the first cone will be

$$V_1 = \tfrac{1}{3}\pi s^3 \tan^2 \alpha.$$

If the base KL of the second cone intersects the axis SM at an angle SNK $=\beta$, and at a distance SN $= s_1$ from the apex S, the longer axis of the base will be

$$2a = \mathrm{KL} = \mathrm{KN} + \mathrm{NL} = \frac{s_1 \sin \alpha}{\sin(\beta+\alpha)} + \frac{s_1 \sin \alpha}{\sin(\beta-\alpha)}$$
$$= \frac{s_1 \sin 2\alpha \sin \beta}{\sin(\beta+\alpha)\sin(\beta-\alpha)},$$

and the shorter axis

$$2b = 2s_1 \tan \alpha.$$

Since the height of this cone is SR $= s_1 \sin \beta$, its volume will be

$$V_2 = \pi ab \frac{s_1 \sin \beta}{3} = \frac{\pi}{3} s_1^3 \frac{\sin^2 \alpha \sin^2 \beta}{\sin(\beta+\alpha)\sin(\beta-\alpha)} = \frac{\frac{1}{3}\pi s_1^3}{\cot^2 \alpha - \cot^2 \beta}.$$

The volume contained in each vessel will now be given by

$$V = V_1 - V_2.$$

In the tympanum with spiral partitions the vertical section of the body of water in each compartment is a segment, and its volume will therefore be approximately expressed by

$$V = \tfrac{2}{3} abl,$$

when a, b, and l respectively denote the height, width, and length of the mass of water contained.

EXAMPLE.—A scoop-wheel has twelve conical receptacles of the following dimensions: radius at the bottom $r = 0{\cdot}16$ m. [0·52 ft.], radius at the top $r_1 = 0{\cdot}08$ m. [0·26 ft.], height of receptacle $a = 0{\cdot}64$ m. [2·1 ft.]; the inclination of the axis of the vessel when the latter emerges from the water is $\beta = 25$ degrees. How much water does the scoop-wheel deliver per minute, and how much power does it absorb, when running at $u = 5$ revolutions per minute and raising the water to a height of 4 m. [13·12 ft.]?

For the angle 2α at the apex of the cone we have

$$\tan \alpha = \frac{r - r_1}{a} = \frac{0{\cdot}16 - 0{\cdot}08}{0{\cdot}64} = \tfrac{1}{8};$$

further, the total height of the cone (Fig. 17) is

$$\mathrm{MS} = s = r \cot \alpha = 0{\cdot}16 \times 8 = 1{\cdot}28 \text{ m. [4·2 ft.]},$$

and the length of the axis NS of the supplementary cone is

$$s_1 = r_1 (\cot \alpha + \cot \beta) = 0{\cdot}08 \times 10{\cdot}1445 = 0{\cdot}812 \text{ m. [2·666 ft.]}$$

Consequently we have for the volume of the whole cone

$$V_1 = \tfrac{1}{3}\pi r^2 s = \frac{\pi}{3} 0{\cdot}16^2 \times 1{\cdot}28 = 0{\cdot}0328 \frac{\pi}{3},$$

and for that of the supplementary cone

$$V_2 = \frac{\pi}{3} \frac{s_1^3}{\cot^2 \alpha - \cot^2 \beta} = \frac{\pi}{3} \frac{0{\cdot}812^3}{8^2 - 2{\cdot}1445^2} = 0{\cdot}0090 \frac{\pi}{3}.$$

The volume of water in each vessel will therefore be

$$V = V_1 - V_2 = (0{\cdot}0328 - 0{\cdot}0090)\frac{\pi}{3} = 0{\cdot}0249 \text{ cub. m.} = 24{\cdot}9 \text{ litres } [0{\cdot}879 \text{ cub. ft.}]$$

The quantity of water raised per minute is

$$nuV = 12 \times 5V = 60V = 1{\cdot}494 \text{ cub. m. } [52{\cdot}762 \text{ cub. ft.}],$$

and the work required per second

$$L = \frac{nu}{60} V\gamma 4 = 24{\cdot}9 \times 4 = 99{\cdot}6 \text{ m. kg. } [720{\cdot}4 \text{ ft. lbs.}],$$

or, when the friction at the journals is included, about $1\frac{1}{2}$ horse-power.

§ 7. **Chain-Pumps.**—In place of securing the receptacles for the water directly to a revolving wheel, an *endless chain* is sometimes employed to which they are attached. The lower end of the chain is submerged in the water and its upper end is passed around a sprocket-wheel; when the latter revolves, water is thus raised and discharged at the top. Pistons or discs are frequently used in place of actual receptacles or buckets, a tube being then required for the discs to ascend in. Apparatus of this class are commonly termed *chain-pumps* (*pater-noster-pumps*), and the special types are called *chain-bucket-pumps*, when the water is elevated in buckets; *chain-disc-pumps*, when discs or pistons are made use of; and *rosary-pumps* (or *chaplets*), when the chain is provided with pallets or buttons approaching the spherical shape.

A description of the simplest form of *chain-bucket-pump*, or the so-called *noria*, may be found in the volume on *Hydraulics*. It is there represented as a prime mover, the machine ABD, Fig. 18, being assumed to derive its motion from the weight of the water admitted to the buckets at the top and allowed to descend on one side. If, however, the upper wheel be revolved by some other force in the opposite direction and the apparatus be immersed in the water to a sufficient depth, the bucket D will be filled with water at the lower end and will discharge it when it arrives at the top, in passing over the wheel C. This may, at first sight, appear to be a very perfect mode of raising water, but the objections to its application are numerous. For instance, when the buckets are dipped into the water vertically, the air contained in them opposes the entrance of the water, and, besides, it is difficult to empty the receptacles without losing a large portion of their contents. It is also necessary to lift part of the water to a considerable height above the point where it is collected, and the machine must be run at a very slow rate of speed in order to effect properly the filling and emptying of the buckets and cause the chain-links to occupy their correct positions on the wheels or drums.

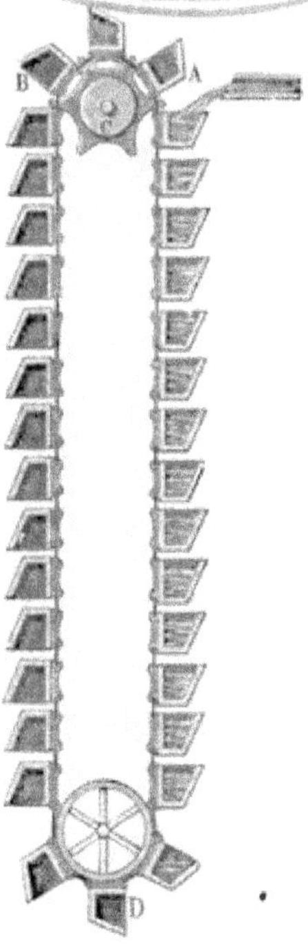

Fig. 18.

The manner of driving a noria by means of a revolving wheel provided with studs, and the means employed for collecting the discharging water, are evident from Fig. 19, I. and II., where AB is the wheel revolving with the shaft C, D are the driving studs projecting from the side of the former, E the box-shaped receptacles, and F the trough where the water is gathered, while G represents the chute for its further conduction.

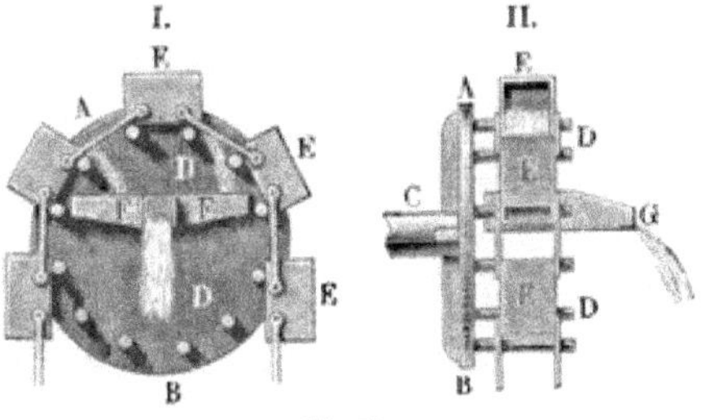

Fig. 19.

Spouts made of sheet-iron are sometimes introduced between the driving studs or the arms of the wheel for the purpose of collecting the discharging water and conducting it to a chute at the side.

The upper end of such a scoop-machine is illustrated in Fig. 20. The apparatus consists of two chains carrying the vessels A, B, etc. between them, the joint-pins being supported by the forked ends of the arms D, E, F of the wheels. Between each adjacent pair of arms, and extending between the two wheels, sheet-iron spouts are placed, which receive and carry off the water.

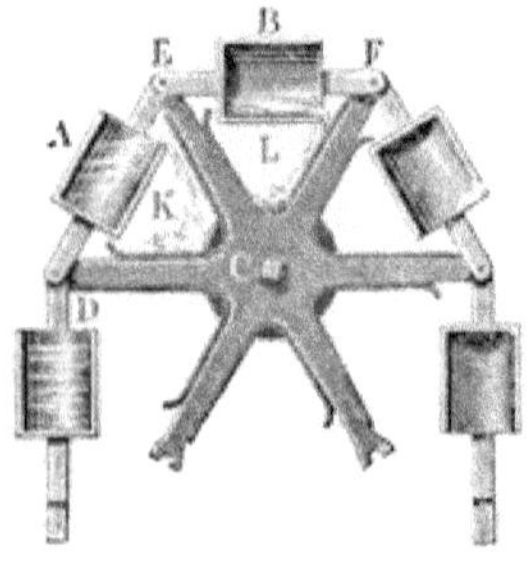

Fig. 20.

In Fig. 21 another form of noria is shown, which was first constructed by *Gateau* and is very commonly used in France. It consists of sheet-iron buckets A, etc., 0·3 m. [1 ft.] high, 0·15 m. [6 ins.] wide, and 0·24 m. [9½ ins.] long, the upper end being somewhat oblique and provided with a large opening *a* at the side for the admission and discharge of the water, and the lower end having a small hole *b* fitted with a flap-valve and serving as an in- and out-let for the air. When the receptacles ascend, the valves are, of course, closed; but when the former have passed over the top of the wheel, the valves open by the action of their own weight and admit the air required to allow of the efflux of the water. Moreover, when the buckets are dipped into the water at HR at the lower end of the chain, the air contained in them is forced out through the valves by the pressure of the entering

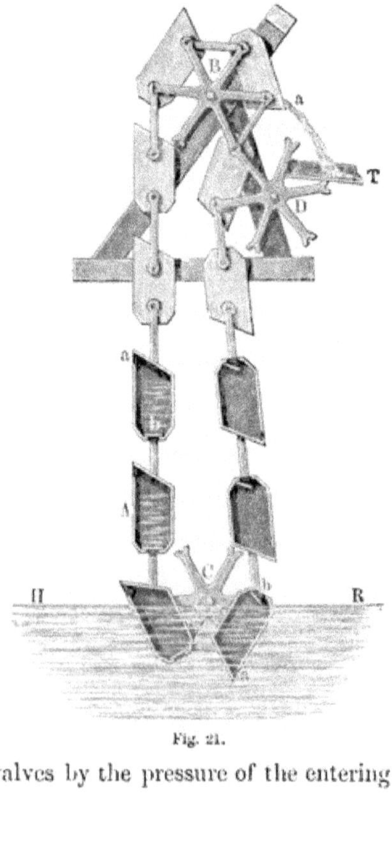

Fig. 21.

water. In order that the discharging water may be collected more readily, a guide-wheel D is located below the upper wheel B, the chain being thereby pushed back sufficiently to allow of the chute being placed almost directly beneath the discharging bucket.

NOTE.—The application of bucket-chains to the hoisting of grain or other loose material in *elevators* for flour mills, as well as to raising sand, slime, etc., in *dredging-machines*, is treated of in the volume on *Hoisting Machinery*.

§ 8. **Chain-Pumps** (*continued*).—As pumps of this class work satisfactorily also with muddy water and are easily made portable, they are frequently made use of for moderate lifts of from 2 to 3 m. [6 to 10 ft.] One form of water-elevator of this kind consists essentially of a double, endless chain to which, at the middle of the links, are secured rectangular boards of 20 to 40 mm. [$\frac{3}{4}$ to $1\frac{1}{2}$ in.] in thickness, 0·3 to 0·4 m. [1 to 1·3 ft.] long, and 0·15 to 0·20 m. [6 to 8 ins.] high. The length of the chain links, and consequently also the distances apart of the boards, are likewise 0·15 to 0·20 m. [6 to 8 ins.], and the chain wheels over which the double chain runs usually have either 6 studs or 8 or more radial prongs on which the links are supported. The ascending part of the chain travels through a trough of rectangular section, the boards having a clearance at the top and sides of about 8 to 12 mm. [$\frac{5}{16}$ to $\frac{1}{2}$ in.]; for supporting the descending part either a plain board or an open trough is used, having a length sometimes reaching 10 m. [33 ft.], and an inclination to the horizon of from 10 to 30 degrees. Motion is given to the apparatus through the upper chain-wheel shaft, usually by the turning of cranks. If it is desired to operate the machine by horse-power or by a water-wheel or a windmill, this may be done by the use of an upright shaft provided with a bevel-gear engaging another bevel on the chain-wheel shaft.

An apparatus of the above description is shown in Fig. 22, partly in elevation and partly in longitudinal section. A is the ascending, B the descending chain, and C shows the chain wheel provided with only four studs, and operated by means of a crank D. There are further visible at EE the rising chute,

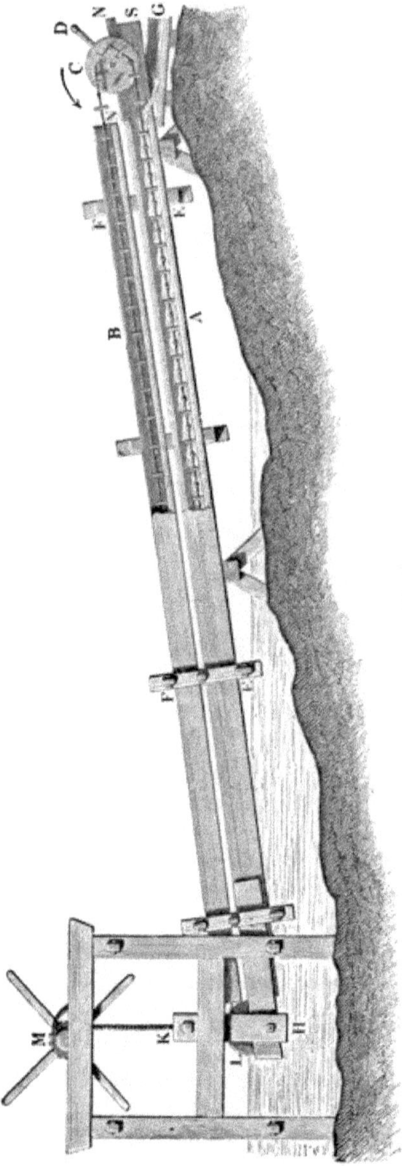

Fig. 22

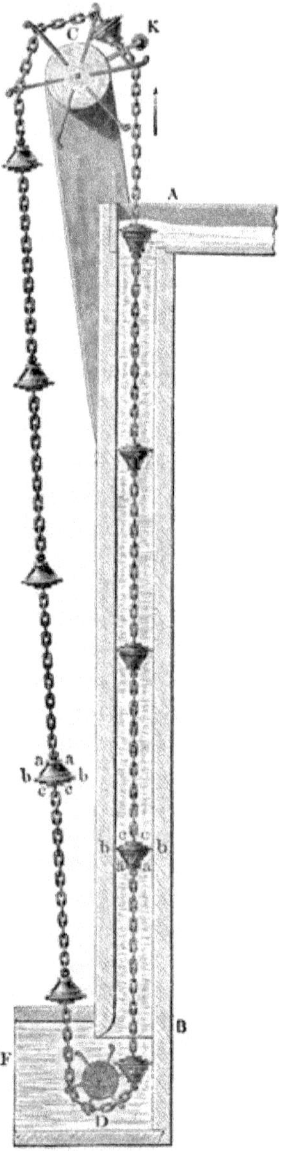

Fig. 23.

and at FF the carrying trough, also at G the discharge conduit, and at EF the frames for coupling the two inclined troughs to each other. At the lower end a frame HK is applied which receives the bearings for the lower chain pulleys, and is suspended by means of a rope or chain from a drum M provided with a ratchet, in order that the inclination of the machine may be adjusted to suit the position of the water-level. To provide a means of giving the proper tension to the chain, the bearings of the upper chain wheels are arranged in a sliding piece NN, which is connected to the head S of the chute E by means of a rack and pawl.

Such chain-pumps are sometimes used for *dredging* slimy bottoms. A very good example of a steam dredge of this kind, built at *Waltjen's* works in Bremen for the harbour of Geestemünde, is found in Wiebe's *Skizzenbuch f. d. Ing. u. Maschinenbauer*, parts 32 and 33.

In the *chain-disc-pump*, shown in Fig. 23, the chain is fitted with circular discs or pistons, and ascends in a vertical pipe AB. The discs consist of a wooden piece *a*, a packing-ring *b* of leather soaked in a mixture of tallow, oil, and tar, an iron plate *c*, and a spindle or bolt passing through the whole, and attached to the chain at each end. Both at the top and bottom the disc chain is held by the forked ends of the arms of the wheels C and D, of which the upper one is rotated by means of a crank K. Motion is thus imparted to the chain in the direction of the arrow, and the water raised by the discs from the box F through the vertical pipe to the point of discharge A. The vertical pipe or barrel is given a diameter of 0·12 to 0·15 m. [$4\frac{3}{4}$ to 6 ins.], the leather discs being made about 3 mm. [$\frac{1}{8}$ in.] smaller, so as to reduce the friction, and located at intervals of 0·80 to 1 m. [$2\frac{1}{2}$ to $3\frac{1}{4}$ ft.] In order to prevent the water, as far as it is possible, from leaking by the discs, without materially increasing the frictional resistances in the barrel, the latter is frequently made of iron instead of wood, or at least its lower end is formed by an iron tube in which the leather discs are made to fit accurately.

§ 9. **Mechanical Effect of Chain-Pumps.**—The effect of a chain-pump may be computed as follows. Let V be the volume of water carried by each bucket or disc, etc., let further

n denote the number of studs, arms, or forks of the upper wheel for the support of the links of the chain, and let u be the number of revolutions of this wheel per minute; then, the volume of water raised by the machine per second will be

$$Q = \frac{nu}{60} V.$$

If the vertical height to which the water is raised be denoted by h, and all hurtful resistances be omitted, the work required per second will be

$$L = Qh\gamma = \frac{nu}{60} Vh\gamma.$$

As a rule, the volume V may be calculated as that of a body of prismatic shape. In the case of the pump with rectangular lifting boards, the volume depends to a great extent on the inclination of the machine. Letting d denote the height AH (Fig. 24) of each board, b the width, and e the distance AB between the boards, then, for an angle of inclination ACH = RHD = a of the chute to the horizon, we have the volume of water carried by each board

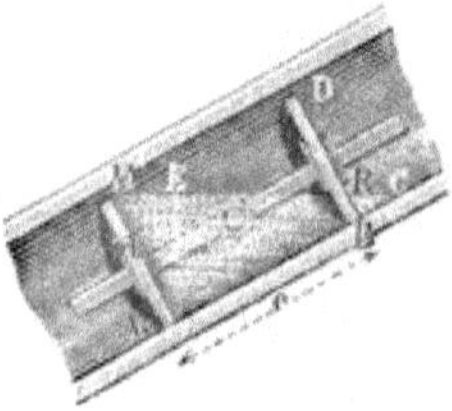

Fig. 24.

$$V = b(AB \,.\, AH - \tfrac{1}{2} HD \,.\, DR)$$
$$= b(d \,.\, e - \tfrac{1}{2}e \,.\, e \tan a) = be(d - \tfrac{1}{2}e \tan a).$$

This formula, however, is only applicable when the water-line HR intersects both boards; were it to terminate, on one side, at the bottom wall of the chute (at C), that is, were AC < AB or $d \cot a < e$, then the section of the volume of water between the boards will no longer be a trapeze AHRB, but a triangle AHC having the base AC = $d \cot a$, and consequently we should have

$$V = \tfrac{1}{2} d^2 b \cot a.$$

In a chain-disc-pump with a vertical barrel, the volume will be V = Ge, if G is the area of each disc and e the distance between the discs.

The actual volume elevated, however, is always somewhat smaller, since we must deduct in the first place the space V_1 occupied by the chain links between the two boards or discs, and in the second place the volume which leaks through in the clearance space between the latter and the chute or barrel. This latter loss is to be computed for one board or disc only, namely that at the top, as the water lost from the spaces below is replaced by that flowing down from above.

For the chain-disc-pump this loss may be calculated as follows: If r is the radius of the disc and r_1 the inside radius of the tube, then the area of the clearance space will be

$$F = \pi(r_1^2 - r^2),$$

in place of which, however, we can put approximately

$$F = 2\pi r s,$$

if $s = r_1 - r$ denotes the width of the clearance space.

Let z designate the varying height of water above the discharging disc (at the top), and μ the coefficient of efflux (0·7) corresponding to the area F, then the quantity of water passing through during an element of time dt will be

$$dW = \mu F \sqrt{2gz} \,.\, dt.$$

If the chain moves with a velocity c, and the distance between the discs is e, we have

$$z = e - ct, \text{ hence } dz = -c dt, \text{ and } dt = -\frac{dz}{c}.$$

From this we obtain

$$dW = \mu F \sqrt{2gz} \,.\, \left(-\frac{dz}{c}\right) = -\frac{\mu F}{c} \sqrt{2g} \,.\, z^{\frac{1}{2}} dz,$$

and hence

$$W = -\frac{\mu F}{c} \sqrt{2g} \,.\, \tfrac{2}{3} z^{\frac{3}{2}} + \text{const.},$$

or, since z decreases gradually from e to 0, the total leakage for each disc will be

$$W = \frac{\mu F}{c} \sqrt{2g} \,.\, \tfrac{2}{3} e^{\frac{3}{2}} = \tfrac{2}{3} \frac{\mu F e}{c} \sqrt{2ge}.$$

Now is, however, $\frac{e}{c}$ equal to the time of discharge t for each disc, and consequently we can place

$$W = \tfrac{2}{3}\mu F t \sqrt{2ge},$$

and the leakage per second

$$\frac{W}{t} = Q_1 = \tfrac{2}{3}\mu F \sqrt{2ge}.$$

In a chain-pump with rectangular lifting-boards we can assume that leakage takes place only at the two sides of the latter, as the water-line reaches the top of the lowest board only, and the bottom of each board slides on the bottom of the chute. For an angle α of the chute to the horizon, the variable head of the water above the top board will be

$$z \sin \alpha,$$

and the variable height of the orifice

$$z \tan \alpha.$$

If therefore the clearance between the board and the chute at each side is denoted by s, we shall have

$$dW = \mu \,.\, 2sz \tan \alpha \sqrt{2gz \sin \alpha} \,.\, dt$$

$$= -2\mu \frac{s}{c} \tan \alpha \sqrt{2g \sin \alpha} \,.\, z^{\frac{3}{2}} dz,$$

and hence

$$W = -2\mu \frac{s}{c} \tan \alpha \sqrt{2g \sin \alpha} \,.\, \tfrac{2}{5} z^{\frac{5}{2}} + \text{const.}$$

The definite integral between the limits $z = e$ and $z = 0$ will then give

$$W = 2\mu \frac{s}{c} \tan \alpha \sqrt{2g \sin \alpha} \,.\, \tfrac{2}{5} e^{\frac{5}{2}}$$

$$= 2 \,.\, \tfrac{2}{5}\mu \frac{e}{c} se \tan \alpha \sqrt{2ge \sin \alpha},$$

and consequently the water which returns through leakage per second:

$$Q = \frac{W}{t} = \frac{Wc}{e} = \tfrac{4}{5}\mu se \tan \alpha \sqrt{2ge \sin \alpha}.$$

Friction of the boards at the bottoms of the two troughs also materially increases the force required for moving the chain. If R is the weight of the chain and boards, and ϕ the coefficient of friction, this frictional resistance will be $= \phi R \cos \alpha$ (we may here assume $\phi = \frac{1}{4}$).

To this must be added the journal friction of the two wheels, and the friction generated when the chain winds on and off the latter. Both resistances may be computed from the formulæ deduced in Weisbach's *Mechanics*, vol. i. and vol. iii., 1 [Machinery of Transmission].

If the hurtful resistances were neglected, the force required in the axis of the chain would be

$$P = \frac{L}{c} = \frac{nu}{60} \frac{Vh}{c} \gamma,$$

and consequently, since the distance

$$e = ct = \frac{60}{nu} c, \text{ or } \frac{nu}{60} = \frac{c}{e} \text{ and } \frac{L}{c} = \frac{Vh\gamma}{e},$$

when the friction of the boards is taken into account, and further the length of the crank is assumed $= a$ and the radius of the driving wheel $= b$, the turning force necessary at the crank

$$P = \frac{b}{a}\left(\frac{L}{c} + \phi R \cos \alpha\right) = \frac{b}{a}\left(\frac{Vh\gamma}{e} + \phi R \cos \alpha\right).$$

NOTE.—For a given length l of the machine there is always a certain inclination α which gives a maximum useful effect, that is, makes the product of the water delivered Q into the head $h = l \sin \alpha$ a maximum. Placing, in accordance with the above, the quantity of water delivered per second equal to

$$Q = \frac{nu}{60} V = \frac{nu}{60} bc(d - \tfrac{1}{2}e \tan \alpha) = bc(d - \tfrac{1}{2}e \tan \alpha),$$

we shall obtain

$$Qh = bcl \sin \alpha (d - \tfrac{1}{2}e \tan \alpha).$$

This expression obtains its greatest value when $d \sin \alpha - \frac{1}{2}e \tan \alpha \sin \alpha$ becomes a maximum, and the corresponding value of α is therefore, according to the calculus, obtained from the equation

$$d \, . \cos \alpha - \frac{e}{2}\left(\tan \alpha \cos \alpha + \frac{\sin \alpha}{\cos^2 \alpha}\right) = 0,$$

or

$$\tan^3 \alpha + 2 \tan \alpha = \frac{2d}{e}.$$

Hence, if $d=e$, for instance, we have $\tan^3 a+2\tan a=2$, and consequently $a=37°\ 38'$.

EXAMPLE.—In an inclined chain-pump the lifting-boards have a width $b=0\cdot3$ metre [0·98 ft.], a height $d=0\cdot15$ metre [0·49 ft.], and are placed $e=0\cdot20$ metre [0·66 ft.] apart; further, the inclination of the chain to the horizon is $a=20°$, the height to which the water is lifted is $h=1\cdot5$ metre [4·92 ft.], the clearance of the lifting-boards at each side in the chute is $s=10$ mm. [0·39 in.], the number of driving studs or arms in the upper wheel is $n=6$, and the number of revolutions of the latter per minute is $u=40$. Find the volume of water which can be lifted by the machine and the power required to drive it.

Neglecting all losses, we should have the volume of water raised per second:

$$Q=\frac{nu}{60}bc\left(d-\frac{1}{2}c\tan a\right)=\frac{6\times40}{60}\times0\cdot3\times0\cdot2(0\cdot15-0\cdot10\tan20°)$$
$$=0\cdot24(0\cdot15-0\cdot0364)=0\cdot02726 \text{ cub. metre [0·9697 cub. ft.]}$$

Under the assumption that the portion of chain included between two adjoining lifting-boards displaces 0·2 litre [0·0071 cub. ft.] of water, the volume raised per second will be decreased by $0\cdot0002\frac{nu}{60}=0\cdot0008$ cub. metre [0·0283 cub. ft.], and if we further take into account the leakage—

$$Q_1=\frac{4}{5}\mu sc\tan a\sqrt{2ge\sin a}=\frac{4}{5}0\cdot7\times0\cdot010\times0\cdot2\tan20°\sqrt{2\times9\cdot81\times0\cdot2\sin20°}$$
$$=0\cdot00047 \text{ cub. metre [0·0166 cub. ft.]},$$

we shall obtain the volume of water actually raised—

$$Q=0\cdot02726-0\cdot0008-0\cdot00047=0\cdot026 \text{ cub. metre [0·918 cub. ft.]}$$

The velocity of the chain is given by $c=\frac{nu}{60}e$, in which formula e represents the distance from centre to centre of the lifting-boards, *i.e.* the distance between the boards 0·2 metre [0·656 ft.] plus the thickness of board 0·025 metre [0·082 ft.], or in all 0·225 metre [0·738 ft.] Accordingly we have

$$c=4\times0\cdot225=0\cdot9 \text{ metre [2·95 ft.]}$$

The work consumed is

$$L_0=Qh\gamma=0\cdot02726\times1\cdot5\times1000=40\cdot9 \text{ metre kilograms}$$
$$[L_0=0\cdot9627\times4\cdot92\times62\cdot5=296 \text{ ft. lbs.}],$$

and consequently the force

$$P=\frac{L_0}{c}=\frac{40\cdot9}{0\cdot9}=45\cdot5 \text{ kg.}\left[=\frac{296}{2\cdot95}=100\cdot3 \text{ lbs.}\right]$$

It would be sufficient to make the length of each part of the chain equal to $\frac{h}{\sin \alpha} = \frac{1 \cdot 5}{\sin 20^\circ} = 4 \cdot 38$ metres [14·37 ft.], but let us assume that this length be made 6 metres, or the whole chain 12 metres [39·37 ft.]; then the total number of lifting-boards will be

$$12 : 0 \cdot 225 = \sim 54.$$

If the weight of each board together with the piece of chain belonging to it is 2 kg. [4·41 lbs.], the weight of the whole chain will be $R = 2 \times 54 = 108$ kg. [222·26 lbs.], and the corresponding friction at the bottoms of the chutes

$$\phi R \cos 20^\circ = \tfrac{1}{4} \times 108 \times 0 \cdot 940 = 25 \cdot 4 \text{ kg. [56 lbs.]}$$

The total load is therefore

$$45 \cdot 5 + 25 \cdot 4 = 70 \cdot 9 \text{ kg. [156·3 lbs.]},$$

and if the ratio $\frac{b}{a}$ of the radius b of the driving-wheel to the length a of the crank is $= \frac{3}{4}$, there will be required at the crank a force

$$P = \frac{3}{4} 70 \cdot 9 = 53 \cdot 2 \text{ kg. [117·24 lbs.]}$$

Since the velocity of the crank is

$$\frac{a}{b} c = \frac{4}{3} 0 \cdot 9 = 1 \cdot 2 \text{ metre [3·94 ft.]},$$

the total work expended will be

$$L = 53 \cdot 2 \times 1 \cdot 2 = 63 \cdot 8 \text{ m. kg. [461·5 ft. lbs.]},$$

and the efficiency

$$\eta = \frac{L_0}{L} = \frac{40 \cdot 9}{63 \cdot 8} = 0 \cdot 641.$$

§ 10. **Archimedean Screw** or **Water-Snail.**—This is one of the oldest water-raising machines known. It consists essentially of a tube wound spirally around an inclined shaft AB (Fig. 25), and taking part in the rotation of this shaft. The pitch of the screw and the inclination of the shaft are so chosen that a portion of each thread will always point downwards; a certain quantity of water V may then be contained in the lower part of each thread, the number of pockets or receptacles thus formed being equal to the total number of threads. The volumes of water contained in the latter may be regarded as an equal number of nuts of the screw, and hence it follows that for each revolution of the shaft all these volumes of water are elevated, in the direction of the axis, a distance equal to the pitch s of the screw. If, therefore, the screw is so placed that the lower orifice A of the tube, which is submerged to a certain depth in the water HR, can, for each revolution, admit the volume V required to fill one of the pockets, it is evident that when the shaft is rotated there must be discharged at the upper orifice B a quantity of water, which for each revolution equals the volume V contained in each pocket. The value of V depends, not only on the sectional area F of the tube or worm, but also on the length l of the arc in which the water is contained; when the diameter

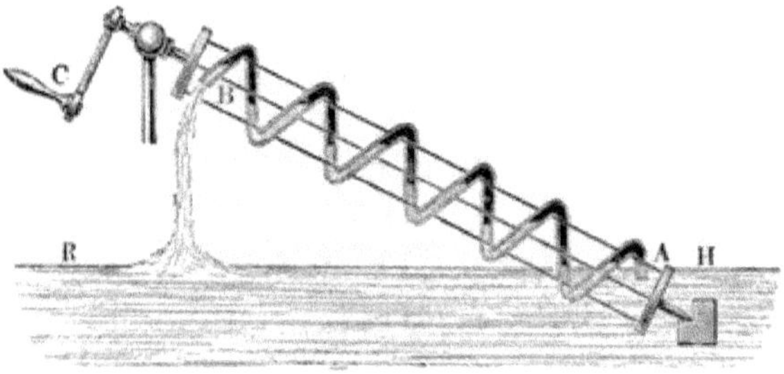

Fig. 25.

of the tube is small, as compared with that of the cylinder around which it is wound, it is sufficiently accurate to place $V = Fl$.

If the machine consists of one helix only, and makes u revolutions per minute, the volume elevated per second will be

$$Q = \frac{u}{60} V = \frac{u}{60} Fl,$$

while for a screw with n spiral tubes or threads the volume is

$$Q = \frac{nu}{60} V = \frac{nu}{60} Fl,$$

and the corresponding work expended, theoretically,

$$L = Qh\gamma = \frac{nu}{60} Vh\gamma = \frac{nu}{60} Flh\gamma.$$

Here h denotes the total lift, $h = zs \tan \beta$, when z is the number of threads of each helix, s the pitch, and β the angle of inclination of the shaft to the horizon.

To determine the length l of the arc in which the water is contained, let r be the radius of the helix described by the axis of the tube, and a the constant angle which the thread makes with a plane normal to the shaft (plane of rotation). If the screw were placed vertically this angle a would represent the inclination of the thread to the horizon at every

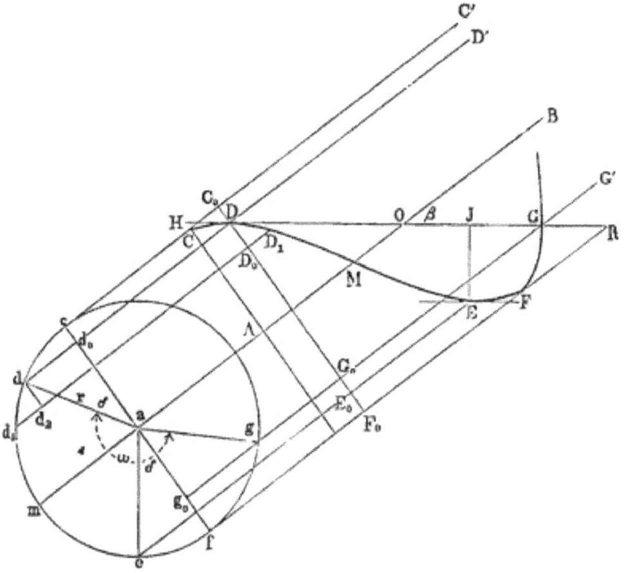

Fig. 26.

point. Inclining the screw until its axis AB (Fig. 26) makes an angle $BOR = \beta$ with the horizon HR will alter the inclination of the helix to the latter to a different degree at different points. At a certain point D, the projection of which in the plane of cross-section may be designated by d, the helix can therefore assume a horizontal direction, and we will now determine under what conditions such is the case. Let δ denote the angle cad, showing how far the point D is removed from the line CC′ imagined at the upper side of the cylinder, and let DD_1 be an infinitely small element of the screw-line, the projection of which in the plane of cross-section is $dd_1 = r \,.\, \partial\delta$. Assuming now through the point D a plane DD_0 at right angles to the axis of the cylinder, it is evident that the rise D_0D_1 axially of the element is

$$D_0D_1 = dd_1 \tan \alpha.$$

But on account of the assumed horizontal position of the element DD_1 we also have

$$D_0D_1 = dd_2 \cot \beta = dd_1 \sin \delta \cot \beta,$$

as may be easily seen from the figure. From the equality of these two expressions for D_0D_1 we obtain

$$\tan \alpha = \sin \delta \cot \beta,$$

or

$$\sin \delta = \tan \alpha \tan \beta \qquad (1).$$

For a given screw, *i.e.* for a given angle α of the thread, and an assumed inclination β to the horizon, this expression will give two adjacent angles δ, provided that $\tan \alpha \tan \beta < 1$. In the limiting case $\tan \alpha \tan \beta = 1$ it is apparent that $\delta = 90°$, *i.e.* the horizontal direction of the helix will be obtained in the point M. As, therefore, the helix would ascend on one side of this point M and descend on the other, it is easily understood that for the limit $\tan \alpha \tan \beta = 1$, or for $\alpha + \beta = 90°$, the screw is unsuited for the elevation of water, since the pockets serving to receive the water then entirely disappear. The first requirement of a water-screw is, therefore, that the *sum of the angle of the thread* and *the angle of inclination of the shaft should be less than* 90°.

Assuming then that this requirement is fulfilled, we obtain from $\tan \alpha \tan \beta = \sin \delta$ two angles δ and $180° - \delta$, at which the tangents of the helix are horizontal. These two points are designated by D and E in the figure, and it is evident that at these places the screw line *rises* on one side and *falls* on the other. From these considerations it follows that when the screw is submerged up to the point D the helix will form a bag-shaped arc for the reception of the water, the arc beginning at D and having its lowest point at E. If C_0F_0 represents the lower end of the screw, we obtain for the vertical height of the extremity C_0 above the water-level HR the expression

$$C_0D \cos\beta = r(1 - \cos\delta)\cos\beta.$$

The arc in which the water is contained extends to the point G where the level HR of the water intersects the helix. To compute the length l of this arc DMEFG, let the corresponding angle *dag* at the centre be designated by ω. As above in the case of D_0D_1 we now obtain two expressions for the rise G_0G of the arc axially, namely,

$$G_0G_1 = deg \tan\alpha = r\omega \tan\alpha,$$

in consequence of the angle of the thread, and

$$G_0G_1 = d_0g_0 \cot\beta = r[\cos\delta - \cos(\omega + \delta)]\cot\beta,$$

on account of the inclination of the shaft.

From the equality of these values, and by paying attention to (1) we therefore obtain

$$\cos\delta - \cos(\omega + \delta) = \omega \tan\alpha \tan\beta = \omega \sin\delta \qquad (2).$$

For any given screw, *i.e.* for a known value of δ, the angle ω may be determined by approximation from this equation, and, referring to the figure, the length l of the arc DEG will then be found to be

$$l = \frac{r\omega}{\cos\alpha}.$$

During the rotation of the screw around its axis both the length and the shape of this arc will remain unchanged, since the latter gradually advances upward in such a manner that all its points, and consequently its extremities as well, move along lines DD′, GG′ . . . which are parallel to the shaft AB. As of the total length $\frac{2\pi r}{\cos\alpha}$ of each thread only the portion $\frac{\omega r}{\cos\alpha}$ is filled with water, the air will occupy the remainder of the arc

$$l_1 = \frac{2\pi r}{\cos\alpha} - \frac{\omega r}{\cos\alpha} = \frac{2\pi - \omega}{\cos\alpha} r.$$

In each revolution the orifice of the tube describes an arc $2cd = 2\delta r$ only in the air, and therefore it takes in a volume of air corresponding to the arc $\frac{2\delta r}{\cos a}$ only. After the orifice has re-entered the water this volume can expand only at the expense of the density of the enclosed air. Such expansion or decrease in density is necessarily combined with a reduction in the pressure, and therefore the state of equilibrium of the water arc is disturbed and a portion of the water must return through the air space. To avoid such disturbances in the regular ascent of the water, the tube is either provided with a large number of small holes along its whole length for the admission of the air needed to fill the arc l_1, a small quantity of water being hereby lost, however, through the arc l, or it is made of such a width that all the air spaces will communicate with each other, and accordingly the air which is lacking below may be supplied from above. In this case the width d must be at least equal to the perpendicular height EJ of the water arc DEG, and from the figure it is thus found to be

$$d = \mathrm{DE}_0 \cos\beta - \mathrm{E}_0\mathrm{E}\sin\beta$$
$$= 2r\cos\delta\cos\beta - r(\pi - 2\delta)\tan a\sin\beta.$$

When this or a greater width is chosen, it is evident that the sectional area F of the water arc must not be placed equal to $\frac{\pi d^2}{4}$, but a mean value must be deduced from which the capacity may be computed. In this calculation *Simpson's* rule may be utilised to advantage after the surface of the cylinder has been rectified; the water arc DEG and the horizontal water line DJG may then be transferred to the paper and the distances between these two lines measured at various points.

In order that the arc DEG may actually become filled with water, it is necessary to revolve the screw very slowly.

EXAMPLE.—If a water-snail be placed at an angle of $\beta = 35°$ and the angle of thread be $a = 30°$, then the most advantageous angle of submersion δ will be obtained from

$$\sin\delta = \tan a \tan\beta = \tan 30° \tan 35° = 0{\cdot}4043,$$

which gives $\delta = 23° 51'$; for the angle ω at the centre, corresponding to the water arc, we have

$$\omega \sin \delta + \cos (\delta + \omega) = \cos \delta,$$
$$\omega \sin 23° 51' + \cos (23° 51' + \omega) = \cos 23° 51',$$

that is, $0{\cdot}4043\omega + \cos (23° 51' + \omega) = 0{\cdot}91461$. Hence follows $\omega° = 211° 4'$, and the length of the water arc

$$l = \frac{\omega r}{\cos a} = \frac{3{\cdot}6839r}{\cos 30°} = 4{\cdot}254r.$$

The angles δ and ω may also be easily obtained by a construction, if the surface of the cylinder, the elliptic outline of the water-level, and the helical axis of the tube be developed on the paper.

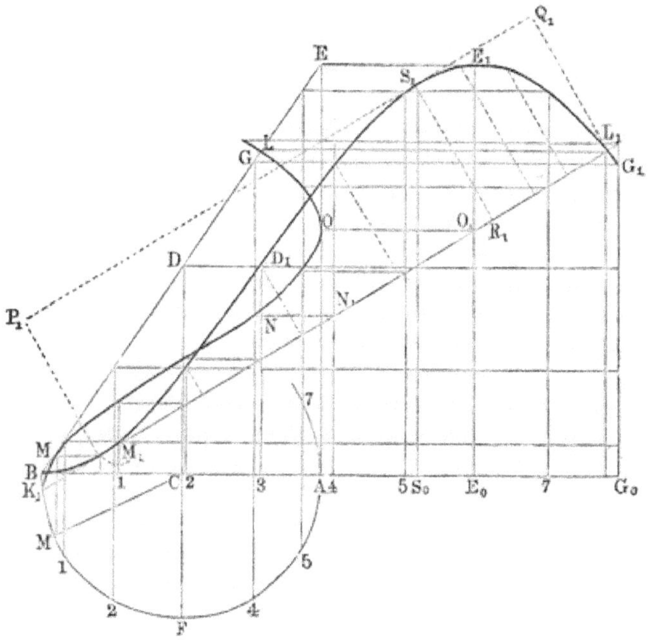

Fig. 27.

In Fig. 27, let AB represent the base, CD the axis of the cylinder, and BE the elliptic section formed by the water-level in cutting the cylinder around which the tube is wound (this section making with the base an angle $ABE = 90° - \beta = 90° - 35° = 55°$). Now, make $BE_0 = \pi r$ equal to half the circumference AFB of the base of the cylinder, further, divide both the semicircle AFB and the straight line BE_0 in (6) equal parts and erect perpendiculars from the points of division (1, 2, 3 . . .) of both lines, and finally, through the points of intersection of the first system of lines (1, 2, 3 . . .) and the section BE draw horizontal lines; then, the intersections of the latter with the second system of perpendiculars will form a curve $BM_1D_1E_1G_1$, which will represent the development of the elliptic water-line, and the shape of which is that of the sinus curve. If we now draw a straight line K_1L_1 intersecting the base AB at the given angle of thread a (30°), and touching the curve in the point M_1, this line will be the desired development of the helix, from which the projection MNOL of the helix is determined by drawing horizontal lines M_1M, N_1N, O_1O, L_1L . . . from the intersections M_1, N_1, O_1, L_1 . . . of the line K_1L_1 and the developed system of perpendiculars to where they intersect the system of perpendiculars drawn from the periphery of the cylinder AFB. The line M_1L_1, extending from the point of tangency M_1 to the point of intersection L_1 of the developed water-line and the developed helix, is the true length of the arc in which the water is contained, and therefore a rectangle drawn on this line with a height $M_1P_1 = L_1Q_1 =$ the diameter of the tube will represent the development of the vertical section through the axis of the latter, while the space between $M_1N_1O_1L_1$ and $M_1D_1E_1L_1$ is the development of the vertical section, axially, through the body of water enclosed in the tube. The distances of the points D_1, E_1, etc., from M_1L_1 are the heights of the circular segments forming the cross-sections of this body of water. We are therefore able to determine the volume of the latter by taking the mean of all the cross-sections and multiplying it by the length $M_1L_1 = l$ of the water arc. From the drawing, which is made accurately to scale, we thus get

$$l = M_1L_1 = M_1R_1 + R_1L_1 = 3{\cdot}30r + 0{\cdot}95r = 4{\cdot}25r,$$

which is approximately the same value as was obtained by calculation. The greatest depth S_1R_1 of the screw below the surface of the water is $R_1S_1 = 1{\cdot}1r$. In order to establish communication between all the air spaces, we give the same width to the screw, consequently making $d = M_1P_1 = L_1Q_1 = 1{\cdot}1r$. Dividing the arc M_1R_1 in four and the arc R_1L_1 in three equal parts, and through the dividing points thus obtained drawing additional ordinates between $M_1D_1E_1L_1$ and $M_1N_1O_1L_1$ parallel to R_1S_1, we shall have the necessary data for the determination of the volume of water contained in each thread of the screw. The inserted heights are:

$$\overbrace{0{\cdot}20r;\ 0{\cdot}55r;\ 0{\cdot}90r}^{\text{between } M_1 \text{ and } R_1S_1}\qquad \overbrace{1{\cdot}04r;\ 0{\cdot}80r}^{\text{between } R_1S_1 \text{ and } L_1},$$

or, if we take as unit the radius of the tube $r_1 = \dfrac{d}{2} = \dfrac{1{\cdot}1r}{2} = 0{\cdot}55r$,

$$0{\cdot}364r_1;\ 1{\cdot}000r_1;\ 1{\cdot}636r_1 \text{ and } 1{\cdot}891r_1;\ 1{\cdot}454r_1.$$

The areas of the segments having these heights will be found to be respectively

$$0{\cdot}375r_1^2;\ 1{\cdot}571r_1^2;\ 2{\cdot}751r_1^2 \text{ and } 3{\cdot}075r_1^2 \text{ and } 2{\cdot}446r_1^2.$$

By using *Simpson's* rule we now get the volume of water in the tube M_1R_1,

$$3{\cdot}30r(0+4\times0{\cdot}375+2\times1{\cdot}571+4\times2{\cdot}751+3{\cdot}142)\frac{r_1^2}{12}=5{\cdot}167rr_1^2,$$

and in the tube R_1L_1,

$$0{\cdot}95r(3{\cdot}142+3\times3{\cdot}075+3\times2{\cdot}446+0)\frac{r_1^2}{8}=2{\cdot}340rr_1^2,$$

and consequently the total volume of water in each thread,

$$V=5{\cdot}167rr_1^2+2{\cdot}340rr_1^2=7{\cdot}507rr_1^2=7{\cdot}507(0{\cdot}55)^2r^3$$
$$=7{\cdot}507\times0{\cdot}3025r^3=2{\cdot}27r^3.$$

If the screw has $n=4$ threads and is revolved 20 times per minute the volume of water elevated per second will be

$$Q=\frac{nu}{60}V=\frac{4\times20}{60}2{\cdot}27r^3=3{\cdot}03r^3$$

Assuming, for instance, that the screw has a radius of $r=0{\cdot}15$ metre [0·459 ft.], we should have $Q=0{\cdot}01023$ cub. m. [0·36128 cub. ft.] If the axis of the screw has a length of 6 metres [19·685 ft.], which for an angle of inclination $\beta=35°$ corresponds to a vertical lift of $h=6\sin 35°=3{\cdot}441$ metres [11·289 ft.], the theoretical work absorbed per second would be

$$L=Qh\gamma=10{\cdot}23\times3{\cdot}441=35{\cdot}2 \text{ metre kilograms}$$
$$[L=0{\cdot}36128\times11{\cdot}289\times62{\cdot}5=254{\cdot}9 \text{ ft. lbs.}]$$

§ 11. **Water-Screws.**—As it is difficult to produce a snail having circular cross-sections, threads of rectangular section are almost exclusively employed at the present day in water-screws, being wound on the spindle and enlosed by a cylindrical casing. When the casing is fixed to the screw the machine assumes the appearance of a long barrel or drum. It is more common, however, as in the so-called *Dutch water-screw*, to make the casing stationary, in the shape of a trough, and allow it to surround the screw only at the bottom side. The trough may be made either of iron or wood, or it may be constructed of brick and cement. In order to reduce the leakage through the clearance space to a minimum, it is not only necessary to make the latter as small as possible (from

2 to 4 mm. [0·08 to 0·16 ins.]), but the whole trough should be constructed very accurately and substantially. In Holland, where water-screws are used for draining of low lands, the troughs are constructed with great care of vitrified brick or clinker. The water-screws with revolving casings are usually portable and operated by hand, while the Dutch water-screws are rotated by windmills. In the latter machine the screw shaft is provided with a gear which is driven by another gear placed at the lower end of the upright windmill shaft.

Fig. 28 shows the arrangement of a water-screw with revolving drum and triple thread. The journals C and D of

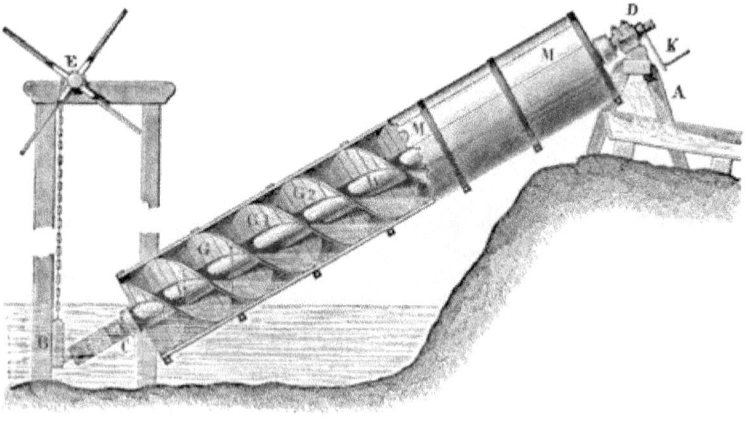

Fig. 28.

the shaft are supported in a wooden frame which at the top rests on a horse A, and at the bottom is held by a guide beam B suspended by chains from a windlass E by means of which the screw can be given the proper inclination. The crank K is not rotated directly by hand, but is turned by means of rods attached to the wrist pin and pulled back and forth by four or six labourers. In the figure the lower half of the machine is shown in section; at G, G_1, G_2 are seen the threads which wind around the shaft WW, while at M in the upper portion the casing, bound with iron hoops, is visible.

Fig. 29, I, represents a Dutch water-screw also having a triple thread, and in which the trough is made of wood, the upper half of the latter being cut away in the engraving. The trough BB rests on the cross timbers E, E . . . between the posts F, F . . ., which are joined into the former, and the whole apparatus is supported by a portable framing GHK. Also here the crank K at the upper bearing C is put in motion by means of push rods; the step for the pivot D at the lower end is secured to a cross girt which is inserted between two posts. A cross-section of the screw together with its trough and the supports for the latter is shown in Fig. 29, II, where W is the shaft and S_1, S_2, S_3 sections

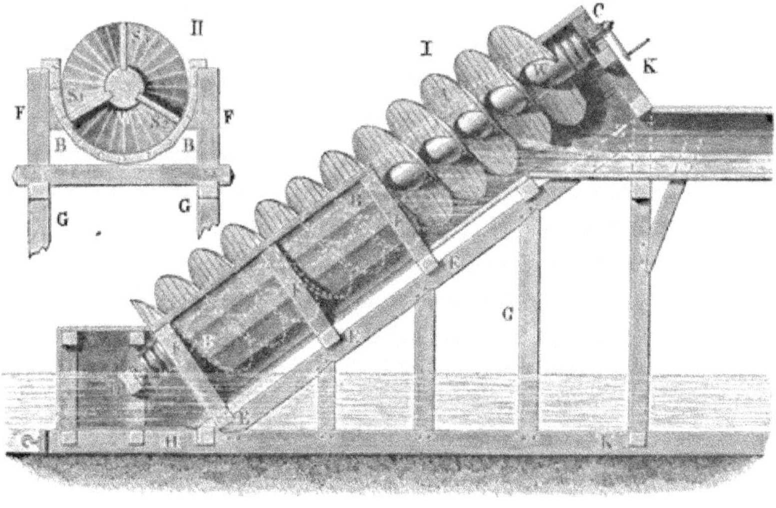

Fig. 29.

through the three threads. After the water has been elevated, it is removed through the discharge conduit Z, which is secured to the trough and also rests on the framing.

NOTE.—The Dutch screw is occasionally used for transporting loose material, as, for instance, in flour mills for taking the mixture of meal and bran to the elevator; in this case the screw is, as a rule, placed horizontally.

It is quite common in practice to make the water-screws entirely of wood; the threads may, however, just as easily be made of iron-plate sectors as of wooden ones. When made of wood they have a thickness of about 25 mm. [1 in.], and the shaft or stem of the screw is provided with helical grooves about 25 mm. [1 in.] deep for the reception of the boards. The adjoining boards are tied together firmly by irons, and no further fastening to the shaft is necessary. When the casing is to be secured to the screw, the former is also provided with grooves for the boards to fit into, and iron hoops are driven on the outside at intervals of about 0·6 metre [2 ft.] Usually the whole machine is coated with tar.

When the threads are put together of sections made of iron plate, a cast-iron shaft is used, which in place of the grooves is provided with screw-shaped flanges to which the sections are riveted. If the casing also is of iron plate it is most suitably fastened to the thread by means of angle irons.

To get the shape of the sections, a prismatic block of wood ACD_1B_1 (Fig. 30) is cut out, the base $AB_1 = CD_1$ of which is a sector of a circle forming the projection of each section in a plane at right angles to the axis of the screw. The outlines BD and B_1D_1 of the thread are then marked on the two cylindrical ends AC and A_1C_1, and finally the block is sawed through along these curves with the blade of the saw always kept at the same height a, a_1 or b, b_1 at both ends; the side surface $BabDD_1b_1a_1B_1$ will then give the curvature of each section.

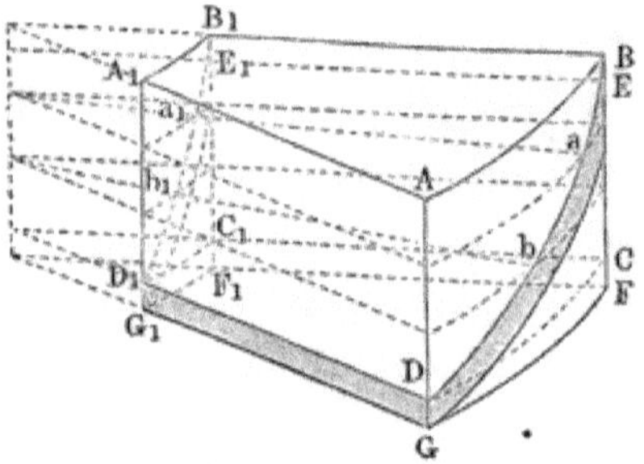

Fig. 30.

If now the concave surface of the block $ABDD_1B_1A_1$ be placed on the plate from which the sections are to be made, the outline of the latter is easily obtained. When the thread is to be made of wood it is necessary to increase the height of the wooden prism by the thickness $CF = DG$ of the thread in the direction of the axis of the screw. By making a second cut EGG_1E_1 parallel to the concave surface BDD_1B_1 and at a distance $BE = B_1E_1$ from it, the desired section BGG_1B_1 will be produced.

In place of curved sections, plane sectors may be used, if they are made to overlap each other and placed at right angles to the axis. A step-shaped thread is then obtained, like a winding stairway.

The ordinary water-screws have a length of from 3 to 6 metres [10 to 20 ft.], the diameter of the shaft is from 0·15 to 0·30 metre [6 to 12 ins.], and that of the screw from $\frac{1}{2}$ to 1 metre [1·6 to $3\frac{1}{4}$ ft.] The angle of the thread at the outside circumference is usually from 10° to 30°, and an inclination of from 30° to 35° is given to the shaft. At the shaft the angle of the thread is evidently much greater, since, according to the formula $\tan a = \frac{s}{2\pi r}$, the tangent of the angle a varies inversely as the distance r from the axis of the screw. If, for instance, the diameter of the shaft is $\frac{1}{3}$ of that of the screw, and the angle of the thread at the outside circumference were 15°, then the angle at the shaft would be determined by $\tan a = 3 \tan 15° = 0·80385$, which gives $a = 38\frac{3}{4}°$.

In order to admit and discharge the water as uniformly as possible, the distance between the threads is usually made only 0·15 or 0·20 metre [6 or 8 ins.], and from two to four separate threads are instead employed. If s is the pitch of the screw, n_1 the number of separate threads, and s_1 the distance from centre to centre of the latter measured in the direction of the axis, then we have

$$s = 2\pi r \tan a = n_1 s_1,$$

and hence

$$n_1 = \frac{2\pi r \tan a}{s_1};$$

for instance, if $r = 0·5$, $s_1 = 0·18$ metre [$r = 1·64$ and $s_1 = 0·59$ ft.], and $a = 12\frac{1}{2}°$,

$$n_1 = \frac{2 \times 0·5}{0·18} \pi \tan 12\tfrac{1}{2}° = \backsim 4.$$

For a length l of the screw each thread will wind around the shaft

$$n = \frac{l}{s} = \frac{l}{n_1 s_1}$$

times; for instance, if $n_1 = 4$, $s_1 = 0{\cdot}18$, and $l = 6$ metres [$s_1 = 0{\cdot}59$, $l = 19{\cdot}685$ ft.], we have

$$n = \frac{6}{4 \times 0{\cdot}18} = 8{\cdot}33.$$

NOTE.—Water-screws may be classed among the most perfect types of water-raising machines, their efficiency being at least 0·75. According to *Mallet's* observations, nine men operating a water-screw of the drum type, having triple thread, can raise 1480·2 cubic feet of water per hour to a height of 10·8 feet, the screw making thirty-five revolutions per minute. The corresponding hourly effect is therefore $1480{\cdot}2 \times 10{\cdot}8 \times 62{\cdot}5 = 999150$ foot pounds = 138130 m. kg.; or, for each man, 111017 foot pounds = 15350 m. kg., which, with an actual working day of six hours, is equal to a daily effect of $L = 6 \times 111017 = 666102$ foot pounds = 92100 m. kg. In France L is estimated at 100,000 m. kg.

§ 12. **Water-Screws** (*continued*).—It is a matter of some complication to determine geometrically the volume of water raised in each revolution of a water-screw, because we no longer have a water arc to deal with, but instead solids of peculiar shape, being bounded by cylindrical and helical surfaces and one horizontal plane. The development method is the quickest, the screw lines being transformed into straight lines and the elliptic outlines of the water-level into sinus curves.

In Fig. 31 let CX be the axis of the screw placed vertically, let AFB be a half transverse section of the screw shaft, and $A_1F_1B_1$ one of the screws, both unfolded in the vertical plane of the drawing. Further, let AO represent half the pitch of a thread, and BMGOH the vertical projection of the thread on the shaft or stem, as constructed in accordance with well-known rules, and let KMDE be the tangent of the curve representing the vertical projection of the level of the water in the thread. Making BL equal to the semicircle $AFB = \pi \, . \, CB$, and LO_1 = half the pitch, we shall have the developed screw line represented by the straight line BO_1, while KM_1D_1NP is the development of the ellipse along which the surface of the water intersects the stem of the screw. The two lines M_1D_1NP and M_1P enclose the surface of contact between the stem of the screw and the water contained in one division; this surface is one of the elements necessary to determine the desired volume, and its area may easily be calculated by means of *Simpson's*

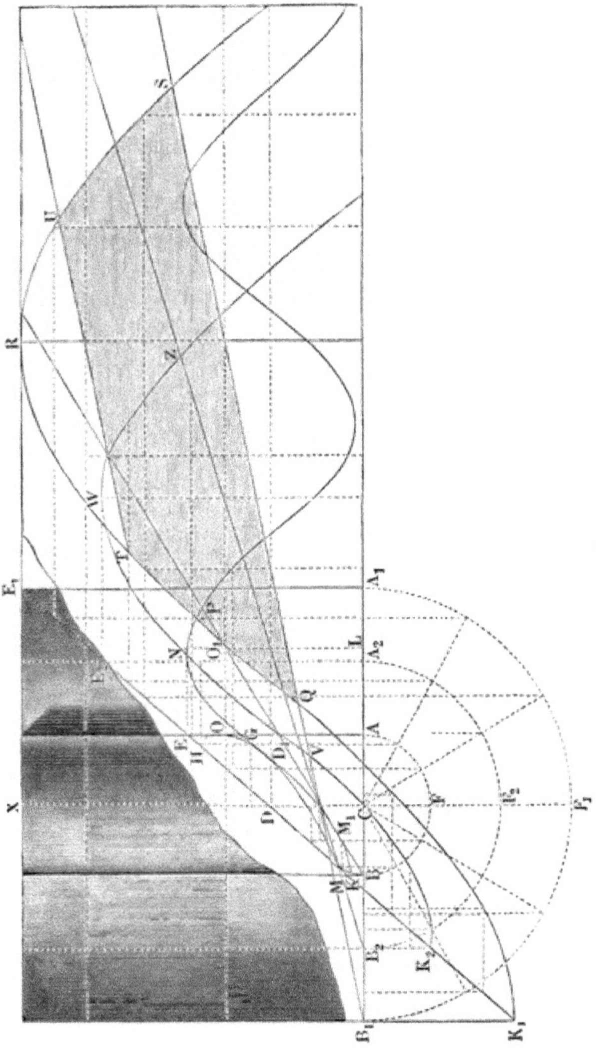

Fig. 31.

rule. The section KE corresponding to the water-level in the thread intersects the outside surface of the water-screw in a second ellipse whose transverse axis is $K_1E_1 = 2DE_1 = 2DK_1$, and which may be developed into a sinus curve K_1QTRUS intersected at two points Q and S by the developed helix B_1QS at the outside surface of the screw. Were the screw provided with single thread only, the area QTRUS between the sinus curve QTRUS and the straight line QS would now represent the surface of contact between the inside of the casing or drum and the water contained in one thread. On the other hand, if the screw has double thread, then the sinus curve is intersected by a second line TU which is parallel to QS and at a distance from it equal to half the pitch AO; the zone QTUS between these two parallel lines will then be the desired element for the determination of the volume of water in the thread.

If now other cylinders be inserted between the stem, above AFB, and the outer surface, above $A_1F_1B_1$, and the same proceeding be repeated as was applied to the outside cylinder, additional surface elements for determining the volume of water will be obtained, as, for instance, VWZ, which corresponds to the cylinder that has a radius $CA_2 = CB_2$ equal to the mean of the radius of the stem and that of the screw. After computing the areas of all these elements by the use of *Simpson's* rule, the required volume of water in each thread may be obtained by taking the mean of all these surface elements, also by the aid of *Simpson's* rule, and multiplying it by the normal distance between the casing or drum and the shaft or stem.

Example.—In the water-screw shown in Fig. 31, let the radius of the shaft be $CA = CB = r = 0{\cdot}15$ metre [0·49 ft.], the radius of the drum $CA_1 = CB_1 = r_1 = 0{\cdot}45$ metre [1·48 ft.]; let half the pitch be $AO = \frac{1}{2}s = 0{\cdot}3$ metre [0·98 ft.], the number of individual threads be $n = 2$, and the angle of inclination of the axis of the screw to the horizon be 40°, which must be <EDX. We then have for the angle of the inside helix (at the stem)

$$\tan \alpha = \frac{s}{2\pi r} = \frac{\frac{1}{2}s}{\pi r} = \frac{2}{\pi} = 0{\cdot}6366\text{ ; hence } \alpha = PBC = 32^\circ\ 29'\text{ ;}$$

for the outside helix (at the drum) we have, further,

$$\tan \alpha_1 = \frac{\frac{1}{2}s}{\pi r_1} = \frac{2}{3\pi} 0{\cdot}2122\text{ ; hence } \alpha_1 = QB_1C = 11^\circ\ 59',$$

and for an intermediate angle $ZB_2C = \alpha_2$,

$$\tan \alpha_2 = \frac{\frac{1}{2}s}{\pi r_2} = \frac{1}{\pi} = 0{\cdot}3183\text{, consequently } \alpha_2 = 17^\circ\ 39'.$$

Drawing, in accordance with these values, the straight lines BP, B_1QS, and B_2VZ to represent the developed screw lines, and constructing in addition the sinus curves KNP, K_1QRS, and K_2VWZ corresponding to the developed elliptic water-lines in a thread, also, finally, lying off at a distance $\frac{1}{2}s = 0{\cdot}3$ metre [0·98 ft.] above QS the developed helix TU of the next thread, then we obtain the surface elements M_1NP, QTUS, and VWZ, and it is now an easy matter to calculate the volume of water in a thread.

We shall find the surface

$$F_0 = M_1NP = 0{\cdot}0289 \text{ sq. m. } [0{\cdot}311 \text{ sq. ft.}]$$
$$F_1 = QTUS = 0{\cdot}3358 \text{ sq. m. } [3{\cdot}613 \text{ sq. ft.}]$$
$$F_2 = VWZ = 0{\cdot}1609 \text{ sq. m. } [1{\cdot}731 \text{ sq. ft.}]$$

and consequently the mean of these

$$F = \frac{F_0 + 4F_2 + F_1}{6} = \frac{0{\cdot}0289 + 0{\cdot}6436 + 0{\cdot}3358}{6} = 0{\cdot}1681 \text{ sq. m. } [1{\cdot}809 \text{ sq. ft.}]$$

As the width $r_1 - r = AA_1 = BB_1$ of the water space is 0·3 metre [0·98 ft.], we get the desired volume

$$V = F(r_1 - r) = 0{\cdot}1681 \times 0{\cdot}3 = 0{\cdot}0504 \text{ cub. m.} = 50{\cdot}4 \text{ litres } [1{\cdot}78 \text{ cub. ft.}]$$

The screw being double threaded will therefore deliver $2V =$ 0·101 cub. m. [3·56 cub. ft.] of water per revolution. For a length $l = 6$ metres [19·68 ft.] of the screw the water will be lifted to an approximate height

$$h = l \sin \beta = 6 \sin 40° = 6 \times 0{\cdot}6428 = 3{\cdot}857 \text{ metres } [12{\cdot}65 \text{ ft.}].$$

and consequently the work absorbed per revolution will amount to

$$Vh\gamma = 101 \times 3{\cdot}857 = 389{\cdot}6 \text{ m. kg.}$$
$$[Vh\gamma = 3{\cdot}56 \times 12{\cdot}65 \times 62{\cdot}5 = 2815 \text{ foot pounds}].$$

With a speed of $u = 30$ revolutions per minute, the theoretical work expended will be

$$\frac{30}{60} 389{\cdot}6 = 194{\cdot}8 \text{ m. kg.} = 2{\cdot}6 \text{ p.c.}$$

$$\left[\frac{30}{60} 2815 = 1407{\cdot}5 \text{ ft. pounds} = 2{\cdot}56 \text{ h.p.}\right]$$

With reference to journal friction, which is greater for screws of the drum type than for those revolving in a trough, and to the hydraulic resistances, an increase of 20 to 25 per cent must be made in this result, and therefore the amount of power absorbed must be placed at about 3·25 horse-power.

For further information on water-screws and the literature pertaining to the subject, see Rühlmann, *Allgemeine Maschinenlehre*, vol. iv.

CHAPTER II

ELEMENTARY ACTION OF RECIPROCATING PUMPS

§ 13. **Reciprocating Pumps** are the most common water-raising machines in use. They elevate water by means of a reciprocating piston fitting closely in a cylinder to which are attached the requisite pipes and regulating apparatus. The principal parts of a pump of this description are—

(1) The *pump cylinder* or *pump barrel*;

(2) The *piston* moving in this cylinder;

(3) The *pipes* through which the water is conducted to or from the barrel; and

(4) The *valves* by means of which communication between the pump barrel and the pipes is alternately established and interrupted.

As a rule two valves are employed, one admission and one delivery valve, the former governing the entrance of the water into the barrel and the latter regulating the discharge. According as both valves have immovable seats, or one seat only is fixed while the other is arranged in the piston, two different types of pumps may be distinguished—

I. *Pumps having solid pistons*, or *plunger pumps*.
II. *Pumps having hollow valved pistons*, or *bucket pumps*.

The pipe which conducts the water from the barrel is called the *delivery pipe*, and that through which the water is led to the barrel is termed the *supply* or *suction* pipe, according as the reservoir which supplies water to the pump is located above or below the latter. Occasionally one of these pipes is dispensed with, the pump cylinder being placed either directly in the water to be lifted, or just above the upper level to which it is to be delivered.

Pumps having a suction pipe but no delivery pipe are called *suction pumps*, and those having a delivery but no suction pipe are named *force* or *lift pumps*, according as the delivery pipe is attached to the barrel below or above the piston, *i.e.* according as the piston acts with its lower or its upper surface on the water column, thus exerting either a *forcing* or a *lifting* action on the water. In most cases either combined *suction* and *lift* or combined *suction* and *force* pumps are employed.

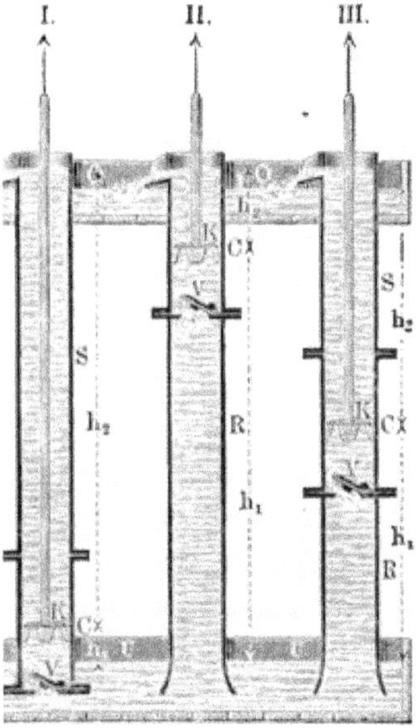

Fig. 32.

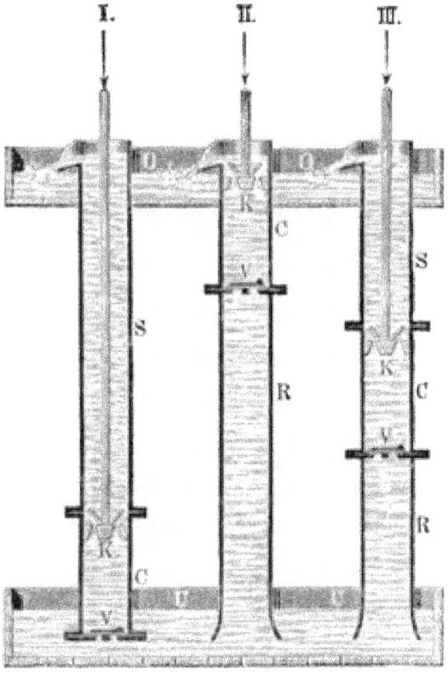

Fig. 33.

§ 14. **Bucket Pumps.**—The manner in which bucket pumps operate can be seen from Figs. 32 and 33, where sections of three pumps are shown, the pistons being indicated in their up strokes in Fig. 32, and in their down strokes in Fig. 33. The sections at I. represent a *lift pump*, those at II. a *suction pump*, and those at III. a *combined suction and lift pump*. In all three pumps C is the barrel, K the valved piston, V the *suction* or *admission valve*, UU the lower and OO the upper level of the water. Moreover in I. and III. S is the *delivery* *pipe*, and in II. and III. R is the *suction pipe.* During the up stroke (Fig. 32) the piston valves are closed and the suction valves (V) are open, owing to the pressure of the atmosphere on the lower level. A portion of the water above the piston is therefore discharged, and the volume below it is increased by the influx from the lower level. On the other hand, during the down stroke (Fig. 33) the piston valves are open and the suction valves closed, and consequently no fresh supply is taken in. The volume below the piston then simply flows through it to the space vacated above, and the amount delivered merely equals the displacement of the piston rod.

Theoretically, these three pumps are identical both as regards the power absorbed and the effect delivered. Letting b denote the height of the water barometer, h_2 the height of the water column above the piston, and h_1 the height of the column below it, measured from the lower level UU. Further, letting F designate the area of the piston, and γ the weight of a unit volume of water, or of the fluid which is to be lifted, then during the *up stroke* of the piston we have:—

(1) For the downward pressure of the air and water acting on the upper side of the piston K,

$$R_2 = F(b + h_2)\gamma, \text{ and}$$

(2) For the upward pressure of the air and water acting on the piston from below,

$$R_1 = F(b - h_1)\gamma.$$

The difference of these pressures gives the force required for lifting the piston—

$$P = R_2 - R_1 = F(b + h_2 - b + h_1)\gamma = F(h_2 + h_1)\gamma, \text{ that is, } P = Fh\gamma,$$

where h is the total vertical lift $h_1 + h_2$ measured from the lower level to the upper surface of water *in the pump.*

It is thus evident that *the force necessary to lift the piston of a bucket pump* is constant, and does not depend either on the position of the piston or on the pressure of the atmosphere, being simply equal to the *weight of a water column* having the piston area F as a base and a lift h for its height.

As the air pressing on the lower level UU can only balance a column of water of the height b (10·336 m. [33·9 ft.]), the water can follow the rising piston *only as long as the height* h_2

of the lower end of the piston above the lower water-level is less than the height b *of the water barometer.*

If s represents the stroke of the piston, the quantity discharged during the up stroke is

$$V = Fs.$$

Neglecting the sectional area of the piston rod and all the hurtful resistances, it is easily seen that the pressure of the water above and below the piston is the same during the down

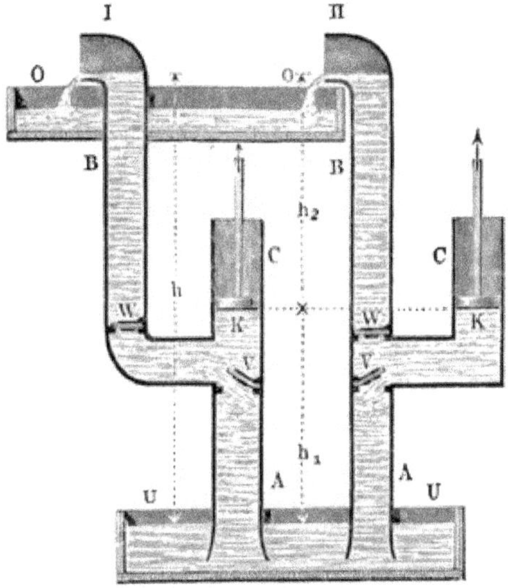

Fig. 34.

stroke, as the piston valves are then open, consequently the force needed to push down the piston and the quantity of water discharged on the down stroke are both equal to zero.

It follows that the work needed to lift the volume of water V to the height h by means of these pumps is

$$A = Ps = Fh\gamma s = Fsh\gamma = V\gamma h\,;$$

or if $G = V\gamma$ represents the weight of the lifted water, the work per stroke will be

$$A = Ps = Gh.$$

§ 15. **Plunger Pumps.**—The action of the plunger pumps can be seen from Figs. 34 and 35, the first representing the up stroke of the plunger, and the second (Fig. 35) the down stroke. In the pump shown at I the suction pipe A is a continuation of the pump barrel C; at II the suction pipe is a continuation of the delivery pipe B. Otherwise the action of the two pumps is the same. During the up stroke of the piston K the suction valve V is open, and the delivery valve W is closed. On the other hand, during the down stroke the former is closed and the latter open. In the first case the empty space left by the piston on its up stroke is filled with

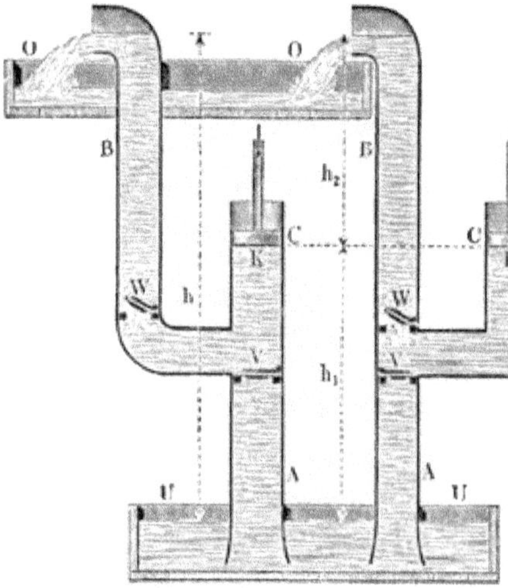

Fig. 35.

water that is driven into the suction pipe A by the atmospheric pressure; in the second case the piston presses this water out of the cylinder into the delivery pipe, and thus causes the discharge of an equal quantity of water into the catch basin OO.

Let F again represent the area of the piston, b the height of the water barometer, and h_1 the height of the piston K above the lower level, then, during the *up stroke* the downward pressure of the atmosphere on the piston is

$$R = Fb\gamma.$$

The upward pressure of the atmosphere and the water acting on the piston from below is

$$R_1 = F(b - h_1)\gamma.$$

From this follows the force needed *to lift the piston* when hurtful resistances are neglected:

$$P_1 = R - R_1 = F(b - b + h_1)\gamma = Fh_1\gamma.$$

Moreover, if h_2 is the height of the upper level of the water in the delivery pipe B above the piston K, the pressure of the water on the piston, from below upward, is

$$R_2 = F(b + h_2)\gamma;$$

consequently the total force *needed to push down the piston is*

$$P_2 = R_2 - R = F(b + h_2)\gamma - Fb\gamma = Fh_2\gamma.$$

As the heights h_1 and h_2 are continually changing during the piston motion, the piston forces P_1 and P_2 are not constant, and therefore

$$P_1 = Fh_1\gamma, \text{ and } P_2 = Fh_2\gamma$$

are only average values when h_2 is the height of the point of discharge from the average piston position (at half stroke), and h_1 the height of the average piston position above the lower water-level. If s is the stroke of the piston, then during the up stroke the force P_1 gradually increases from $F\left(h_1 - \frac{s}{2}\right)\gamma$ to $F\left(h_1 + \frac{s}{2}\right)\gamma$, and during the down stroke the force P_2 varies from $F\left(h_2 - \frac{s}{2}\right)\gamma$ to $F\left(h_2 + \frac{s}{2}\right)\gamma$.

The work that must be expended per double stroke is again

$$A = P_1 s + P_2 s = (Fh_1\gamma + Fh_2\gamma)s = Fs(h_1 + h_2)\gamma = Fsh\gamma = Vh\gamma = Gh,$$

where again $h = h_1 + h_2$ is the total lift, $V = Fs$ is the volume discharged per double stroke, and $G = V\gamma$ the weight of the water discharged.

Consequently in plunger pumps the work is distributed over both piston strokes, while in bucket pumps the work is only performed during the up stroke of the piston.

§ 16. **Plunger Pumps** (*continued*).—When the open end of the pump barrel points downward, the piston rod must either be guided through a *stuffing box* or it must also be pointed downward. The former arrangement is shown in Fig. 36 and the latter in Fig. 37. The first figure represents a combined suction and lift pump, the solid piston causing the work to be distributed over the up and down stroke, which is not the case with the lift pumps shown in Fig. 32. R is the suction pipe with the suction valve V, and S the delivery pipe with the delivery valve W. During the down stroke of the piston K water is sucked in through R, and during the up stroke it is lifted through S; in the first case, of course, the

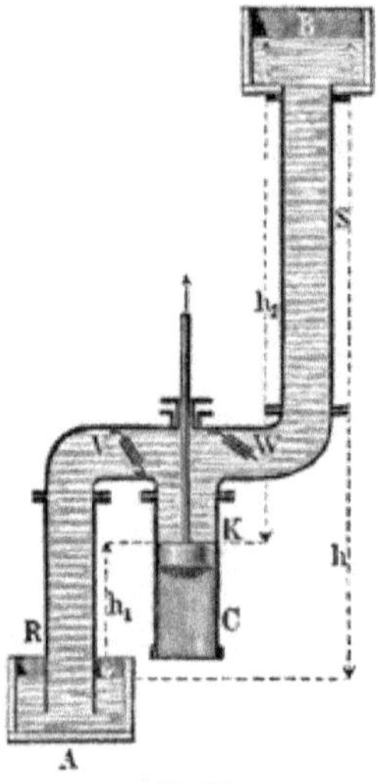

Fig. 36.

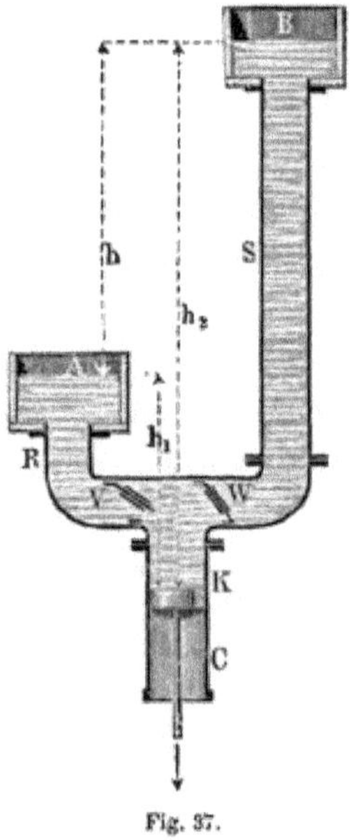

Fig. 37.

valve V opens, and in the second, which is shown in the figure, the valve W opens. In Fig. 37 a force pump having the supply pipe R is represented in its down stroke. The peculiarity of this arrangement is that the piston operates below the lower water-level, while in pumps with suction pipes its place of action is above this level; consequently in pumps with supply pipes, the perpendicular distance h_1 between the lower water-level and the average piston position will be negative if it be regarded as positive in pumps with suction pipes. If h_2 again represents the height of the upper water level B above the average piston position, and F the area of the piston, then for the pump with suction pipe (Fig. 36) *pushing down* the piston will require a force

$$P_1 = Fh_1\gamma\,;$$

on the other hand, in the pump with supply pipe (Fig. 37) the same force will be

$$P_1 = -Fh_1\gamma,$$

while the force needed to *lift* the piston in both cases is

$$P_2 = Fh_2\gamma.$$

The work needed per double stroke is, therefore, in the first pump,

$$A = P_1 s + P_2 s = Fs(h_1 + h_2)\gamma = V(h_1 + h_2)\gamma,$$

and in the second pump

$$A = P_1 s + P_2 s = Fs(h_2 - h_1)\gamma = V(h_2 - h_1)\gamma\,;$$

but as the difference of level in the first case is

$$h = h_1 + h_2,$$

and in the second case

$$h = h_2 - h_1,$$

the work in each of the pumps per double stroke will be expressed by

$$A = Vh\gamma.$$

The pump barrels are not always vertical, but frequently horizontal or inclined; it is quite evident, however, that this does not alter the mode of action nor the driving force.

§ 17. **Double Pumps.**—In order to obtain a continuous discharge of water we may employ either *one double-acting* pump, or a combination of *two single-acting* pumps.

A double-acting suction and force pump is shown in Fig. 38. It has a suction pipe R and a delivery pipe S, like most single-acting pumps; but the pump barrel C is connected at each end with the two pipes R and S, and there are, therefore, four communicating pipes leading off from the barrel C, the pipes containing two suction valves V and V_1, and two delivery valves W and W_1. As the suction valves open inward and the delivery valves outward, it is easy to see that during each stroke a suction valve will be open on one side of the piston and a delivery valve on the other. The figure shows the piston on the up stroke, during which the valves V and W_1 are open. On the other hand, when the piston is on the down stroke the valves V_1 and W are open; but in both cases the water is sucked in through R and driven up through S.

The same object is attained by a combination (Fig. 39) of

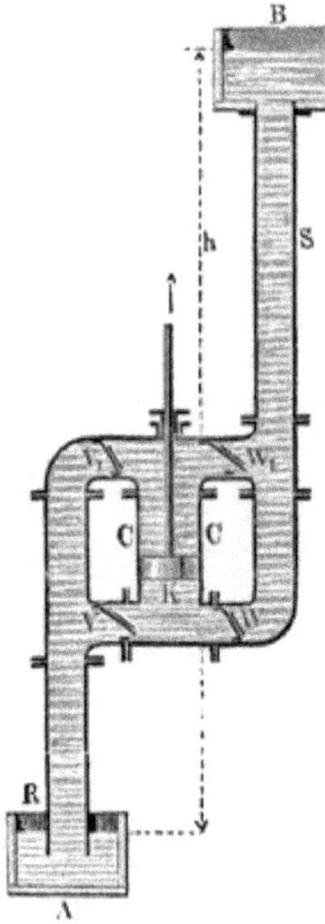

Fig. 38.

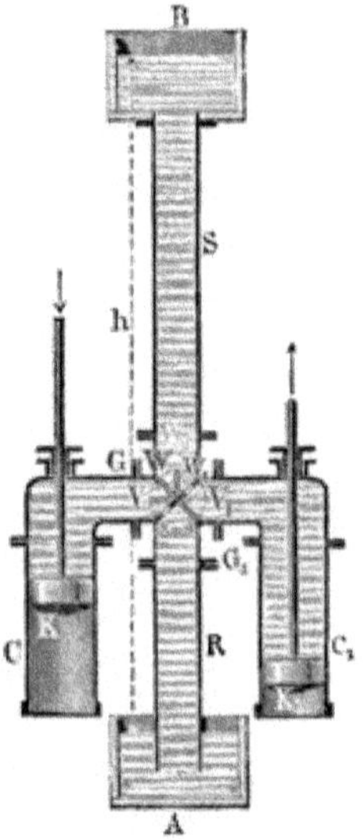

Fig. 39.

two single-acting pumps whose pistons K and K_1 alternately move up and down. In the pump shown the four valves V and V_1, W and W_1 are in the same valve chamber GG_1, and their seats are in two planes at right angles to each other. The figure shows the machine with the left piston K moving downward and the right piston K_1 moving upward, consequently the valves V and W_1 are open while V_1 and W are closed, and as a result the water passes from A through R to C, and from C_1 through S to B. The two systems of pumps

(Fig. 38 and Fig. 39), besides a continuous discharge, possess the advantage that the force to be exerted on the piston K is the same for the up and the down stroke, namely, $P = Fh\gamma$, where h again represents the total lift or difference of level and F the area of the piston.

In bucket pumps (Fig. 32 and Fig. 33) the same condition can also be brought about by using a *piston rod* of correspondingly *large diameter*. Let F_1 be the sectional area of this rod, then during the down stroke it displaces the volume of water $F_1 s$, and consequently during the up stroke only the volume $(F - F_1)s$ is discharged; therefore, if the same quantity of water is to be discharged during the down as during the up stroke, we must make

$$F_1 = F - F_1, \text{ that is, } F_1 = \tfrac{1}{2}F.$$

Accordingly, if d represents the diameter of the piston and d_1 that of the piston rod, we have

$$d_1^2 = \tfrac{1}{2}d^2, \text{ therefore } d_1 = d\sqrt{\tfrac{1}{2}} = 0{\cdot}707d.$$

Such distribution of the discharge is of course accompanied by a corresponding distribution of driving force. In this case the force necessary to lift the piston is

$$P_1 = Fh_1\gamma + (F - F_1)h_2\gamma = F(h_1 + h_2)\gamma - F_1h_2\gamma = Fh\gamma - F_1h_2\gamma;$$

on the other hand, the force needed on the down stroke, since the water acts with a head h_2 on the lower surface F and with the same head on the upper surface $F - F_1$, will be

$$P_2 = [F - (F - F_1)]h_2\gamma = F_1h_2\gamma,$$

or for the case that $F_1 = \frac{1}{2}F$

$$P_1 = F(h - \tfrac{1}{2}h_2)\gamma \text{ and } P_2 = \tfrac{1}{2}Fh_2\gamma.$$

In case we should have $h_2 = h$, that is, $h_1 = 0$ (Fig. 32, I.), we should get $P_1 = P_2 = \frac{1}{2}Fh\gamma$. In general, if we wish to expend the same force for the up and the down stroke, we place $P_1 = P_2$ and obtain the ratio of the cross-sections $\frac{F_1}{F} = \frac{h}{2h_2}$; in this case the discharge is of course unequal for the two strokes.

In order to maintain a practically uniform discharge from

the delivery pipe, as is necessary in fountains and fire-engines, for instance, *air chambers* are introduced, the mode of action of which will be described hereafter.

§ 18. **Pump Cylinder and Pipes.**—The *pump cylinder* or *barrel* is the vital part of every pump, and its proportions are of prime importance to the performance of the machine. Pump cylinders are usually made of cast-iron, though occasionally for small diameters brass or even zinc is employed; wooden pump barrels are only used for small lifts in hand pumps for agricultural or building purposes; in such cases oak or maple is usually the material chosen. Occasionally, in the crudest forms of pumps employed for building purposes, the barrels are made square, being put together of four boards; with this exception, however, pump cylinders are always made of circular cross-section. This form is not only most easily produced, but also best suited for making the piston water-tight. With the latter object in view, the cylinder must always be bored accurately, whereas the inside of the water pipes is never finished. The diameter of the pump barrel, of course, ranges within very wide limits, according to the quantity of water to be lifted. For example, in pumps which supply hydraulic presses or feed accumulators or steam boilers, the plungers are often only 20 mm. [0·79 in.] in diameter, while in those used for draining low lands they may reach 2 m. [6·56 ft.] or more.[1] The length of the cylinder is, as in steam engines, at least equal to the length of stroke plus the depth of piston; the stroke must be determined for each particular case, in accordance with the prescribed piston speed and the desired number of strokes. As a rule, the thickness of the walls of the pump barrel cannot be determined according to the formulæ for strength from the pressure of the fluid, for by this method such slender dimensions would be obtained in most cases that the practical execution would be impossible. But, in order to furnish ample resistance to the great torsional stress accompanying the boring process, and to make re-boring possible when they are worn, the walls of cast-iron cylinders are seldom made less than 20 mm. [0·79 in.] thick; for greater pressures and diameters we can use the formula given

[1] The pumps used in draining the low lands near Bremen have cylinders 8 ft. in diameter; see Berg, *Die Entwässerung des Blocklandes*, 1864.

in Weisbach's *Mechanics*, vol. ii., under the head of Water-Pressure Engines, for the thickness of cylinder walls:

$$e = 0.0025\,pd + 32 \text{ mm.} \quad [e = 0.00017\,pd + 1.26 \text{ ins.}]$$

where p is the pressure in atmospheres [lbs. per sq. in.], d the diameter of the cylinder, and e the thickness of the cylinder wall.

In all the better pumping arrangements the suction and delivery pipes are of metal, either cast-iron, wrought-iron, copper, or lead. It is only in small wells that we find pipes made of wood. The diameter of the pipes depends on the size of the pump barrel and on the piston speed, which latter is commonly chosen at 0·3 or 0·4 metre [1 to $1\frac{1}{4}$ ft.], although in mining pumps with long strokes the piston velocities may reach 1 m. [3·28 ft.] or more.[1] Usually, the pipe diameters are so proportioned as to give the water a velocity of about 1 m. [3·28 ft.] per second. The pipes are made in lengths of from 3 to 4 m. [12 to 16 ft.], and either provided with flanges for bolting the sections together or with a socket at one end to receive the adjoining pipe, the joint in this case being made by pouring in molten lead and then calking.

Special care should be taken to make the joints as staunch and tight as possible, particularly in the suction pipes, as a very small opening in the latter may endanger the entire suction by admitting air; in delivery pipes the only effect of poor joints is the loss of water due to leakage. It is, of

[1] In modern steam pumping machinery these speeds have been considerably exceeded, experience having proved that advantages can be gained by increasing the speed of the plunger. The principal advantage is that smaller steam cylinders can be used for a given amount of work to be done, thus reducing the losses due to radiation, and increasing the efficiency of the machine, which latter will occupy less space and be cheaper to procure. In an article entitled "A Few Examples of High Grade Pumping Engines," by E. D. Leavitt, in the *Journal of the New England Waterworks Association*, a number of modern pumping engines produced by able builders both in Europe and America are enumerated, in which the velocity of the plunger ranges from 1·28 to 2·29 m. [4·2 to 7·5 ft.] per second. The latter is the plunger velocity in a large pumping engine built by I. P. Morris and Co. of Philadelphia for the Calumet and Hecla Mines. Velocities of more than 2 m. [6·56 ft.] are likewise used in pumping engines built by the Quintard Iron Works of New York for Boston, Mass.; by the Hannov. Maschinenbau A. Gesellschaft for Barmen, Germany; by Ehrhardt and Schume of Saarbrücken for Eschweiler, Germany; and by L. Lang of Buda-Pesth for the latter city.—The Translator.

course, evident that suction pipes are from without subjected to a pressure that tends to collapse them, whereas delivery pipes are acted upon from within by a pressure which strives to burst them.

The thickness of metal in pipes is not generally determined from considerations of strength, but with reference to the processes of manufacture. For cast-iron pipe it varies from 10 to 20 mm. [0·4 to 0·8 in.], according to the diameter of the pipe, and for drawn wrought-iron pipe from 3 to 5 mm. [0·12 to 0·2 in.] for diameters varying from 0·04 to 0·2 m. [1·6 to 8 ins.] Pipes of larger diameter are sometimes made of riveted plates like steam boilers, but are not in favour because of their rapid corrosion when used with pumps. The smallest pipes are of sheet-copper soldered, or are pressed out of lead. Wooden suction pipes have a thickness varying from 50 to 80 mm. [2 to $3\frac{1}{4}$ ins.]; their inside diameter rarely exceeds 0·10 m. [4 ins.] When the pressure is considerable the thickness δ can be determined from the following empirical formulæ:

For cast-iron $\delta = 0{\cdot}0025\,pd + 12$ mm. [$\delta = 0{\cdot}00017\,pd + 0{\cdot}47$ ins.]

" wrought-iron $\delta = 0{\cdot}0009\,pd + 4$ mm. [$\delta = 0{\cdot}000061\,pd + 0{\cdot}16$ ins.]

" lead $\delta = 0{\cdot}0040\,pd + 8$ mm. [$\delta = 0{\cdot}00027\,pd + 0{\cdot}315$ ins.]

" wood $\delta = 0{\cdot}0050\,pd + 50$ mm. [$\delta = 0{\cdot}00034\,pd + 1{\cdot}97$ ins.]

where p is expressed in atmospheres [lbs. per sq. in.], and d is the inside diameter expressed in millimetres [inches].

In order to facilitate the influx of the water from the *sump* or lower level into the suction pipe, the inlet of the latter is rounded so as to diminish the contraction of the entering water; this orifice is also provided with a strainer to prevent impurities and foreign bodies from entering the pump. The total area of the holes in the strainer is usually equal to two or three times the cross-section of the suction pipe. When the delivery pipes discharge into an open basin their upper end is provided with a special discharging piece or nozzle of suitable form.

§ 19. **Pump Valves.**—Pump valves are placed in special valve chambers, which communicate directly with the suction and delivery pipes, and are provided with lateral openings ordinarily closed by bungs, plates, or doors, in order that the valves may be examined or repaired. In simple suction pumps only one chamber is needed, for the suction valve, since the delivery valve, which is here located in the piston, can be seen and repaired from the discharging orifice; in the *suction and lift* pumps, on the other hand, there must be a second chamber directly above the pump barrel, in order that the piston may be reached when repairs are to be made.

Pump valves are either simple or compound; the latter are used chiefly in large pumps in order to obtain a large inlet area together with prompt action. The *simple valves* are either *clack valves* or *lift valves*. The clack valves are fastened at one edge, and, when opening and closing, work like a trap-door on its hinges; lift valves move along their geometrical axis when they open and close; to this class belong *cone* (*poppet*) *valves* and *ball valves*. Cone valves have the form of a low frustum of a cone, and ball valves are perfect spheres. In the former case the valve seat is of course conical, and in the latter it is spherical. *Balance* valves have been proposed by *Belidor*; these operate like balanced sluice gates, being divided into two unequal parts by their horizontal axis, and when they open one part is raised above the seat while the other recedes below it.

Valves must be so arranged as to offer the least possible resistance to the passage of the water. For this reason the valve opening must be of a size corresponding to the cross-section of the valve chamber, and the lift of the valve must correspond to its diameter. Let r represent the radius CD of the valve chamber or pump barrel N (Fig. 40), r_1 the mean radius CB of a lift valve BB, this being equal to

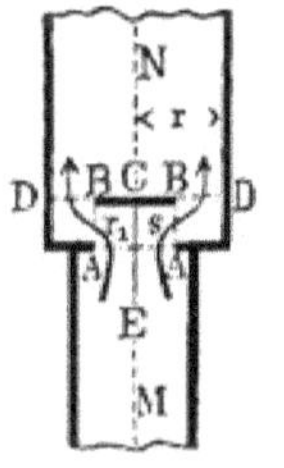

Fig. 40.

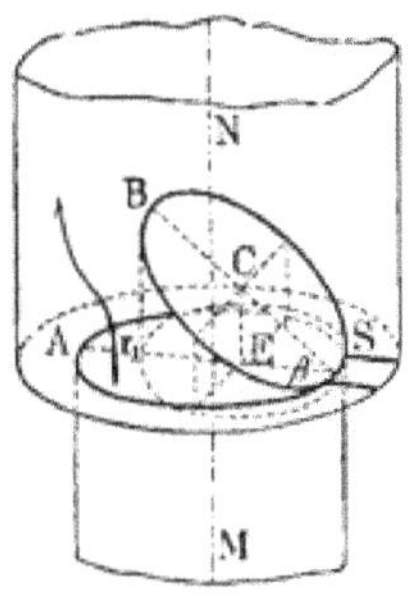

Fig. 41.

the mean radius of the passage AA, and let s be the lift AB of the valve, then the area of the circular orifice AA is

$$\pi r_1^2,$$

the area of the annular space between BB and DD is

$$\pi(r^2 - r_1^2),$$

and that of the cylindrical surface ABBA

$$2\pi r_1 s.$$

Now, in order that the water may pass through these apertures with equal velocity, we must have $\pi r_1^2 = \pi(r^2 - r_1^2) = 2\pi r_1 s$, which gives

$$r_1^2 = \frac{r^2}{2} \text{ and } s = \frac{r_1}{2},$$

i.e.

$$r_1 = r\sqrt{\tfrac{1}{2}} = 0{\cdot}707r \text{ and } s = r\sqrt{\tfrac{1}{8}} = 0{\cdot}35r.$$

If the clack valve BS (Fig. 41) is open and makes the angle ASB $= \beta$ with the valve seat, its projection on to a plane at right angles to the axis of the pipe is an ellipse with the semi-axes r_1 and $r_1 \cos\beta$, and the average height of the opening is EC $= r_1\beta$, which gives for the cross-sections of the current

$$\pi r_1^2,\ \pi r^2 - \pi r_1^2 \cos\beta, \text{ and } 2\pi r_1^2\beta.$$

By placing

$$\pi r_1^2 = \pi(r^2 - r_1^2 \cos\beta) = 2\pi r_1^2\beta$$

we now obtain

$$\beta = \tfrac{1}{2},\ i.e.\ \beta = 28\tfrac{1}{2}^0 \text{ and } r_1^2(1 + \cos\beta) = r^2,$$

or

$$r_1 = \frac{r}{\sqrt{1 + \cos\beta}},$$

or approximately

$$r_1 = \frac{r}{\sqrt{2 - \frac{\beta^2}{2}}} = \left(1 + \frac{\beta^2}{8}\right)\sqrt{\tfrac{1}{2}} \times r = 0{\cdot}730r.$$

Hence we can derive the following constructive rules: *the mean radius of the valve and its orifice should be fully $\frac{7}{10}$ of the diameter of the valve chamber or pump barrel*; moreover, *the* *lift of a lift valve should be equal to half the radius of the valve;* and finally, the *angular motion* of the *clack should be fully* 28°.

§ 20. **Pump Valves** (*continued*).—The arrangement of two simple *lift valves* is shown in Figs. 42 and 43. In both cuts A is the valve seat, B the conical valve disc, M the pipe through which water is led to the valve, and N the valve

Fig. 42.

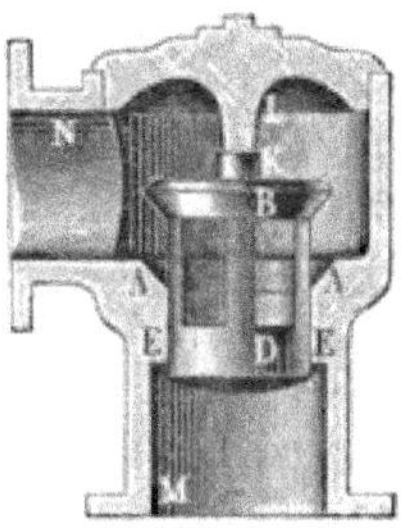

Fig. 43.

chamber. In order that the valve may shift itself exactly in its geometrical axis it is either provided with a *stem* CD (Fig. 42), or with three or four *radial wings* as in safety valves, or it receives a *cylindrical* guiding surface with *perforations* or passages, as at D (Fig. 43).

In the first case the stem is guided through two lugs, C and D, rigidly united to the valve seat by the arms E and

Fig. 44.

Fig. 45.

F; in the other two types the cylindrical surface EE under the valve seat serves to guide the valve. To prevent the valve from lifting too high it is provided with a projection K, which strikes against a stop; for example, in Fig. 42 against the yoke EE, and in Fig 43 against a lug L on the cover of the valve chamber.

Clack valves are shown in Figs. 44 and 45. The simple clack valve in Fig. 44 is of leather, so cut out as to form a circle, with a radial lapping piece B, so that it not only covers the orifice A of the pipe M, but can be fastened between M and N by means of this lap. In order that the leather clack may have the necessary stiffness, it is covered with two iron plates, the upper C being the larger, and, like the leather disc, projecting 20 to 30 mm. [0·8 to 1·2 ins.] beyond the edge of the orifice A. The lower plate D is somewhat smaller than the orifice, so that it can enter the latter without obstacle. One or more rivets or bolts firmly unite the plates and the intermediate leather.

In very large pipes, instead of a single circular plate, two semicircular or segment-shaped plates are employed, as in Fig. 45. Here the leather clack CC rests on a diametral bar, to which it is fastened by a second cross bar SS, the clack as before being covered above and below by iron plates.

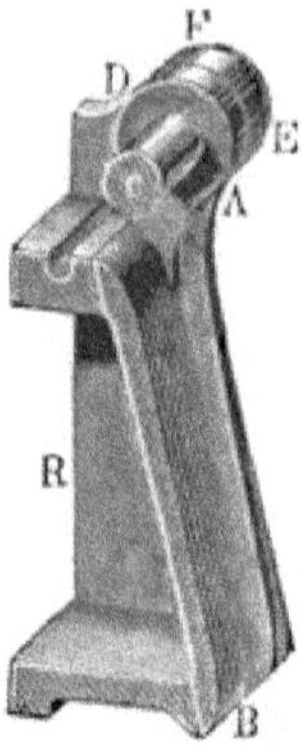

Fig. 46.

To facilitate the opening of heavy valves, they are placed in an *inclined* position, particularly if they cover orifices in horizontal pipes. Such valves are often provided with special hinges, and then work exactly like trap doors. The cross-section of such a valve for closing a rectangular pipe is shown in Fig. 46. Here AB is the clack, AD an ear attached to it, EF an ear attached to the pipe, and C the pin passing through both, and constituting the axis of rotation.

In large pumping engines *double-seated valves*, resembling double-seated steam valves, have recently come into use. The wide end of the bell-shaped valve BDE (Fig. 47) rests on the ring-shaped seat A, and the contracted end on the disc-shaped seat F. Consequently, when the valve opens, the water can flow through two cylindrical orifices ABBA and FEEF. Let r and r_1 represent the radii CD and CE of the valve orifices, and $s = AB = FE$ the lift of the valve, then the area of the passages will be given by

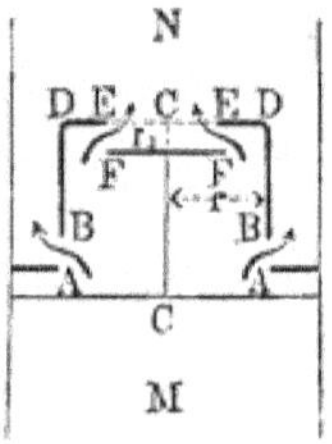

Fig. 47.

$$F_1 = 2\pi rs + 2\pi r_1 s = 2\pi s(r + r_1).$$

Also let z represent the height of a water column measuring the difference between the pressure of water on the upper and lower surface of the valve, then the force with which the water tends to lift the valve is given by

$$\pi(r^2 - r_1^2)z\gamma,$$

and if this force is greater than the weight G of the valve, *i.e.* if

$$\pi(r^2 - r_1^2)z\gamma > G,$$

the lifting will actually take place.

In suction valves $z = b - h_2$, that is, equal to the height b of the water barometer less the height of suction h_2 measured from the valve seat to the lower water-level.

The arrangement of such a double-seated pump valve is shown in Fig. 48. Here also BDE is the movable bell or valve proper, A the wide ring-shaped valve seat, and F the narrow disc-shaped seat. The ring-shaped surfaces of contact must be made to fit accurately; sometimes the valves are allowed to strike against rings a and f which are of wood or some soft metal. The wings L support the disc F and guide the valve, the latter purpose being also served by the cylinder C resting on F, the valve enclosing this cylinder with a ring; the disc k bolted to C prevents too great a lift of the valve.

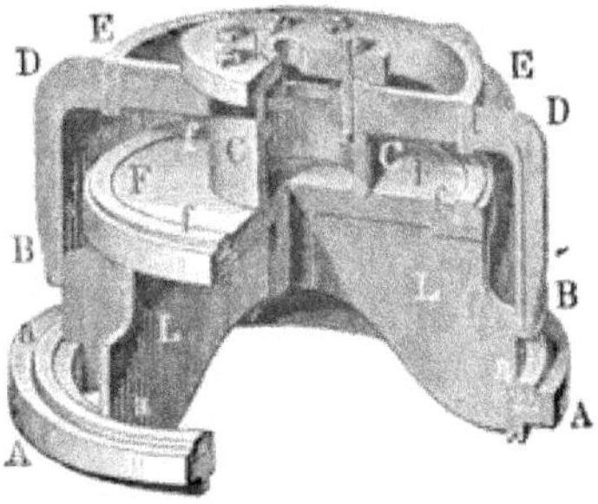

Fig. 48.

Ordinary double-seated valves require considerable excess of pressure in order to open, and unless their lift is small they strike hard against their seat at the end of the strokes. To overcome these objections and also lessen the resistance to the passage of the water, other valves have more recently been designed. Among these are those designed by *Hosking, Jenkin, Simpson*, and others, and provided with several passages.

Hosking's valve, of which a vertical section is given in Fig. 49, deserves special attention. This valve consists of a series of ring-shaped valve seats A, B, C . . ., which lie over each other in the form of a pyramid, and are covered with ring-shaped valve clacks *a, b, c.* . . . The clacks are of leather or rubber, either in the form of complete rings or sections of rings. Each clack is held in position by the valve seat above it, and all the valve seats are clamped together to a rigid whole by means of a bolt DE. Instead of clacks, rubber balls *a, b, c* have been employed,[1] the balls resting in conical seats which are enclosed by chambers of the kind shown in Fig. 50. In the large

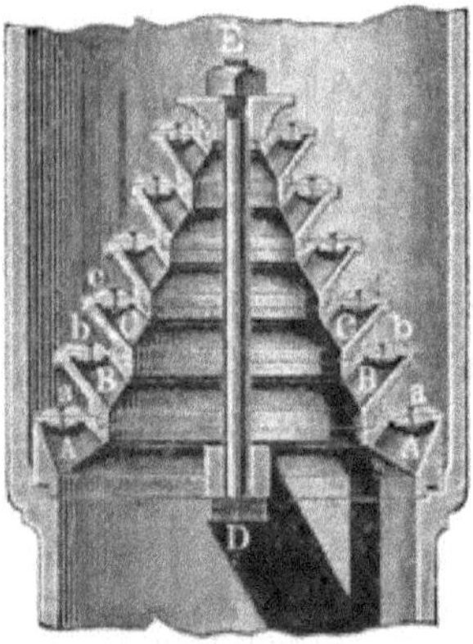

Fig. 49.

Fig. 50.

box pumps described below (§ 24), a series of valves are likewise employed for the admission and discharge of the water.

NOTE.—The manner in which the excess of pressure necessary to open the valves is obtained is explained in Weisbach's *Mechanics*, vol. iii. 1, ch. ix.

§ 21. **Pump Pistons.**—The solid pump pistons, like the driving pistons in water-pressure engines, consist of a low cylinder called the piston *head*, with its surrounding *packing*; or they have the form of a long cylinder, without packing, called the *plunger*, the requisite tightness being then obtained by a stuffing box on the pump barrel. The piston head is either made of wood which has been boiled in oil, or it is cast of iron or bronze. In ordinary pump pistons the packing is simply

[1] See article on "Improvements in Pump Valves," in the *Artizan*, vol. xvi., May 1, 1859.

a strip or disc of leather, or several strips sewed together; in air and hot-water pumps of steam engines, where leather would be rapidly injured by the heat, hempen packing is employed. The width of the packing ring may be taken equal to $b = 50$ mm. $+ 0{\cdot}1d$ [$b = 1{\cdot}96$ in. $+ 0{\cdot}1d$], where d is the diameter of the piston. Evidently pistons for double-acting pumps must have two packing rings. Such a piston is shown in Fig. 51. Here A is the piston head proper, BC its piston rod, D one packing ring or *cup leather*, and F the other; E the upper cover of the packing, and G the lower; and S a nut at the lower threaded end of the piston rod, by means of which all these parts can be pressed against the collar R on the

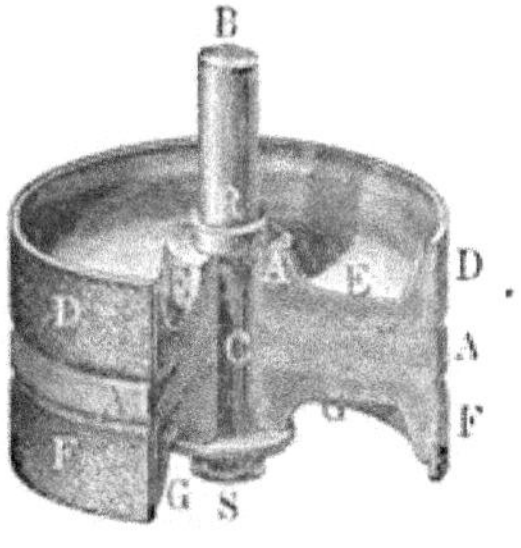

Fig. 51.

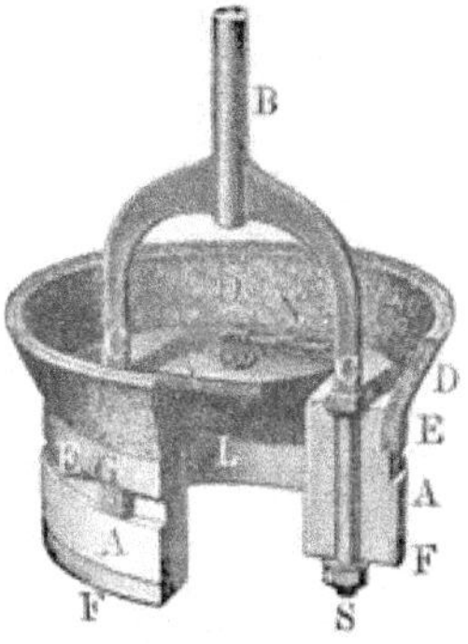

Fig. 52.

piston rod, and thus become firmly united. In *valved pistons* the heads are provided with rectangular holes for letting the water through, and are fitted with clack valves covering the upper ends of these holes. In order to facilitate as much as possible the passage of the water, the latter are made as large as is practicable, and are rounded off at the end where the water enters. When the end of the piston rod is straight and not forked, the head receives two passages and two clack valves.

A *wooden valved piston*, with forked rod and one passage, is shown in Fig. 52. Here A is the piston head, CBC the fork constituting the end of the piston rod, and CS the tines of the fork passing through the piston head and bolted to the latter as at S. Moreover, D is the sewed leather packing fastened to the piston head by an iron ring E driven on from below; F is a second iron hoop which, together with the first, protects the piston head from injury. Finally, V is the leather clack covered with sheet-iron, and secured by nails or screws to the piston head at N. To prevent the ring E from slipping off, wooden or leather pieces G are nailed into the groove occupied by the ring before it was driven in place.

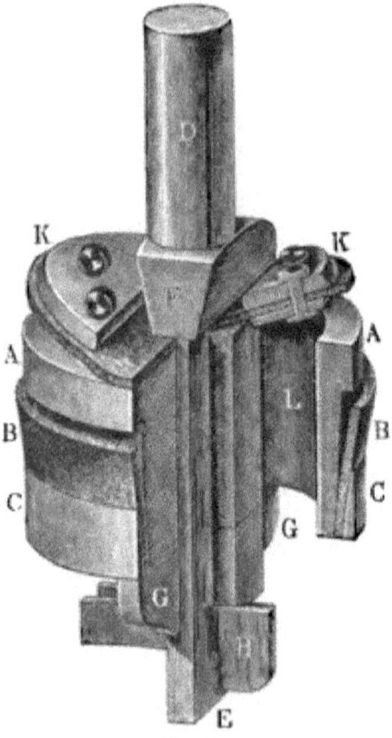

Fig. 53.

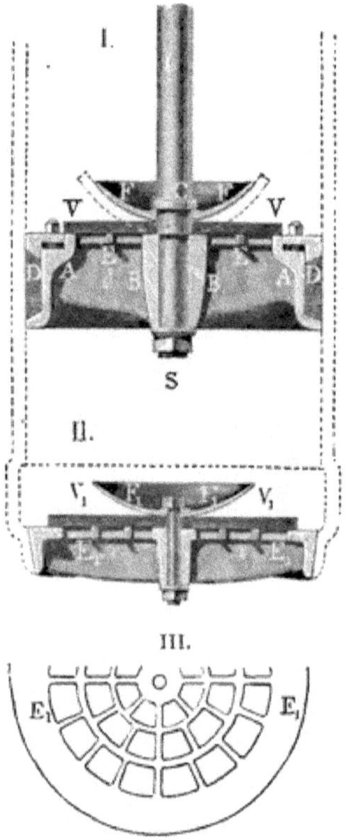

Fig. 54.

An *iron valved piston*, with straight rod and two passages, is shown in Fig. 53. Here A is the cast-iron piston head, B the packing which is secured to the conical outer surface of the head by the bevelled wrought-iron ring C. Moreover, DE is the piston rod with the cross-piece F forged on, and GG is a slotted piece slipped on the end of the piston rod, while H is a key by means of which the head is wedged in firmly between the two cross-pieces, and thus rigidly united to the piston rod. Finally, KK are two leather clack valves stiffened by iron plates, and covering the upper ends of two segment-shaped passages in the piston.

In the air and hot-water pumps of steam engines, *rubber clack valves* have proved most satisfactory, since they close more perfectly than metal valves. A pump piston with such rubber valves is shown in Fig. 54, I. The piston head here consists of a ring A, connected by four arms to the hub B, through which the end of the piston rod CS passes. Hempen packing D is employed as in the earlier forms of steam pistons. The rubber disc V, constituting the valve, rests on a gridiron frame E, supported by the arms of the head, and, when lifting, strikes against a curved stop or disc F attached to the piston rod. In Fig. 54, II., is shown the foot or suction valve V_1, which is arranged exactly like the piston valve; and in Fig. 54, III., the gridiron plate E_1 is visible. To prevent the valve discs from stretching, a layer of cloth is embedded in the rubber.

Pistons for air and hot-water pumps sometimes have metallic double valves, as in Fig. 55, where the piston head A is formed like that in Fig. 54, but the valve V is a dished cast-iron or brass plate, with a hub C movable along the turned portion DE of the piston rod, up to the shoulder D. The valve seats are formed by two brass rings and inserted in the piston head, and ground to make a tight joint with the dished plate.

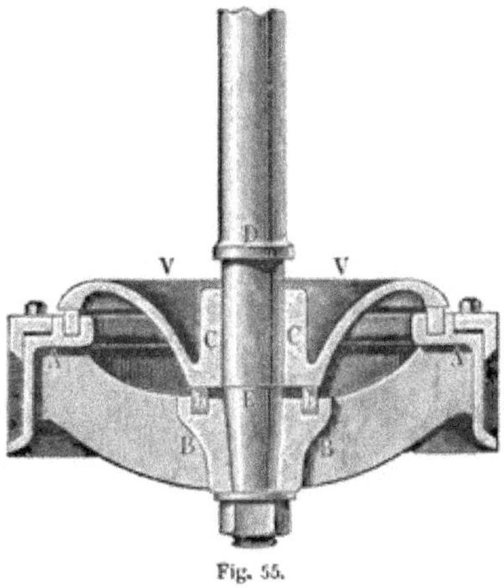

Fig. 55.

Metallic piston packing, such as is now generally used in steam pistons, is never applied to pumps, because experience has shown that it is rapidly destroyed by the sand and other solid particles carried along by the water. On this account rubber and leather valves are to be preferred to metallic valves when dirty water is pumped, as that which collects in excavations, for instance; but for high pressures or temperatures neither rubber nor leather valves can be used.

§ 22. **Suction and Lift Pumps.**—The arrangement of a suction pump, as used in mines, is shown in Figs. 56 and 57.

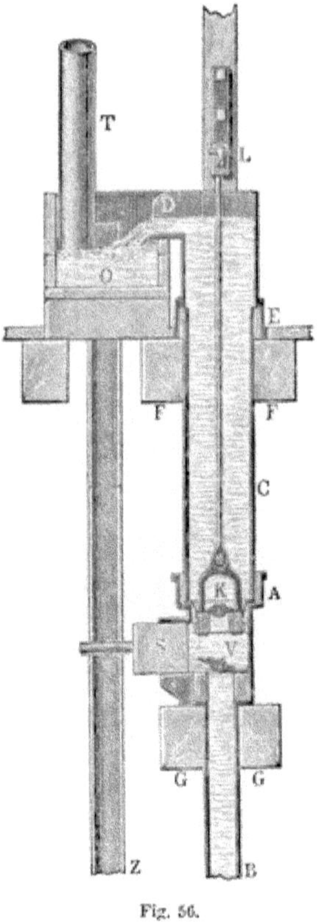

Fig. 56.

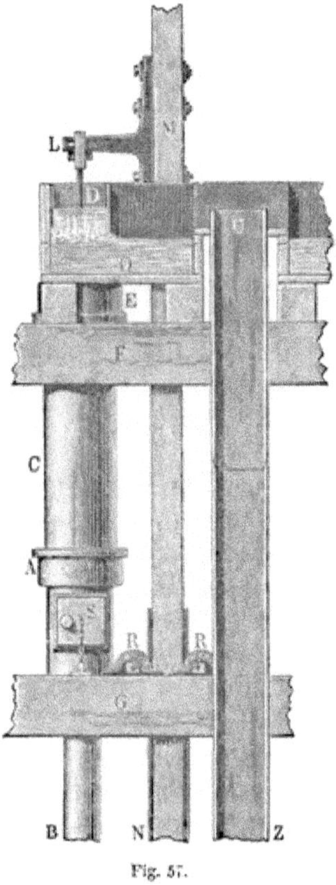

Fig. 57.

The pump barrel here consists simply of a cast-iron pipe resting in a socket on the valve chamber AB, which is at the top of the suction pipe B, and at the upper end fitting into another socket, which is a part of the discharge pipe DE. Both valve chamber and discharge pipe are supported by timbers F and G, and the hand hole through which the suction valve V can be examined and repaired is closed by a wooden bung S; the piston rod KL is attached to an arm LM, which is bolted to the main rod MN, that passes down the shaft. To prevent this rod from bending in consequence of the eccentric action of the pump, it is guided between two rollers RR. The lifted water is discharged at D into the tank O, into which the next higher suction pipe T enters. In case the pump MCB discharges more water into the tank O than the next pump is able to lift, the surplus is returned through an overflow pipe UZ to the lower tank, from which the suction pipe B draws its supply.

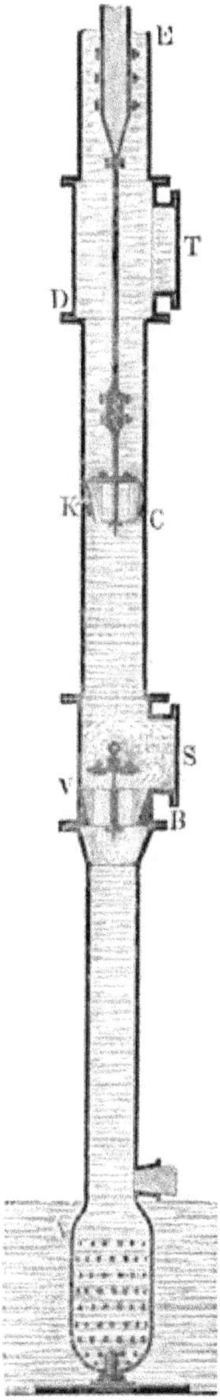

Fig. 58.

In some mines pumps are employed which differ from that just described, chiefly in that a series of pipes are placed on the top of the barrel, thus together with the latter forming a very long delivery pipe. The vertical section of such an arrangement is shown in Fig. 58. Here C is the pump barrel, AB the suction pipe, DE the delivery pipe, B the chamber for the suction valve V, and S the plate covering the hand hole. In order that the piston and its valve may be readily repaired, a second chamber D is placed over the pump barrel, and has a lateral opening which is likewise covered by an iron plate T. The lower end of the suction pipe is enlarged and provided with a number of small perforations, so as to exclude solid matter.

In bucket pumps the delivery pipe may also be connected to the side of the barrel, provided the piston rod is passed through a stuffing box. A vertical section of such an arrangement is shown in Fig. 59, where C is the pump barrel, with a lateral pipe D on which the delivery pipe is placed, and B is the valve chamber to which is attached the pipe A that connects with the suction pipe. The whole rests on the cast-iron base E supported by the beams S. The valve seat consists of two rectangular frames G, which are inclined at an angle of 45° to the vertical partition BF; VV are brass clack valves with lugs cast on that strike against the vertical sides of the chamber when the valves are open. The piston head K shown in half section is a short brass cylinder provided with a groove for the packing L, and with a vertical partition M, in which the semicircular valve plates WW are hinged. To render the discharge as uniform as possible, the piston rod OP terminates in a hollow plunger NO, to which the bucket is attached by means of a yoke and a bolt H. If the sectional area of the plunger is equal to half that of the barrel, the pump will discharge as much water on the down as on the up stroke; and if at the same time the height of suction is small compared with the height of delivery, the driving force needed during the up stroke of the piston will not be much greater than that required on the down stroke (see § 17).

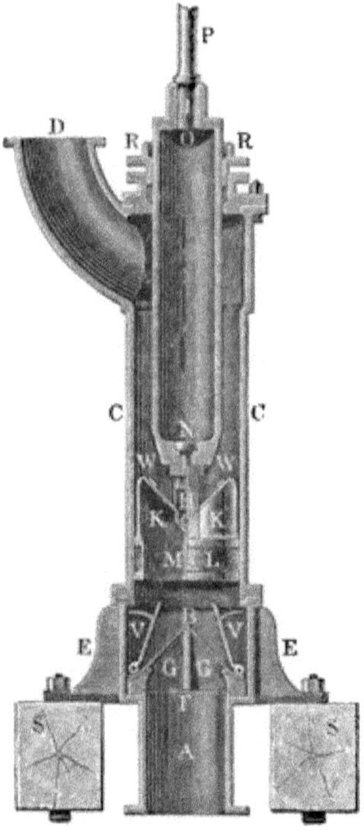

Fig. 59.

In place of the ordinary valved piston there may be substituted an externally finished pipe provided with an internal valve, and guided by stuffing boxes located in the ends of the suction and delivery pipes.

The *telescope pumps* designed by *Althans* and *Rittinger* are of this class. The former were employed in the mine Pfingstwiese near Ems, and were driven by a water-pressure engine. Their peculiarity consists in having a piston composed of two pipes, one within the other, so arranged that, according as the service may require, either both pipes or the inner alone may be reciprocated. The arrangement of a *Rittinger* pump as employed at the mines near Joachimsthal and Schemnitz[1] is shown in the vertical section given in Fig. 60. Here V is the suction valve at the end A of the suction pipe; W the delivery or piston valve in the tubular piston CD; B the pump barrel proper; E the delivery pipe; C is the stuffing box on the pump barrel B, which guides the lower end of the tubular piston; and D is the stuffing box at the upper end of the latter, surrounding the turned end of the delivery pipe E. The manner in which the tubular piston CD is connected to the pump rod is clearly shown by the sections I. and II., Fig. 61. Here also W is the piston valve, G the pump rod, and K a flanged bracket cast on the valve chamber and bolted to the pump rod. The pet-cock H serves to let the water out of the valve chamber.

Fig. 60.

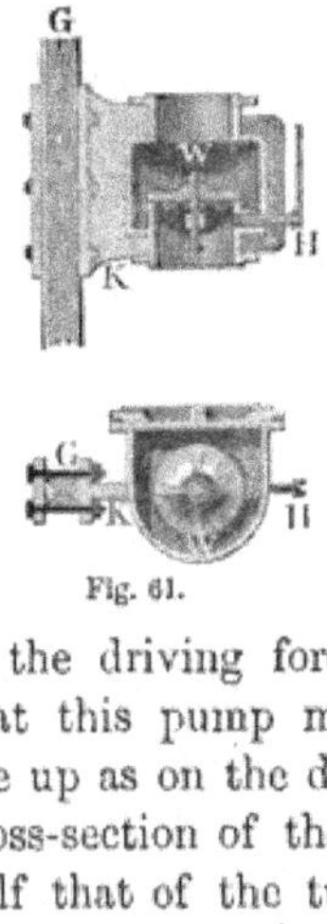

Fig. 61.

In *Althans'* pumps the tubular piston is connected to the pump rod by means of two brackets and auxiliary piston rods, thus rendering the action of the driving force perfectly central.[2] In order that this pump may discharge as much water on the up as on the down stroke, the area of the outer cross-section of the delivery pipe is made equal to half that of the tubular piston, *i.e.* the diameter of the former $= \sqrt{\frac{1}{2}} = 0{\cdot}707$ of the diameter of the latter. In common with the plunger pumps, these constructions possess the advantage of being easily taken care of and lubricated, and besides they are well adapted for pumping water containing slime or sand.

An excellent suction and lift pump with *solid piston* is shown in Fig. 62. It is in operation at Huelgoat in Brittany, and is driven by a water-pressure engine at

[1] See *Polytechn. Centralblatt*, 1851; also Rittinger's *Erfahrungen im Berg- und hütten- männischen Maschinenwesen*, etc., 1856.

[2] See Prechtl, *Technolog. Encyclopädie*, Bd. 11, Art. on Pumps.

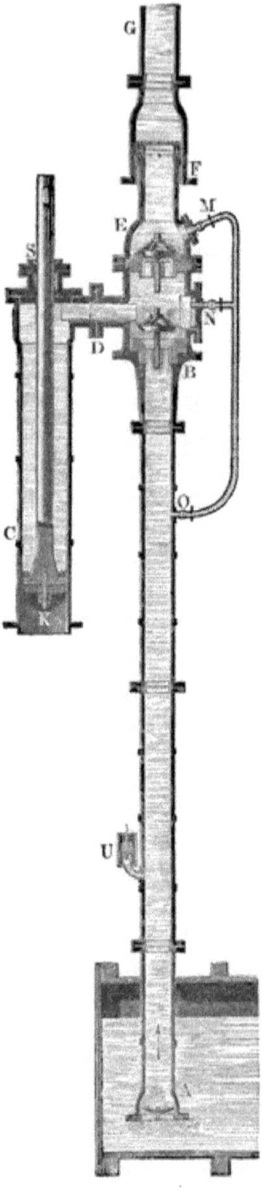

Fig. 62.

the rate of five and a half double strokes per minute, the length of stroke being 2·3 m. [7·54 ft.] The aforesaid water-pressure engine has a fall of 61 m. [200 ft.], and a driving piston 1·2 m. [47·24 ins.] in diameter, while the pump driven by it has a piston 0·422 m. [16·61 ins.] in diameter, and lifts the water to a height of 230 m. [750 ft.] In the figure, AB is the suction pipe, which has a diameter of 0·275 m. [10·83 ins.], B the chamber for the suction valve, E that for the delivery valve, and FG the lower portion of the delivery pipe. The entrance to the suction pipe is about $\frac{1}{2}$ m. [1·64 ft.] below the lower water-level, and is provided with two clacks A, acting as foot valves. The height of suction is about 6 m. [19·7 ft.], and consequently the height of delivery is 230 − 6 = 224 m. [734·7 ft.] Through the short pipe D the valve chamber B communicates directly with the pump barrel CD, which is open at the bottom, and, like the piston head and rod KS, is made of brass. In order to make both piston and piston rod air- and water-tight, the former has leather packing both at the top and bottom, and the latter is guided through a stuffing box S, filled with leather discs. The connection between the delivery pipe and the valve chamber E is also effected by a leather-packed joint, so that, after unscrewing the bolts, the distance pipe E may be pushed upward, and the chamber B removed whenever it is necessary to put in new valves.

The pump is also provided with a small pipe MNO, by means of which, without opening the pump valves, the space between the latter can be connected with the delivery pipe as well as with the suction pipe, and thus the whole apparatus be filled with water before starting. If the delivery pipe is full, and the cocks M and O, as well as the pet-cock that connects the chamber B with the atmosphere, be opened, then the water will flow by way of MO into the suction pipe AB, and the air enclosed in the latter will lift the valve and flow through the pet-cock into the atmosphere. To fill the pump cylinder and valve chamber BE with water, the cock N is also opened and kept so until water flows out of the pet-cock. Finally, there is attached to the suction pipe a small branch pipe closed by a sort of safety valve by which leakage of the valves is indicated. When the suction valve leaks the suction pipe will be in communication with the delivery pipe during the up stroke of the piston, and, as a consequence, the excess of pressure in the suction pipe will raise the test valve U, and allow water to flow out. The same would be the case when the delivery valve leaks, provided the cocks N and O are opened when the pump is at rest.

As long as it is perfectly filled with water, the pump runs smoothly, the piston speed at the beginning of each up and down stroke being equal to zero, then gradually increasing until about half stroke, and finally diminishing until it again becomes zero. But if air has collected in the valve chamber BE in consequence of leakage through the packing, then the delivery valve fails to open until the air has been compressed to a certain degree and the piston has assumed a considerable velocity. The whole mass of water in the delivery pipe must then suddenly assume this velocity, with the result that a severe shock is experienced by the apparatus.

§ 23. **Suction and Force Pumps.**—The *force pumps* press the water upward during the down stroke of the piston, and consequently they cannot be directly driven by long piston rods, which would be apt to bend; *they are therefore employed only when the pump is near the prime mover, or when the water can be pressed upward by the weight of the descending rod.* At the present day *plungers,* as at A, Fig. 63, are mostly employed in force pumps, owing to the ease with which their packings can be maintained. In the pump here shown the delivery valve C is composed of one, and the suction valve B of two, inclined clacks. During the up stroke of the plunger the pressure in the pump barrel is below that of the atmosphere, and, as a consequence, the water drawn in is partially

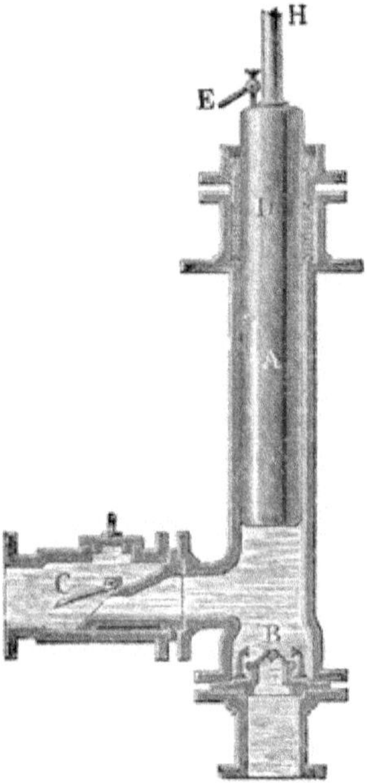

Fig. 63.

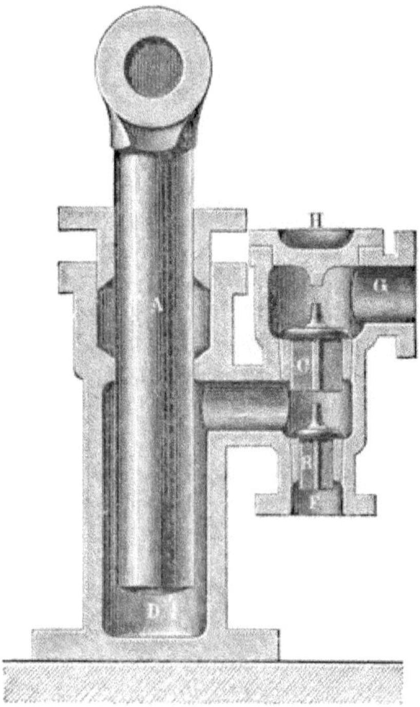

Fig. 64.

freed from the air it contains. To prevent the air from accumulating in the cylinder, which would materially reduce or perhaps destroy the suction, the diameter of the pump barrel is made no larger than that of the plunger. As, however, a small quantity of air is liable to collect under the stuffing box, the plunger is, for its removal, provided with a narrow passage D, which connects the interior of the pump with the atmosphere whenever the cock E at the upper end is open. Evidently this cock must be opened only during the down stroke of the piston if the air in the cylinder is to be removed, for in case it were open during the up stroke merely air would be sucked in instead of water. Occasionally an air-cock of this description is placed on the pump barrel, and it is then possible by opening the cock to destroy the action of the pump without interrupting the motion of the piston. A simple means of preventing air from collecting in the barrel consists in placing the valve chest close to the stuffing box, as in the boiler feed-pump shown in Fig. 64. Here no space exists for the air to collect in, for as soon as it is freed it passes through the delivery valve C to the delivery pipe G. The water drawn from the suction pipe F through the suction valve first fills the cylinder D, and then, during the down stroke of the piston, is forced through the delivery valve to the delivery pipe. Consequently the piston A must not exactly fill the cylinder D, since there must be sufficient space between the piston and the wall of the cylinder to enable the water to pass from the latter to the valve chamber. A special advantage of the construction just given is, that by removing the cover H both valves are at once accessible, the delivery valve C being so large that the suction valve B below it can be taken out through the seat of C. Also in leading the suction pipes from the pump-well it is necessary to carefully avoid leaving high places in the pipes where air can collect, the best plan being to *so arrange the suction pipe that it shall everywhere slope upward toward the pump* in order that the air may be continually removed by the latter.

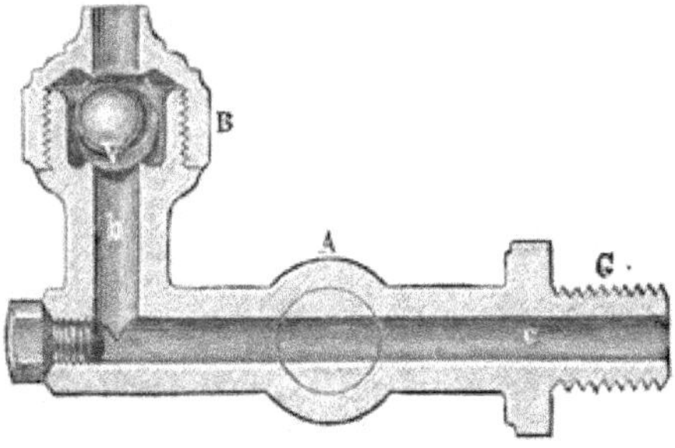

Fig. 65.

For the automatic and continual removal of air from a pump barrel, the *air-escape* designed by *Reuleaux* can be used to advantage. This device is shown in Fig. 65, and consists essentially of a double-seated ball valve V, having a lift of but 3 to 4 mm. [0·12 to 0·16 in.] The apparatus is attached to the pump by means of the screw C, and its cock A is left open. On its inward stroke the piston drives the air from the pump barrel into the pipe *cb*, and this air lifts the ball valve V from its lower seat, the ball falling back to the seat during the return stroke. On the following inward stroke the ball is again lifted, and thus receives an oscillating motion, which continues until the pipe *cb* becomes filled with water, which

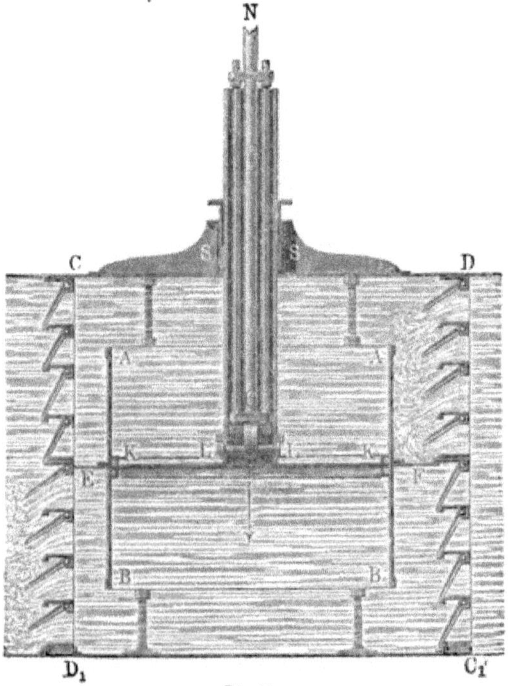

Fig. 66.

will keep the ball pressed against its upper seat during both strokes of the piston.

§ 24. **Double-Acting Pumps.**—The so-called *box pumps*, designed by the Dutch engineer *Fynje*, have been employed to advantage for draining low lands. They differ from ordinary pumps in having their valves located outside the barrel on a separate box. The vertical section of such a double-acting box pump is given in Fig. 66. Here AB is the pump barrel, open above and below, and resting on deck beams, KK is a solid piston packed with rings of wood, and CDC_1D_1 is a sheet-iron box surrounding the pump barrel, and divided into two chambers by a partition EF fitting around the barrel. The piston rod NO is connected to the piston by a pin LL, and is surrounded by a pipe rigidly attached to the piston head, and passing through a stuffing box S in the cover CD of the pump box. The open sides CD_1 and C_1D of this box are provided with cast-iron frames, on which the cast-iron valves, covered with wood, are hung inclined. On one side C_1D the space enclosed by the box is in communication with the lower water-level, and on the other CD_1 with the upper level, consequently the valves hanging at the first-mentioned side are made to open inward, while those on the other side open outward. This construction offers the great advantage of a large valve area for admission and discharge of the water, the velocity of which in the valve passages need be no greater than that of the piston. The speed of the latter can therefore be taken much greater than in ordinary cylinder pumps, in which the sectional area of the valve passages is only a small fraction of the piston area.

The piston therefore is allowed to work with an average velocity of 1·5 m. [5 ft.] per second.[1]

Another advantage peculiar to this type of pumps, and one which renders them particularly adapted for draining low lands, which always involves the raising of large quantities of water to small heights, lies in the fact that they always lift the water to a height h, representing the exact difference between the upper level A (Fig. 67) and the lower B, without being affected by the variations of the two levels. This is accomplished by placing the pump box K in a break in the dam enclosing the low land to be drained, and at a sufficient depth to cause the lowest position of the level B to remain above the valves. In this manner all unnecessary lift is avoided, since the loss which is caused in some pumps by elevating the water into a conduit placed at sufficient height above the upper water-level to discharge into it does not here exist. This advantage is of particular importance in drainage works, where every unnecessary lift, no matter how small, even that due to the depth of the discharging jet or stream, causes a noticeable loss of work. An

[1] See article by Krüger in Erbkam's *Zeitschrift für das Bauwesen*, 1858, "Die Trockenlegung von Ländereien und die Kastenpumpen"; also *Polytechn. Centralblatt*, 1858.

excellent plant of above description, constructed by *Berg* for draining the Bremer Blocklands, is described in his pamphlet: *Die Entwässerung des Blocklandes.*

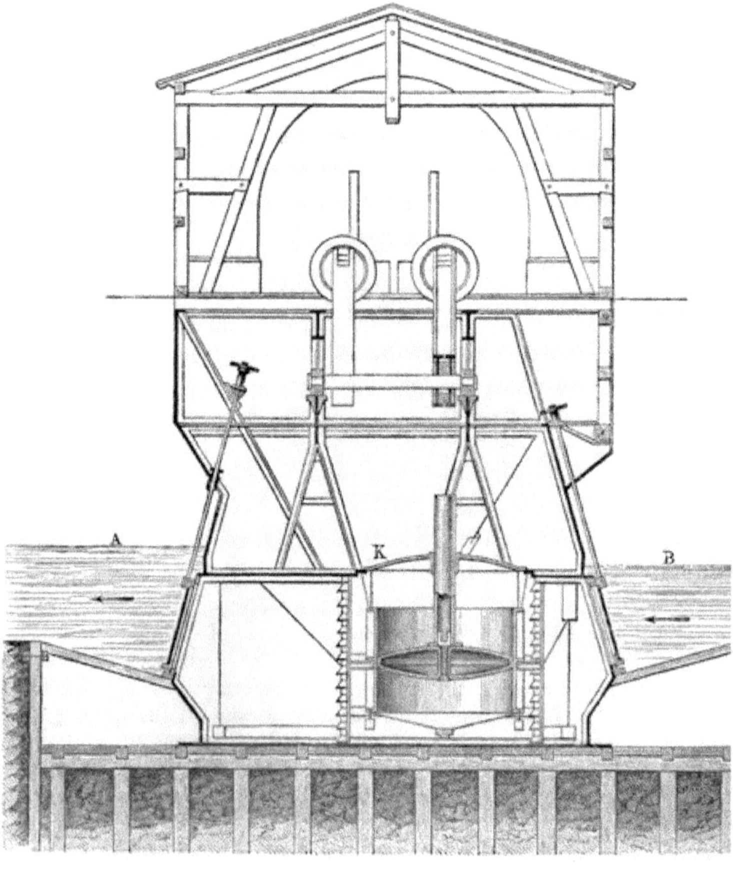

Fig. 67.

A double-acting pump of more ordinary construction is shown in Fig. 68. The pump barrel C has a diameter of 0·135 m. [5·30 ins.], and is cast in one piece with the valve chests and the short pipes B and D for admitting and discharging the water. The piston rod is operated by a rotating crank by means of a connecting rod L, the crank having a length of 0·1 m. [3·94 ins.], and consequently the piston K a stroke of 0·2 m. [7·88 ins.] According to the position of the two-way cock H, the water can be led to the pump either through the pipe A or another pipe A_1 leading from some other reservoir. The valve chambers are accessible from above,

Fig. 68.

through hand holes provided with removable covers. During the up stroke of the piston the valves a and d are open, and during the down stroke the valves a_1 and d_1; water is therefore drawn in at each stroke and forced into the delivery pipe DS.

CHAPTER III

THEORY OF RECIPROCATING PUMPS

§ 25. **Quantity of Water Delivered.**—If F designates the area and s the stroke of the piston, then the theoretical quantity of water lifted by a *single-acting pump* per double stroke is

$$V = Fs;$$

consequently when there are n double strokes per minute the quantity discharged per second will be

$$Q = \frac{n}{60} Fs = \frac{Fns}{60} = \frac{Fv}{2},$$

where $v = \frac{2ns}{60} = \frac{ns}{30}$ represents the average piston velocity.

On the other hand, in a *double-acting* pump we have

$$Q = \frac{n}{60}\, 2Fs = F\,\frac{2ns}{60} = Fv.$$

But the quantity of water actually lifted is much smaller than the theoretical quantity corresponding to the space swept through by the piston (sometimes called *piston displacement*), because even in the most perfect pumps an assignable quantity of water runs back during each stroke. This is partly due to leakage at the packing and valves, and partly to the gradual and not instantaneous closing (slip) of the valves. If a small leak exists in the packing or valves, the water will flow back through it with a velocity $w = \sqrt{2gh}$, where h is the head or lift; if f is the area of the opening, the quantity of water thus lost per second will be

$$q = fw = f\sqrt{2gh},$$

and the loss relatively to the quantity discharged

$$\frac{q}{Q} = \frac{f\sqrt{2gh}}{Fv}.$$

Consequently the loss of water due to leakage at the packing and valves becomes greater the smaller the area F and velocity v of the piston, and the greater the lift h. For this reason pumps which are less perfectly made are run faster than the more perfect ones, and several smaller pumps are employed in preference to a single machine that lifts the water to the same height as the several combined. In plunger pumps leakage at the piston has the additional disadvantage that during suction air is drawn into the pump barrel through the packing, thus preventing the cylinder space from being filled with water.

The quantity of water that runs back during the closing of a valve can be determined approximately as follows. If V_1 is the volume and ϵ the density of a body, and γ the weight of a unit volume of water, then the body will fall under water with the acceleration

$$p = \frac{\text{force}}{\text{mass}} = \frac{V_1\epsilon\gamma - V_1\gamma}{V_1\epsilon\gamma} g = \frac{\epsilon - 1}{\epsilon} g,$$

and such is also the case with an open valve which on both sides is acted on by water of the same pressure. If s_1 is the vertical drop of the valve and t_1 the time of its fall, we have

$$s_1 = \frac{pt_1^2}{2} = \frac{\epsilon - 1}{\epsilon} \frac{gt_1^2}{2},$$

and therefore

$$t_1 = \sqrt{\frac{2\epsilon}{\epsilon - 1} \frac{s_1}{g}}.$$

Now if F_1 is the sectional area of the passage when the valve is open, and if we assume its average value for the time t_1 to be $\frac{1}{2}F_1$, then the quantity of water running back in this time will be

$$V_1 = \tfrac{1}{2}F_1 w t_1 = \tfrac{1}{2}F_1 \sqrt{2gh}\sqrt{\frac{2\epsilon}{\epsilon - 1} \frac{s_1}{g}} = F_1\sqrt{\frac{\epsilon}{\epsilon - 1} h s_1},$$

and its ratio to the total quantity discharged will be

$$\frac{q_1}{Q}=\frac{V_1}{V}=\frac{F_1}{Fs}\sqrt{\frac{\epsilon}{\epsilon-1}hs_1}=\sqrt{\frac{\epsilon}{\epsilon-1}}\,\frac{F_1}{F}\,\frac{\sqrt{hs_1}}{s}.$$

This loss increases directly with the ratio of the areas $\frac{F_1}{F}$, with the head h and with the lift of the valve s_1, but inversely as the piston stroke s. For this reason it is well to employ small valve openings, smaller heads, and greater piston strokes, and above all the valve lift should not be unnecessarily large. On this account therefore multiple valves, for instance, as shown in Figs. 47 to 50, are to be recommended, for with these the requisite valve lift is small.

In the best pumping plants executed these losses amount to 5 per cent, in fairly good pumps to 10 per cent, and in many cases they reach 15 per cent and more of the theoretical or geometrically determined quantity of delivery $V = Fs$. It is therefore advisable, in order to provide a safe margin, to place $V = \mu Fs = 0{\cdot}85Fs$, or in other words to assume a *coefficient of discharge* $\mu = 0{\cdot}85$. Under this supposition we obtain for the quantity discharged per second by *single-acting* pumps

$$Q=\frac{\mu nFs}{60}=\frac{\mu Fv}{2}=0{\cdot}425Fv,$$

and consequently for the area of piston corresponding to a required discharge Q

$$F=\frac{2Q}{\mu v}=2{\cdot}353\,\frac{Q}{v},$$

and the necessary diameter of piston, therefore, is

$$d=\sqrt{\frac{4F}{\pi}}=1{\cdot}128\,\sqrt{F}=1{\cdot}731\sqrt{\frac{Q}{v}}\ \text{m. [ft.]}$$

On the other hand, in *double-acting* pumps we have

$$Q=\mu\frac{2nFs}{60}=\mu Fv=0{\cdot}85Fv,$$

therefore

$$F=\frac{Q}{\mu v}=1{\cdot}176\,\frac{Q}{v},$$

and

$$d=\sqrt{\frac{4F}{\pi}}=1\cdot128\sqrt{F}=1\cdot224\sqrt{\frac{Q}{v}}\text{ m. [ft.]}$$

The average piston speed v and the number n of double strokes depend on various items, namely, on the ratio of the sectional area of the pipes to that of the pump barrel, and on the height of suction, concerning which further information will be given in the following paragraph. Usually the velocity of the piston is not over 0·4 m. [1·31 ft.], the common value ranging between 0·2 and 0·3 m. [0·66 and 1 ft.], though piston velocities of 1 m. [3·28 ft.] per second sometimes occur, for instance in mine pumps.[1]

EXAMPLE 1.—If a pump valve of brass whose specific gravity is $\epsilon=8\cdot5$ has a lift of 0·03 m. [0·0984 ft.], the stroke of the piston being $s=1\cdot2$ m. [3·936 ft.], the head $h=12$ m. [39·36 ft.], and the ratio of the sectional areas $\frac{F_1}{F}=\frac{1}{3}$, then the quantity of water running back through the valves is

$$q_1=\sqrt{\frac{\epsilon}{\epsilon-1}}\,\frac{F_1}{F}\,\frac{\sqrt{hs_1}}{s}\,Q=\sqrt{\frac{8\cdot5}{7\cdot5}}\times\frac{1}{3}\times\frac{\sqrt{12\times0\cdot03}}{1\cdot2}\,Q=0\cdot106\ Q,$$

i.e. more than $10\frac{1}{2}$ per cent of the theoretical discharge.

EXAMPLE 2.—If a single-acting pump has an average piston velocity $v=0\cdot3$ m. [0·984 ft.], and is to lift the quantity $Q=25$ litres [0·883 cub. ft.], it will require a diameter of piston

$$d=1\cdot731\sqrt{\frac{Q}{v}}=1\cdot731\sqrt{\frac{0\cdot025}{0\cdot3}}=0\cdot5\text{ m. [1·64 ft.]},$$

and if the water is to move with a velocity of 1·2 m. [3·94 ft.] in the suction and discharge pipes, the necessary diameter of these pipes will be

$$d_1=d\sqrt{\frac{v}{v_1}}=d\sqrt{\tfrac{1}{4}}=\frac{d}{2}=0\cdot25\text{ m. [0·82 ft.]}.$$

§ 26. **The Suction.**—To ensure the ascent of the water in the suction pipes it is not only necessary that the height of suction should not exceed a certain limit, but the sectional area of the suction pipes should not fall below a certain minimum value. To determine these relations let b again represent the height of the water barometer and h_1 the height of suction or the distance between the lower water-level and

[1] Compare note on p. 57.

the average piston position. This suction height h_1 must, of course, always be less than the height of the water barometer, partly because of the resistances to motion of the water in the suction pipe and partly because a certain head must be available to accelerate the water continually entering the suction pipe, and cause it to follow the motion of the piston.

Let F be the area of the piston, and let us assume the usual case, that the piston is driven by a crank of length r, which makes the piston stroke equal to $2r$, and let this crank make n revolutions per minute. Then the velocity of the crank pin is given by $u = \frac{n}{60} 2\pi r$, and that of the rising piston at any instant by $v = u \sin a$, where a is the angle of rotation of the crank estimated from the lower dead centre. The piston begins its motion at the lower end of the stroke with a velocity $v = 0$, and its acceleration for any crank angle a is expressed by (see Weisbach's *Mechanics*, vol. iii., 1, ch. 6)

$$p = \frac{\partial v}{\partial t} = u \cos a \frac{\partial a}{\partial t} = \frac{u^2}{r} \cos a,$$

consequently the acceleration at the beginning of the up stroke is

$$p_0 = \frac{u^2}{r}.$$

Now if f is the sectional area of the suction pipe, the velocity and consequently the acceleration in the pipe at any instant must be greater than that of the piston in the ratio $\frac{F}{f}$; hence the velocity v_1 of the water in the suction pipe is expressed by

$$v_1 = \frac{F}{f} u \sin a,$$

and the acceleration by

$$p_1 = \frac{F}{f} \frac{u^2}{r} \cos a,$$

if the condition is to be satisfied that the water shall always remain in contact with the piston. A driving force must therefore act upon the water sufficiently great to impart to it at every instant this acceleration. Otherwise the piston will move ahead of the water at the beginning of the motion, when its acceleration $p_0 = \frac{u^2}{r}$ is greatest, *i.e.* a *separation* will take place of the piston and the water. As a consequence, a shock will occur during the upper half of the up stroke when the water has caught up with the retarded piston. Such shocks would interfere with the smooth running of the pump and cause severe stresses in all its parts, and therefore they must be carefully avoided. In order that the piston may not separate from the water at the dead centre when it has its maximum acceleration $\frac{u^2}{r}$, the force driving the water up the suction pipe must always be at least sufficient to generate the acceleration $\frac{F}{f}\frac{u^2}{r}$. Now if l_1 represents the whole length of the suction pipe, and consequently $fl_1\gamma$ is the weight of water in this pipe, then the acceleration imparted to this mass at the first instant of motion by the excess $f(b-h_1)\gamma$ of atmospheric pressure is given by

$$\frac{\text{force}}{\text{mass}} = \frac{f(b-h_1)\gamma}{fl_1\gamma}g = \frac{b-h_1}{l_1}g.$$

The limit at which no separation will occur at the dead centre is therefore given by the equation

$$\frac{b-h_1}{l_1}g = \frac{F}{f}\frac{u^2}{r} = \frac{F}{f}\left(\frac{n \times 2\pi}{60}\right)^2 r = 0{\cdot}011\,\frac{F}{f}n^2 r \qquad (1),$$

from which we obtain the maximum permissible number of revolutions of the crank per minute,

$$n_{max} = \frac{30}{\pi}\sqrt{\frac{f}{F}\frac{b-h_1}{rl}g} = 9{\cdot}554\sqrt{\frac{f}{F}\frac{b-h_1}{rl}g} \qquad (2),$$

or the minimum sectional area of the suction pipe

$$f_{min} = \left(\frac{\pi}{30}\right)^2 Fn^2 r\frac{l_1}{g(b-h_1)} = 0{\cdot}011 Fn^2 r\frac{l_1}{g(b-h_1)} \qquad (3).$$

If instead of the piston area F we have given the theoretical discharge Q_0 per minute, which in a single-acting pump is

$$Q_0 = nF2r,$$

we have only to substitute $\frac{Q_0}{2nr}$ for F in the above expression and then obtain

$$f_{min} = \left(\frac{\pi}{30}\right)^2 \frac{n}{2} Q_0 \frac{l_1}{g(b-h_1)} = 0{\cdot}0055 n Q_0 \frac{l_1}{g(b-h_1)} \qquad (4).$$

On the other hand, in a double-acting pump we have

$$Q_0 = 2nF2r,$$

and substituting $\frac{Q_0}{4nr}$ for F, we obtain the minimum area of the suction pipe,

$$f_{min} = \left(\frac{\pi}{30}\right)^2 \frac{n}{4} Q_0 \frac{l_1}{g(b-h_1)} = 0{\cdot}00275 n Q_0 \frac{l_1}{g(b-h_1)} \qquad (4^a).$$

From this we see that for a given discharge Q_0, other things being equal, the permissible number of revolutions n is directly proportional to the sectional area f; therefore, the smaller the suction pipe the smaller must be the number of revolutions of the pump, and the greater, of course, the volume $\frac{Q_0}{n} = F2r$ of the pump barrel. If the suction pipe of a single-acting pump has an area of cross-section f, which at least equals the value determined from (3) or (4), there will be *no separation of the piston from the water at the dead centre;* the question then arises whether such a separation may not take place *afterwards during the up stroke* of the piston. As the acceleration $\frac{u^2}{r}\cos a$ of the piston continually diminishes during the first quarter of a turn of the crank, becoming zero for $a = 90°$, a separation of the piston is only possible if the acceleration imparted to the water by the atmospheric pressure *decreases more rapidly* than the piston acceleration. Now, the general expression for the acceleration of the water at the angle a, where the velocity of the water in the suction pipe is already $\frac{F}{f} u \sin a$, is

$$p_w = \frac{\text{force}}{\text{mass}} = \frac{b - h_1 - \phi - \left(\frac{F}{f}\right)^2 \frac{u^2 \sin^2 a}{2g}}{l_1} g;$$

for the excess $b - h_1$ of atmospheric pressure is expended partly in overcoming the frictional resistances in the suction pipe, represented by a head of water ϕ, and partly in imparting the velocity $\frac{F}{f} u \sin a$ (already existing in the suction pipe) to the water continually entering the pipe, the head needed for this purpose being $\left(\frac{F}{f}\right)^2 \frac{u^2 \sin^2 a}{2g}$. On the other hand, the acceleration which the water must possess in order to keep up with the piston is $p_k = \frac{F}{f} \frac{u^2}{r} \cos a$. Therefore, if p_w is always to be greater than p_k, the decrement of p_w, *i.e.* the absolute value of ∂p_w, must always be smaller than the decrement of p_k, or the absolute value of ∂p_k. Hence we have the condition

$$\partial p_w \leqq \partial p_k,$$

or

$$\left(\frac{F}{f}\right)^2 \frac{2u^2 \sin a \cos a}{2l_1} \partial a \leqq \frac{F}{f} \frac{u^2}{r} \sin a \partial a, \textit{ i.e. } Fr \cos a \leqq fl_1 \qquad (5).$$

This condition is always fulfilled under ordinary circumstances, for the quantity of water fl_1 in the suction pipe is always considerably larger than the half cylinder volume Fr. Consequently *no separation of the piston from the water can take place except at the beginning of the stroke*, and no separation whatever will occur if we satisfy condition (4), which requires that the sectional area f of the suction pipe shall increase directly as its length l_1.

When this length is considerable the area of the suction pipe may evidently become inconveniently large. To avoid this difficulty, it is advantageous to apply *an air chamber to the suction pipe*, as close to the pump as possible. The action of such an air chamber will be understood from the following. Let us suppose the chamber W (Fig. 69) to be inserted in the suction pipe S, a certain quantity of air being contained in the space L; then it is evident in the first place that when the pump is at rest the tension of this air must be less by the head $AB = h'$ than that of the outer atmosphere represented by a water column of a height $AC = b$. If therefore b_1 denotes the height of water column corresponding to the air enclosed in the chamber, we have when the pump is at rest,

$$b_1 = b' = BC = b - h'.$$

Now assume the piston K to move upwards from its lowest position, then water from the air chamber will evidently follow it, since the pressure b_1 in the latter will act in the same manner as the atmospheric pressure b when the suction pipe has no air chamber. As a result the air in L is expanded to a pressure BE, for instance, smaller than BC. Owing to this reduction in pressure, water will now enter the suction pipe S, the rate of motion being more rapid the greater the diminution of pressure in L, or in general, the smaller the air chamber in comparison with the volume of water withdrawn from it by the pump piston. In order that the piston shall not separate from the water supplied to it from the air chamber through the pipe H, the condition expressed by equation (4) must again be complied with, *i.e.* we must have

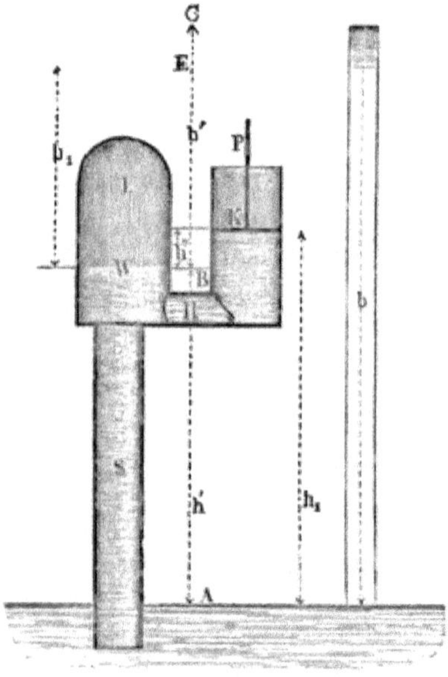

Fig. 69.

$$f'_{\min} = \left(\frac{\pi}{30}\right)^2 \frac{n}{2} Q_0 \frac{l'}{g(b_1 - h'')},$$

where f' is the sectional area and l' the length of the pipe H, while h'' represents the height of suction between the air chamber and pump. Under the supposition that this condition as well as that expressed by (5) is fulfilled, the water will remain in contact with the piston K during the up stroke, and at the end of the stroke the air chamber will have supplied the quantity of water $2Fr$ to fill the cylinder. Now, in the meantime a certain quantity of water must have reached the air chamber from the suction pipe S, though it is easily seen that during the first stroke this quantity will be smaller than that which has flowed out of the air chamber, because at the start the force $b-b_1-h'$, available for acceleration of the water in S, is equal to zero, and it assumes a definite value only when the pressure b_1 in the air chamber becomes smaller, *i.e.* when more water is withdrawn from the latter by the pump than is supplied to it by the suction pipe. After the piston has reached its highest position and reversed its motion no water is withdrawn from the air chamber during the down stroke, but the upward motion already begun in the suction pipe S continues, since the excess $b-h'$ of the atmospheric pressure at the lower water-level is greater than the diminished pressure b_1 in the air chamber. Therefore while the motion of the water in the pipe H between pump and air chamber is *periodical*, the flow into the air chamber through the suction pipe S is *continuous*. It is evident that after a short while a certain normal condition will maintain in the action of the pump, characterised by the relation that the air chamber during a whole revolution of the crank receives the same volume of water $2Fr$ from the suction pipe S as is withdrawn from it by the piston during one half of a revolution. On account of this difference in the supply and discharge of the water, there arise in the air chamber certain periodical fluctuations of pressure, which evidently will be more marked the smaller the volume of the air chamber in comparison with the cubic contents of the pump barrel. By giving a large volume to the air chamber these fluctuations can be reduced, but they cannot be wholly avoided, as that would require the air chamber to be made of infinite size. There is therefore a certain resemblance between the regulating action of air chambers and that of fly wheels.

Assuming that the air chamber is of such large proportions that we may neglect these variations in pressure, we can determine the internal pressure b_1 as follows. Supposing the water to be regularly supplied to the piston so that no separation can take place, then the condition (4) must be fulfilled,

$$f'_{\min}=\left(\frac{\pi}{30}\right)^2\frac{n}{2}Q_0\frac{l'}{g(b_1-h'')} \quad . \qquad (6),$$

where h'' is the height of the piston above the water-level in the air chamber, l' the length, and f' the sectional area of the communicating pipe H.

From this follows

$$Q_0 = \frac{2}{n}\left(\frac{30}{\pi}\right)^2 f'g\frac{b_1 - h''}{l'},$$

and if we here substitute the value for n given by (2), namely,

$$n = \frac{30}{\pi}\sqrt{\frac{f'}{F}g\frac{b_1 - h''}{rl'}},$$

we get

$$Q_0 = \frac{60}{\pi}f'\sqrt{\frac{Fr}{f'}g\frac{b_1 - h''}{l'}}.$$

Now, in order that no separation of the piston from the water may occur, condition (5) $Fr\cos a \leq f'l'$ must be satisfied, and we may therefore place $Fr = f'l'$, then getting

$$Q_0 = \frac{60}{\pi}f'\sqrt{g(b_1 - h'')} \quad . \quad . \quad (7).$$

In the suction pipe S the water moves with a uniform velocity given by $\sqrt{2g(b - b_1 - h' - \phi)}$, when ϕ is the frictional resistance in the pipe, expressed as a head of water; therefore the quantity discharged per minute is also given by

$$Q_0 = 60f\sqrt{2g(b - b_1 - h' - \phi)} \quad . \quad . \quad (8).$$

If we assume the cross-sections of the pipes S and H to be equal, that is, $f = f'$, we get by equating (7) and (8)

$$b_1 - h'' = 2\pi^2(b - b_1 - h' - \phi),$$

from which we deduce the pressure in the air chamber,

$$b_1 = \frac{2\pi^2(b - h' - \phi) + h''}{1 + 2\pi^2} = 0{\cdot}95(b - h' - \phi) + 0{\cdot}05h'' \quad (9).$$

With this value of b_1 we can now determine from (8) the minimum area of the suction pipe,

$$f_{\min} = \frac{Q_0}{60\sqrt{2g \times 0{\cdot}05(b - h' - \phi - h'')}}$$

$$= \frac{Q_0}{13{\cdot}4\sqrt{2g(b - h' - h'' - \phi)}}.$$

As this value represents the *minimum* area of the suction

pipe, it is advisable, for the sake of safety, to choose a larger dimension. If, with *Fink*, we take double the above value, we should have

$$f = \frac{Q_0}{6 \cdot 7 \sqrt{2g(b - h' - h'' - \phi)}} = \frac{0 \cdot 15 Q_0}{\sqrt{2g(b - h_1 - \phi)}} \qquad (10),$$

where, in place of $h' + h''$, the total height of suction h_1 is introduced. The value ϕ of the frictional resistance in the suction pipes (according to Weisbach's *Mechanics*, vol. i. § 7, ch. 3) is given by

$$\phi = \left(0 \cdot 01439 + \frac{0 \cdot 009471}{\sqrt{v}}\right) \frac{l}{d} \frac{v^2}{2g},$$

where d is the diameter of the suction pipe.

When the pump is double-acting, we have from (4^a) the equation

$$f_{\min} = \left(\frac{\pi}{30}\right)^2 \frac{n}{4} Q_0 \frac{l'}{g(b_1 - h'')},$$

which also gives

$$Q_0 = \frac{120}{\pi} f' \sqrt{g(b_1 - h'')} \quad . \qquad (7^a).$$

Equating this and (8), and assuming $f = f'$, we get

$$b_1 - h'' = \frac{\pi^2}{2}(b - b_1 - h' - \phi),$$

from which follows

$$b_1 = \frac{\pi^2(b - h' - \phi) + 2h''}{2 + \pi^2} = 0 \cdot 83(b - h' - \phi) + 0 \cdot 17 h'' \qquad (9^a).$$

With this value of b_1 we now find from (8) the minimum area of the suction pipe, when the pump is double-acting and provided with an air chamber for the suction, to be

$$f_{\min} = \frac{Q_0}{60 \sqrt{2g 0 \cdot 17(b - h' - h'' - \phi)}} = \frac{Q_0}{24 \cdot 74 \sqrt{(b - h' - h'' - \phi) 2g}},$$

or, if here also we double this area,

$$f = \frac{Q_0}{12 \cdot 37 \sqrt{2g(b - h' - h'' - \phi)}} = \frac{0 \cdot 08 Q_0}{\sqrt{2g(b - h_1 - \phi)}} \qquad (10^a).$$

As the air chamber, however, is comparatively small, the

pressure b_1 is not constant, as was assumed in the preceding investigation, but is subject to certain fluctuations, which somewhat modify the action; nevertheless, equations (10) and (10^a), for single- and double-acting pumps respectively, may serve as starting-points when the suction pipe is provided with an air chamber, the volume of which, in accordance with *Fink*, is taken *equal to the volume* F2*r* *swept through by the piston during one stroke.*

On the other hand, when the pump works without an air chamber on the suction pipe, we must determine the diameter of the latter so as to satisfy equations (4) and (4^a) respectively, in order that no shocks may take place during the influx. It should here be noticed that everywhere in the present investigations Q_0 represents the theoretical discharge, $nF2r$ and $2nF2r$ respectively, which, according to § 25, may be placed equal to, say $\frac{Q}{0{\cdot}85} = 1{\cdot}18Q$, where Q is the actual discharge of the pump. Usually this actual discharge Q is given, and then 1·18Q is to be substituted for Q_0 in all the preceding formulæ.

§ 27. **The Suction** (*continued*).—When the piston has attained its maximum velocity $v = u$ at the middle of the stroke, *i.e.* when the crank (always presupposing a very long connecting rod) is 90° from the lower dead centre, the *retardation* of the piston begins; from zero it increases to the value $\frac{u^2}{r}$ at the upper dead centre, and for any crank angle α it is expressed by $\frac{u^2}{r}\cos\alpha$. Now, if the water ascending the suction pipe is to continue to follow the piston without any tendency to *get ahead of* it, it must, by the action of its own weight, be subjected to a retardation not less than $\frac{F}{f}\frac{u^2}{r}$. If the retardation should have a smaller value, p_w for instance, the water will, at a certain crank position, tend to move ahead of the piston. This crank position is given by the equation $p_w = \frac{F}{f}\frac{u^2}{r}\cos\alpha$, where α is the crank angle estimated from the lower dead centre. In consequence of this tendency, the water will exert a pressure on the delivery valve, and cause the latter to open. In this case *water will discharge from the delivery pipe also during the up stroke, and thus the quantity of water lifted will exceed the cubic contents of the pump cylinder.*

In order to investigate under what circumstances this action occurs, we will first assume that the pump is a suction and lift pump, with hollow, valved piston, and that the sectional area f of the delivery pipe is the same as that of the suction pipe. In this case, while the piston is rising, the water in the delivery pipe at every instant moves with the same velocity as that in the suction pipe. Now, if the case of premature opening of the delivery valve occurs, the weight of a water column having a height equal to the *total* head $h_1 + h_2 = h$ will retard the water rising in the pipes, and therefore the retardation will be expressed by

$$p_w = \frac{f(h_1 + h_2)\gamma}{f(l_1 + l_2)\gamma} g = \frac{h}{l} g,$$

where $l = l_1 + l_2$ is the sum of the length l_1 of the suction, and l_2 of the delivery pipe. If this premature opening of the delivery valve occurs at the crank angle a, where the retardation of the piston is $\frac{u^2}{r} \cos a$, we have the equation

$$\frac{F}{f} \frac{u^2}{r} \cos a = -g \frac{h}{l} \quad . \quad . \qquad (11).$$

It is here assumed that no air chamber is applied to the suction pipe; if the reverse were the case, the force retarding the water between the air chamber, where the pressure is b_1, and the point of discharge would be represented by a water column of the weight $f(b - b_1 + h_2)\gamma$. Now, since $b - b_1 = h_1 + \phi$, if ϕ again designates the head due to the resistance in the suction pipe, the retarding force may be expressed by $f(h_1 + h_2 + \phi)\gamma = f(h + \phi)\gamma$, the retarded water being merely $fl_2\gamma$ contained in the delivery pipe. For this case, therefore, the equation corresponding to the premature opening of the piston valve becomes

$$\frac{F}{f} \frac{u^2}{r} \cos a = -g \frac{h + \phi}{l_2} \qquad (11^a).$$

In a force pump with solid piston the delivery valve will not open at the instant when the retardation of the piston has diminished to that of the water in the suction pipe, for at this moment the head of the water h_2 still rests on the valve. The force due to the inertia of the water in the suction pipe will instead push the piston ahead, and the delivery valve will not open until this force is greater than that of the head of water in the delivery pipe.

From equation (11) we get

$$\cos a = -g\frac{frh}{Flu^2},$$

and hence we can determine the angle a at which the piston valve opens. From this moment on the water will pass upward through the piston valve with a retarded motion until its initial velocity $\frac{F}{f}u \sin a$ has been extinguished. The period required for this purpose depends on this initial velocity of the water and the retardation p_w; at all events, the ascent of the water will continue until the crank has passed the upper dead centre, for during the time intervening the retardation of the water will be less than the continually increasing retardation of the piston. Consequently, the suction valve will not at once close when the piston begins its down stroke, but will remain open as long as the ascending motion of the water continues. It may happen that this ascending motion will last beyond the time needed for the crank to move from the upper to the lower dead centre. When this is the case, the *suction valve does not close at all*, since the new up stroke of the piston causes suction. To ascertain under what conditions this state of inactivity of the suction valve will occur, we first determine the time needed, by virtue of the retardation $p_w = \frac{F}{f}\frac{u^2}{r}\cos a$, to destroy the velocity $\frac{F}{f}u \sin a$, and find it to be

$$t = \frac{\text{velocity}}{\text{retardation}} = \frac{u \sin a}{\frac{u^2}{r}\cos a} = \frac{r}{u}\tan a = \frac{30}{\pi n}\tan a,$$

since

$$u = \frac{2\pi r n}{60}.$$

Placing this time equal to that needed by the crank to pass through the angle $360 - a^\circ$, *i.e.* equal to $\frac{360 - a^\circ}{360} \frac{60}{n}$ seconds, we obtain the equation

$$\frac{30}{\pi n} \tan a = \frac{360 - a^\circ}{360} \frac{60}{n};$$

and hence

$$\tan a = \pi \frac{360 - a^\circ}{180} = 2\pi - 0{\cdot}01745a.$$

This equation is satisfied by $a = 102^\circ\ 34'$, and thus we see that *the suction valve of a pump remains inactive* when the retardation p_w of the water caused by the weight of the latter is only $\frac{F}{f}$ times as large as the retardation of the piston when the crank is at the angular distance $a = 102^\circ\ 34'$ from the lower dead centre. If the retardation p_w is greater, the premature opening of the piston valve will not take place until the crank has described a greater angle, and the suction valve will close during the return stroke. No premature opening of the delivery valve will occur when the retardation p_w of the water is equal to or greater than $\frac{F}{f}$ times the piston acceleration $\frac{F}{f}\frac{u^2}{r} g$ at the dead centre.

When there is a premature opening of the delivery valve, a greater volume of water is discharged during every revolution than would otherwise be the case. The quantity discharged can be determined as follows: Let a represent the angle at which the premature opening of the delivery valve begins. While the crank has turned through the angle a, the piston has advanced the distance $r(1 - \cos a)$, and therefore has lifted a volume $Fr(1 - \cos a)$ of water. The water then moves through the open delivery valve with the initial velocity $\frac{F}{f} u \sin a$, under the influence of the retardation $-\frac{F}{f}\frac{u^2}{r}\cos a$.

If we regard this motion as a uniformly retarded one, we find the distance s through which the water moves, according to the law of uniformly retarded motion, to be

$$s = -\frac{\left(\frac{F}{f} u \sin a\right)^2}{2\frac{F}{f}\frac{u^2}{r}\cos a} = -\frac{Fr}{f}\frac{\sin^2 a}{2\cos a}.$$

Therefore the amount of water that passes upward after the delivery valve opens is equal to the volume of a cylinder having the sectional area f of the pipes for base and the length s, that is, equal to the volume $fs = -Fr\frac{\sin^2 a}{2\cos a}$. In this manner we obtain the total quantity discharged during one revolution of the crank:

$$V = Fr\left(1 - \cos a - \frac{\sin^2 a}{2\cos a}\right) \qquad (12).$$

Introducing in this equation the above value $a = 102^\circ\ 34'$, at which the suction valve is entirely out of action, we get

$$V = Fr(1 + 0{\cdot}217 + 2{\cdot}190) = 3{\cdot}407Fr;$$

that is, about 1·7 times as great as the cylinder volume $2Fr$. Of course this increased discharge is accompanied by a correspondingly greater expenditure of work.

The numerical values found for a and V in the preceding investigations are in reality modified by the influence of the frictional resistances of the water in the pipes, which were neglected in the above discussion, as was also the hydraulic resistance due to the imparting of a velocity to the water originally at rest, and entering the suction pipe while the aforesaid upward motion through the delivery valve is taking place. In consequence of these resistances the motion of the water is not uniformly retarded, as was assumed. Nevertheless, the above discussion explains the peculiar result, often observed in practice, that sometimes the *actual discharge of a pump is greater than the theoretical discharge.* It is likewise a fact that has been practically demonstrated that water can be pumped without any suction valve, as is done by the well-known device[1] shown in Fig. 70. It consists of a pipe open at the top and bottom, and provided at the lower end with a delivery valve. By rapidly moving this pipe up and down water can be made to rise and discharge from the spout *c*. During the upward motion the valve is kept closed, and on the downward stroke it is opened by the inertia of the water ascending in the pipe.

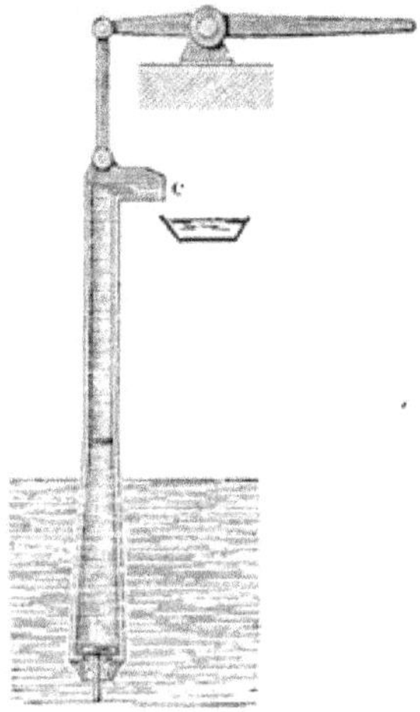

Fig. 70.

A satisfactory investigation and explanation of the occurrences that take place during the suction was, so far as our knowledge goes, first presented by *Fink*,[2] whose investigations have formed the basis of the present article and the following one.

Remark.—The action of the forces of inertia of the water during suction, just analysed, can be clearly shown by a diagram like that employed in Weisbach's *Mechanics*, vol. iii. 1, chap. vi. for the investigation of the crank train. For this purpose let the axis of abscissæ A_1A_2 (Fig. 71) be taken equal to the piston stroke $2r$, and let a parallel B_1B_2 to this axis be drawn at the distance $A_1B_1 = h_1$. Then the constant ordinates included between A_1A_2 and B_1B_2 can be regarded as the load on the piston due to the column of water in the suction pipe. Now if we suppose the force needed to produce the acceleration $\frac{F}{f}\frac{u^2}{r}\cos\alpha$ of the water in the suction pipe to be expressed by the weight of a water column having the height x, we can determine x from

$$fx\gamma = \frac{fl_1\gamma}{g}\frac{F}{f}\frac{u^2}{r}\cos\alpha,$$

which gives

$$x = \frac{Fl_1}{fg}\frac{u^2}{r}\cos\alpha.$$

Laying off this height from every point of the axis A_1A_2, we shall obtain a line C_1C_2, which will represent the forces of inertia of the water in the suction pipe.

[1] Rühlmann, *Allgem. Maschinenlehre*, Bd. 4.

[2] C. Fink, *Theorie und Construction der Kolben- und Centrifugalpumpen*, Berlin; also *Zeitschr. deutsch. Ing.* 1862, p. 177.

It is readily seen that when we assume a very long connecting rod, these forces of inertia will be represented by the straight line C_1C_2 cutting the axis at the middle A_0, and also that the ordinates at the dead centres A_1 and A_2 are given by $\pm \frac{Fl_1}{fg}\frac{u^2}{r}$.

Now, if we suppose the ordinates of the two lines, B for the suction and C for the acceleration, to be combined by drawing through the end B_0 of the ordinate at A_0 the straight line D_1D_2 parallel to C_1C_2, we shall get in the diagram $A_1D_1aD_2A_2$ a graphical representation of the forces acting on the piston, ordinates on opposite sides of the axis A_1A_2, of course, representing forces of opposite directions. At the same time the algebraic sum of the areas of this diagram is a measure of the work which must be performed by the piston during an up stroke, provided the pump has a *solid* piston or *plunger*, and consequently the head h_2 of the

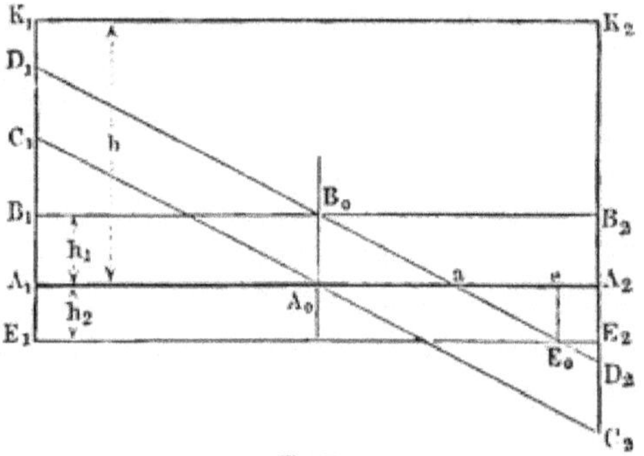

Fig. 71.

pump does not act on the pump piston during the up stroke. Apparently a separation of the piston from the water will take place whenever the ordinate A_1D_1 is greater than the height b of the water barometer. In order that such a separation may not occur, the line D_1D_2, representing the resultant piston pressure, must not cut the atmospheric line K_1K_2 drawn parallel to the base line A at a distance $A_1K_1 = b$. The figure further shows that at the point a, where the base line intersects the line of resultant piston pressures, the driving force exerted by the crank on the piston is equal to zero, and that for the subsequent movement the force of inertia of the water in the suction pipes even exerts a *driving* action on the piston. Now, laying off the head h_2 of the water column resting on the delivery valve equal to A_1E_1, and drawing through E_1 the straight line E_1E_2 parallel to the base line A, we see that at the instant when the piston reaches position e the delivery valve is pressed upward by the force of inertia, so that during the remainder of the movement of the piston from e to A_2 the water rises through the delivery valve, and increases the discharge as calculated above. While the piston travels from a to e it is driven upward by the inertia of the water, the force of which is gradually increasing, and becomes equal to E_0e at the piston position e. This value it retains to the end A_2, while from e onward the excess of the force of inertia is expended in forcing the water through the valve and up the delivery pipe. As the ordinates of the diagram are proportional to the forces exerted on the piston area, we see that the corresponding areas of the diagram can be regarded as measures of the various amounts of work performed. Accordingly, the work transmitted to the piston from the crank during the up stroke is represented by the area A_1D_1a, which may be placed equal to $A_1B_1B_2D_2aA_1$, because the triangles $B_1B_0D_1$ and $B_2B_0D_2$ are equal. Of this work a part

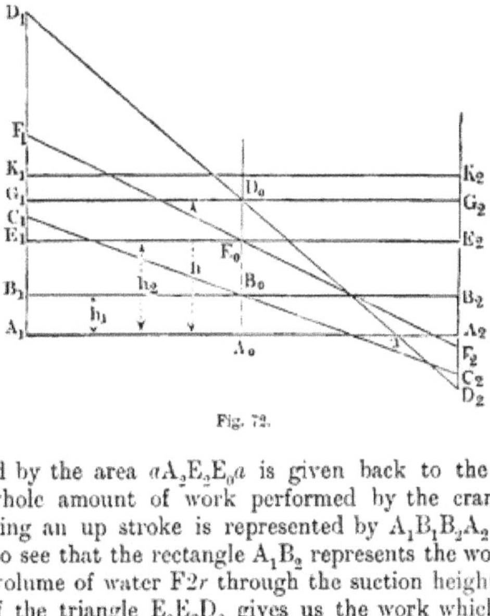

Fig. 72.

represented by the area $aA_2E_2E_0a$ is given back to the crank, so that the whole amount of work performed by the crank on the piston during an up stroke is represented by $A_1B_1B_2A_2 + E_0E_2D_2$. It is easy to see that the rectangle A_1B_2 represents the work needed to lift the volume of water $F2r$ through the suction height h_1, while the area of the triangle $E_0E_2D_2$ gives us the work which must be expended to lift the aforesaid extra water through the whole height $h = h_1 + h_2$. The weight of this extra water is given by $\frac{L_1}{h}$, where L_1 is the work represented by the triangle $E_0E_2D_2$.

The diagram on p. 98 applies to a suction and force pump, *i.e.* to a pump with solid piston; for a suction and lift pump, with hollow, valved piston, the diagram looks somewhat different, for during the up stroke a resistance corresponding to the whole lift $h = h_1 + h_2$ must be overcome. In Fig. 72 is shown the diagram for such a pump, and it will be easily understood after what has preceded. Here also $A_1A_2 = 2r$ represents the base line, and the parallels B_1B_2 and E_1E_2 are drawn at the distances $A_1B_1 = h_1$ and $A_1E_1 = h_2$ from the base. If in the same manner as before we draw the straight lines C_1C_2 and F_1F_2 for the accelerating forces of the water in the suction pipe and in the delivery pipe, so as to have

$$B_1C_1 = \frac{Fl_1}{fg}\frac{u^2}{r} \text{ and } E_1F_1 = \frac{F}{f}\frac{l_2}{g}\frac{u^2}{r},$$

we first get the condition under which a separation of the piston from the water will not take place, namely, that the line C_1C_2 corresponding to the height of suction shall not reach the line K_1K_2 representing the atmospheric pressure. If we also draw the line D_1D_2 to represent the resultant pressures, by making $A_1D_1 = A_1C_1 + A_1F_1$ and $A_2D_2 = A_2C_2 + A_2F_2$, the intersection a of this line with the base will give the piston position at which the premature opening of the delivery valve must occur, and the area aA_2D_2 again represents the work needed to lift the extra quantity of water. The work performed by the crank on the piston is again represented by the area

$$A_1D_1a = A_1G_1G_2A_2 + aA_2D_2,$$

which two surfaces are to each other as the volume of the pump barrel to the volume of the extra quantity of water. For example, if the point a were so situated that the piston travel A_1a corresponded to a crank angle of 102° 34′, then, according to the above determinations,

$$aA_2D_2 = 0{\cdot}7 \times A_1G_1G_2A_2.$$

EXAMPLE.—A single-acting suction and lift pump, whose piston has a diameter of 0·3 m. [11·81 ins.] and a stroke of 0·6 m. [23·62 ins.] makes 30 revolutions per minute with a height of suction of 6 m. [19·68 ft.]; it is required to determine the cross-section of a suction pipe which shall be free from shocks caused by the water, without the use of an air chamber.

If the whole length of the suction pipe is assumed to be l_1 = 10 m. [32·8 ft.] we have from equation (3)

$$f_{min} = 0{\cdot}011 F n^2 r \frac{l_1}{g(b - h_1)} = 0{\cdot}011 F \times 30^2 \times 0{\cdot}3 \frac{10}{9{\cdot}81(10{\cdot}34 - 6)}$$
$$= 0{\cdot}698F = 0{\cdot}698 \times 0{\cdot}0707 = 0{\cdot}0493 \text{ sq. m. [76·4 sq. in.]}$$

This requires a diameter of pipe $d = \sqrt{\frac{4}{\pi}0{\cdot}0493} = 0{\cdot}251$ m. [9·88 ins.] To avoid the use of such a large pipe let us suppose the suction pipe to be provided with an air chamber whose cubic contents is equal to the volume of the cylinder $F \times 2r = 0{\cdot}0707 \times 0{\cdot}6 = 0{\cdot}0424$ cb. m. [15 cub. ft.] Now, if we give to the pipes a diameter $d = \frac{2}{3}D = \frac{2}{3} \times 0{\cdot}30 = 0{\cdot}20$ m. [7·87 ins.], *i.e.* make the area $f = \frac{4}{9}F$

$=0\cdot0314$ sq. m. [48·5 sq. ins.], we obtain a velocity of the water in the pipe of

$$v=\frac{Q_o}{60f}=\frac{F\times2r\times n}{60f}=\frac{0\cdot0707\times0\cdot6\times30}{60\times0\cdot0314}=0\cdot675\text{ m. [2·21 ft.]},$$

and from this follows the head due to the resistance in the suction pipe,

$$\phi=\left(0\cdot01439+\frac{0\cdot00947}{\sqrt{0\cdot675}}\right)\frac{10}{0\cdot20}\frac{0\cdot675^2}{2\times9\cdot81}=0\cdot026\times50\times0\cdot023=0\cdot030\text{ m. [0·1 ft.]}$$

Now, if the water-level in the air chamber is 0·5 m. [1·64 ft.] below the average piston position, that is, $h'=5\cdot5$ m. [18·04 ft.] and $h''=0\cdot5$ m. [1·64 ft.], we obtain from equation (9) the head representing the average pressure in the air chamber:

$$h_1=0\cdot95(b-h'-\phi)+0\cdot05h''=0\cdot95(10\cdot334-5\cdot5-0\cdot030)+0\cdot05\times0\cdot5=4\cdot588\text{ m.}$$
$$\text{[15·05 ft.]}$$

Now, if the height of delivery or distance of the upper level from the average piston position is $h_2=14$ m. [45·92 ft.], the total lift $h=h_1+h_2=20$ m. [65·6 ft.], and if the whole length of the delivery pipe is $l_2=36$ m. [118·08 ft.], we find the crank angle at which the premature opening of the delivery valve takes place from equation (11^a), or

$$\cos\alpha=-g\frac{h+\phi}{l_2}\frac{fr}{Fu^2}=-9\cdot81\frac{20+0\cdot030}{36}\frac{4}{9}\frac{0\cdot3}{(2\pi\times0\cdot3\times\frac{1}{2})^2}=-0\cdot820.$$

This gives $\alpha=145°$, hence if there were no losses of water the quantity discharged per stroke would equal that given by equation (12)—

$$V=Fr\left(1-\cos\alpha-\frac{\sin^2\alpha}{2\cos\alpha}\right)=Fr\left(1+0\cdot820+\frac{0\cdot573^2}{2\times0\cdot820}\right)=2\cdot02Fr=1\cdot01\times F\times2r,$$

i.e. one per cent larger than the cylinder volume. But on account of leakage at the piston and at the valves, we must, in designing a pump, always assume the actual discharge less than the theoretical one $F\times2r$.

§ 28. **The Forcing Action in Pumps.**—The investigations of the preceding articles only referred to the *sucking* action of the pump piston, and therefore only to the motion of the crank from the lower to the upper dead centre. The second half of the rotation of the crank can be discussed in the same way; during this part of the revolution the piston, which we will here suppose to be *solid*, moves downward, forcing the water that has been raised from the pump-well up the delivery pipe to the height h_2. In order that this latter action may take place, it is necessary that the piston be driven by a force not only equal to the weight of the water column pressing against the piston and having a head h_2, but also capable, hurtful resistances neglected, of imparting to the water in the delivery pipe an acceleration corresponding to the motion communicated to the piston by the crank.

The force needed for this acceleration may be represented by a water column of the height $\frac{Fl_2}{fg}\frac{v^2}{r}\cos a$, where a is the angle of the crank estimated from the upper dead centre, f the sectional area, and l_2 the length of the delivery pipe. The resistance experienced by the piston is therefore equal to the pressure of a water column having the height $h_2 + \frac{Fl_2}{fg}\frac{v^2}{r}\cos a$, and this resistance, like the action of suction, can be represented by a diagram (Fig. 73). Again, make $A_1A_2 = 2r$, and draw E_1E_2 parallel to the base line, at the distance $A_1E_1 = h_2$; also make $E_1D_1 = E_2D_2 = \frac{Fl_2}{fg}\frac{v^2}{r}$, then the ordinates of the straight line D_1D_2 again represent the pressures on the piston. As the water is here directly driven by the piston, no assistance being rendered by the atmospheric pressure, as in the case of suction, the initial pressure A_1D_1 is limited by no condition, since no separation of the water from the piston is possible as long as the latter accelerates the water, *i.e.* while the crank is passing from the dead centre through the third quadrant. In other respects the same remarks apply to this diagram as to the diagram in Fig. 71. While the piston is travelling from A_1 to a, the crank acts on the piston with a force that varies from the initial value A_1D_1 down to zero, and when the latter value is reached, the force of inertia of the water set in motion in the delivery pipe exerts a pulling action on the piston in the direction in which it moves. This pull gradually increases from zero at a to the value B_0b at the position b, which just corresponds to the suction height h_1; from this point on the moving mass of water in the delivery pipe exerts a *constant* driving force B_0b on the piston, and at the same time a certain quantity of water passes through the *prematurely opened suction valve*, for, at position b, the driving force of inertia of the water is equal to the weight of the suction column. The work performed by the crank is again represented by the areas $A_1D_1a - aA_2B_2B_0 = A_1E_1E_2A_2 + B_0B_2D_2$. The resemblance between this procedure and that described in the preceding article is readily perceived from the diagrams. We likewise find that the *delivery valve becomes entirely inactive* when the premature opening of the suction valve takes place at a piston position b, for which the crank is at the angular distance $\alpha = 102° 34'$ from the upper dead centre.

It must be noted, however, that in the proceeding just considered, no attention was paid to the fact that when no air chamber is applied to the suction pipe the water in the latter is entirely at rest at the moment when the premature opening

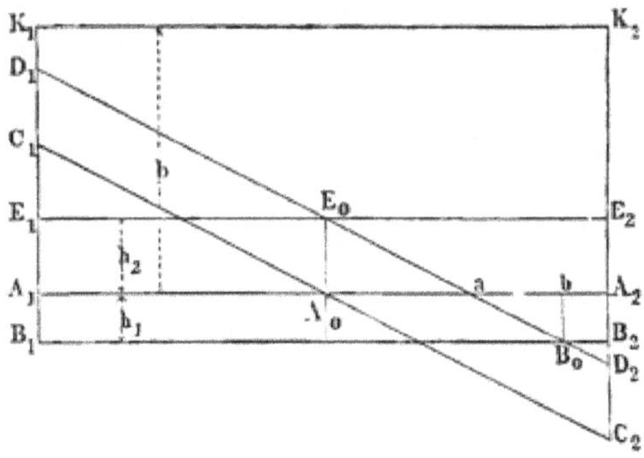

Fig. 73.

of the suction valve takes place. Now, as this water cannot *instantaneously* assume the velocity $\frac{F}{f} u \sin \alpha$ which it must have in order that it may follow the water in the delivery pipe, the latter will move ahead and there will be a separation of the water at the pump. As a result, a shock (water-blow) will occur as soon as the water from the suction pipe catches up with that in the delivery pipe, for the latter, after its living force has been spent, will fall back into the vacuum created by the separation of the water. Such a separation can, of course, only take place when the greatest retardation $\frac{F}{f} \frac{u^2}{r}$ of the piston is greater than that $g \frac{b + h_2}{l_2}$ experienced by the water from the combined action of the delivery head h_2 and the atmospheric pressure on the orifice of efflux.

On the other hand, if the suction pipe is provided with an air chamber, such a separation at the pump, and the accompanying shock, *cannot* take place, provided that, as is generally the case, the quantity of water between the air chamber and the suction valve is comparatively small. In this case there will be an *opening of the suction valve* when the retardation $\frac{F}{f}\frac{u^2}{r}\cos a$ reaches the value

$$g\frac{b+h_2-b_1}{l_2}=\frac{h_1+h_2+\phi}{l_2}=\frac{h+\phi}{l_2},$$

for the water in the delivery pipe is pressed downward by the water column $b+h_2$ and upward by the pressure $b_1=b-h_1-\phi$ in the air chamber, where ϕ again represents the head necessary

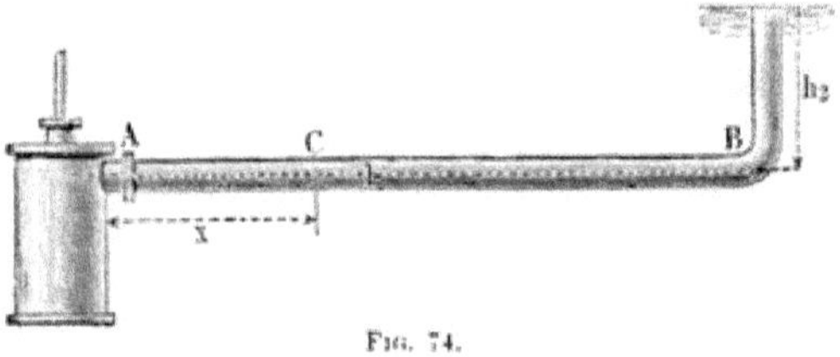

Fig. 74.

to impart to the water taken from the lower reservoir the uniform velocity it possesses in the suction pipe.

But even if a separation of the water be made impossible *at the pump*, for instance by employing an air chamber in the suction pipe, it is liable to occur with the accompanying shock at some point *in the delivery pipe* whenever the retardation at this point, due to the weight of the water and the atmospheric pressure, is smaller than the retardation $\frac{F}{f}\frac{u^2}{r}$ effected by the crank. The possibility of this occurrence depends wholly upon the arrangement of the delivery pipe. In the first place, it is evident that *no such separation can take place* if the delivery pipe is entirely horizontal or entirely vertical. For, in a pump (Fig. 74) with horizontal delivery pipe AB of the length l_2, the retardation at the point A near the pump is

$$p_w=g\frac{b+h_2}{l_2},$$

while for a point C at a distance $AC = x$ from the pump it is

$$p_w' = g\frac{b + h_2}{l_2 - x},$$

which is greater than at A. No separation can therefore take place at the pump, since the condition

$$g\frac{b + h_2}{l_2} > \frac{F}{f}\frac{u^2}{r}$$

is satisfied, and the possibility of separation at any point C will be smaller the farther this point is removed from A. In

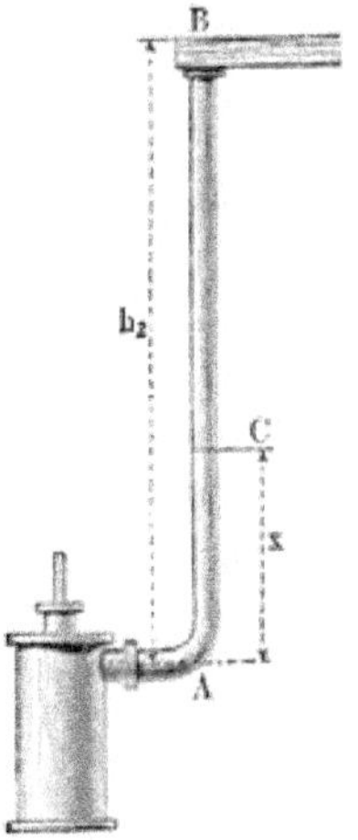

Fig. 75.

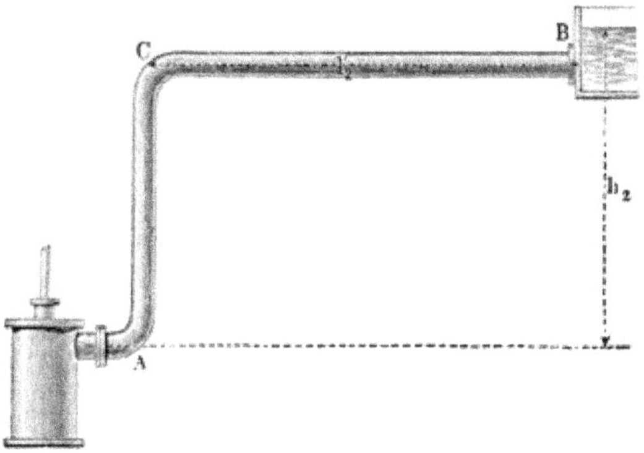

Fig. 76.

like manner we find that with a vertical delivery pipe (Fig. 75) for which $l_2 = h_2$, while at A the retardation is expressed by

$$p_w = g\frac{b + h_2}{h_2},$$

for a point C at the distance x above the pump it is given by

$$p_w' = g\frac{b + h_2 - x}{h_2 - x}.$$

As $\frac{b + h_2}{h_2}$ is an improper fraction, it is easy to see that p_w' is greater than p_w, and that no separation and shock can occur at any point *above* the pump.

On the other hand, if the delivery pipe is inclined or partly horizontal and partly vertical, as in Fig. 76, a shock may occur in the pipe, for instance at the elbow C, if the retardation $p_w' = g\frac{b}{l_2'}$ at this point is smaller than $\frac{F}{f}\frac{u^2}{r}$, which may well happen if the length l_2' of the horizontal portion of the pipe is sufficiently great, even though separation at the pump be impossible because of the fact that the condition

$$g\frac{b+h_2}{l_2} > \frac{F}{f}\frac{u_2}{r}$$

has been satisfied. As the arrangement of the pipes is determined by local considerations, it is always advisable to examine beforehand in each case whether any shock can occur, and, if so, at what point in the pipe conduit.

§ 29. **Air Chamber on the Delivery Pipe.**—Separation of water in the delivery pipe, and the consequent shock, can be *wholly avoided in every case* by placing an *air chamber* on the *delivery pipe* as close to the pump as possible; the pressure b_2 of the air in this chamber maintains an approximately uniform motion of the water in the pipe, here performing the same function as the suction air chamber does at the suction pipe. The periodic motions caused by the reciprocation of the piston are thus, as in the case of the suction, confined to the short piece of pipe connecting the air chamber and the pump. If this air chamber be made sufficiently large, shocks in the delivery pipe, and hence *all shocks* whatever, can be avoided, for we know from what has preceded that separation at the pump can be prevented either by satisfying the fundamental condition

$$g\frac{b+h_2}{l_2} > \frac{F}{f}\frac{u^2}{r}$$

or by employing an air chamber on the suction pipe.

But there are other advantages gained by placing an air chamber on the delivery pipe; one of these is that an approximately uniform discharge is obtained, a matter of importance in fire engines and in pumps that supply fountains; another advantage, due to the regulating action of the air chamber, is constituted by the materially reduced stresses in the machine parts, for example in the piston rod, connecting rod, crank shaft, etc. For, without an air chamber, the resistance opposing the piston at the dead centre would be equal to the pressure of a water column having the height

$$h_2 + \frac{Fl_2}{fg}\frac{u^2}{r},$$

and we see that, when the lengh l_2 of the delivery pipe is considerable, the second term, which depends on the inertia of the water in this pipe, may become many times greater than the delivery height h_2. Consequently all the driving parts of the pump would have to be calculated for this maximum stress at the dead centre, and with long pipes extremely large dimensions would become necessary to prevent fractures. On the other hand, if an air chamber is employed on the delivery pipe, only the inertia of the volume of water contained between this chamber and the pump will resist the motion of the piston during the first part of the down stroke, and this quantity is always very small, inasmuch as the common practice, justified by the above, is always to *place the air chamber as close to the pump as possible.* The water contained between the pump and the air chamber is, of course, removed from the regulating action of the latter, being therefore directly influenced by the motion of the piston.

As in the case of fly wheels, the equalisation of motion attainable by the use of air chambers can never be complete, for, on account of the variable velocity of the pump piston, water alternately enters and leaves the air chamber, thus subjecting the pressure of the air in the latter to continual fluctuations having the same periodicity as the discharge from the pump. It is evident that these fluctuations will be greater the greater the ratio of the volume of water periodically entering and leaving the air chamber to the volume of the air space in the latter. The air chamber will therefore have to be made larger when slight variations of pressure are desired, and the more so the greater the irregularity in the delivery of the pump.

To investigate these relations, let us assume that the velocity v in the delivery pipe beyond the air chamber is *constant*, which assumption, though not strictly correct on account of the variations in pressure of the air, is permissible in the following determinations if the air chamber is of sufficient size.

Taking then the case of a single-acting force pump having the piston area F and the crank radius r, and supposing the crank to have turned through any angle a from the upper dead centre, the piston velocity will be $u \sin a$ and the crank pin velocity $u = \frac{n}{60} 2\pi r$. Now if f is the sectional area of the delivery pipe, then in the element of time ∂t a quantity of water $Fu \sin a . \partial t$ passes from the pump into the air chamber, while the volume $fv . \partial t$ passes from the air chamber into the delivery pipe, and therefore, in the interval under consideration, the difference

$$\partial q = Fu \sin a . \partial t - fv . \partial t$$

of these two quantities is taken up or given out by the air chamber, according as this value is positive or negative. During one complete revolution of the crank, *i.e.* in $\frac{60}{n}$ seconds, the volume $F2r$ is lifted and, when the pump is running normally, passes into the delivery pipe; hence we have

$$fv \frac{60}{n} = F2r,$$

or

$$fv = F \frac{2rn}{60} = F \frac{u}{\pi},$$

which value, substituted in the above formula for water received by the air chamber, gives

$$\partial q = Fu\left(\sin a - \frac{1}{\pi}\right)\partial t.$$

This quantity becomes equal to zero when $\sin a = \frac{1}{\pi}$, *i.e.* for the crank angles

$$a_1 = 18° \ 35' \text{ and } a_2 = 161° \ 25'.$$

While the crank is turning from the dead centre through the angle $a_1 = 18° \ 35'$, water is continually passing from the air chamber into the delivery pipe, as the negative value of ∂q shows, and while the crank is moving from a_1 to $a_2 = 161° \ 25'$ the air chamber receives water; finally, while the angular distance from $161° \ 25'$ to $360°$ is passed over, water again leaves the air chamber. Evidently the latter contains the smallest quantity of water at the crank angle a_1, and the greatest volume at the angle a_2. The quantities in question can be determined from

$$\partial q = \mathrm{F}u\left(\sin a - \frac{1}{\pi}\right)\partial t,$$

by substituting $r\partial a$ for $u\partial t$, and then integrating, which gives

$$q_1 = \mathrm{F}r\int_0^{a_1}\left(\sin a - \frac{1}{\pi}\right)\partial a = \mathrm{F}r\left(1 - \cos a_1 - \frac{a_1}{\pi}\right),$$

and

$$q_2 = \mathrm{F}r\int_0^{a_2}\left(\sin a - \frac{1}{\pi}\right)\partial a = \mathrm{F}r\left(1 - \cos a_2 - \frac{a_2}{\pi}\right).$$

Introducing in these expressions the values $a_1 = 18^\circ\ 35'$, and $a_2 = 161^\circ\ 25'$, we obtain

$$q_1 = -0{\cdot}05\ \mathrm{F}r,$$

and

$$q_2 = +1{\cdot}05\ \mathrm{F}r;$$

hence the quantity of water entering and again leaving the air chamber during each revolution is given by

$$q_2 - q_1 = 1{\cdot}10\ \mathrm{F}r = 0{\cdot}55\mathrm{V},$$

where $\mathrm{V} = \mathrm{F}2r$ is the cubic capacity of the pump barrel.

Knowing this fluctuating quantity of water and the size of the air chamber, we can now determine the variations of pressure in the latter. Let W_2 represent the volume of air in the chamber when it contains the most water, *i.e.* at the crank angle a_2, and let w_2 be the corresponding pressure of the air, then the volume W_1 of the air when the air chamber contains the least water, or at the crank angle a_1, is given by $\mathrm{W}_1 = \mathrm{W}_2 + \nu\mathrm{V}$, where ν is the ratio of the fluctuating volume of water to the volume of the pump cylinder. The minimum air pressure w_1 is then given by *Mariotte's* law

$$w_1 = w_2\frac{\mathrm{W}_2}{\mathrm{W}_2 + \nu\mathrm{V}},$$

since the temperature may here be assumed constant.

When the pump is double-acting the only change to be made in the preceding formulæ is due to the fact that the discharge per revolution is $F \times 4r$, the crank positions for minimum and maximum water contained in the air chamber being now given by the equation $\sin a = \frac{2}{\pi}$, which is satisfied by the angle $a_1 = 39° \ 35'$ and $a_2 = 140° \ 25'$. Substituting these values in the formula

$$q = Fr\left(1 - \cos a - \frac{2a}{\pi}\right),$$

we obtain

$$q_1 = -0{\cdot}210Fr \text{ and } q_2 = 0{\cdot}210Fr,$$

and consequently the quantity of water entering and leaving the air chamber during every *half* of a revolution is given by

$$q_2 - q_1 = 0{\cdot}42 \ Fr = 0{\cdot}21V.$$

If we take the frequently occurring case of two double-acting pumps, with cranks set at right angles to each other, the discharge during a complete revolution is given by $F \times 8r$, and it is evident that the quantity of water that has entered the air chamber, when the crank reaches the angle a from the dead centre, is given by

$$q = Fr\left(1 - \cos a + \sin a - \frac{4a}{\pi}\right).$$

This expression becomes a minimum and a maximum for the respective angles $a_1 = 19° \ 10'$ and $a_2 = 70° \ 50'$, obtained from the equation $\sin a + \cos a = \frac{4}{\pi}$. These angles give

$$q_1 = -0{\cdot}042Fr \text{ and } q_2 = 0{\cdot}042Fr,$$

and hence the fluctuating quantity of water is

$$q_2 - q_1 = 0{\cdot}084Fr = 0{\cdot}042V.$$

This volume enters and leaves the air chamber *four times* during every revolution. The foregoing determinations show that in single-acting pumps the air chamber must be made proportionately larger than in double-acting pumps. *Fink* gives for the volume W of the air space of the chamber, when the air is compressed by the delivery head h_2, the following rules :—

W = 4V for single-acting pumps.
W = 1·6V for double-acting pumps.
W = 0·8V for two double-acting pumps, with cranks at right angles.

When the pump is at rest (the delivery pipe being filled with water) the pressure of the air in the chamber is determined by the sum $h_2 + b$ of the delivery head and the atmospheric pressure. When the pump is at work this pressure is increased by the head ϕ_2, representing the frictional resistances in the delivery pipe. The pressure in the air chamber, when the water is moving uniformly up the delivery pipe, we will call the average pressure, and designate it by w, thus having

$$w = (b + h_2 + \phi_2)\gamma,$$

and the volume of air enclosed at this average pressure we will designate by W. As was shown above, the variable delivery of the pump causes the volume W and its pressure w to change continually, in conformity to the periods of discharge, and therefore the average volume W of the air will alternately increase and decrease. Designating the greatest and smallest volumes of air by W_1 and W_2 respectively, and the corresponding pressures by w_1 and w_2, we have, according to *Mariotte's* law,

$$Ww = W_1 w_1 = W_2 w_2 \quad . \quad (1),$$

and according to the above

$$W_1 - W_2 = \nu V \, . \quad (2),$$

where νV represents the fluctuating quantity of water.

Now the average values of W and w are given by the construction, and enable us to determine approximately the maximum and minimum values as follows. When the maximum pressure w_2 in the air chamber falls to the average w, the work performed while the air expands is expended in *accelerating* the water in the delivery pipe. Neglecting all hurtful resistances due to contraction, etc., this work will, in accordance with well-known laws, be expressed by

$$A_2 = Ww \text{ hyp log } \frac{w_2}{w}.$$

If the pressure is still further reduced from the average w down to w_1, the expanding air will exert a retarding action on the water, which is measured by

$$A_1 = Ww \text{ hyp log } \frac{w}{w_1}.$$

Now, when the pump is running normally, we have $A_2 = A_1$, and this gives

$$\text{hyp log } \frac{w_2}{w} = \text{hyp log } \frac{w}{w_1},$$

or

$$w_1 w_2 = w^2 \quad . \quad . \qquad (3),$$

i.e. *the average pressure* w *may be regarded as a mean proportional of the minimum pressure* w_1 *and the maximum pressure* w_2.

By means of equations (1), (2), and (3) we can now easily determine the maximum and minimum pressures w_2 and w_1 from the given values W, w, and νV.

For, according to these equations we have

$$w_1 = \frac{w^2}{w_2} = w\frac{w}{w_2} = w\frac{W_2}{W} = w\frac{W_1 - \nu V}{W} = w\frac{\frac{Ww}{w_1} - \nu V}{W},$$

from which follows

$$w_1^2 + \nu\frac{V}{W}ww_1 = w^2,$$

and this gives

$$w_1 = -\frac{\nu V}{2W}w + \sqrt{w^2 + \left(\frac{\nu V}{2W}w\right)^2};$$

in like manner we obtain

$$w_2 = +\frac{\nu V}{2W}w + \sqrt{w^2 + \left(\frac{\nu V}{2W}w\right)^2}.$$

We may now proceed, in the manner indicated in the preceding article, to determine whether the minimum air pressure w_1 is sufficient to prevent the water from separating in the delivery pipe near the pump, and, if this is not the case, we must increase the volume W of the air chamber in order to obtain a sufficiently large value for w_1 to exclude the possibility of a waterblow. If the air chamber, as is usually the case, is located as close to the pump as possible, there is generally no danger of a separation of the water.

The values w_1 and w_2, determined above, represent the minimum and maximum air pressures when the pump is *running normally,* but at the moment the pump is *started* a pressure w'' is generated in the air chamber, which exceeds the value w_2, and in the like manner when the pump is *suddenly stopped* the absolutely smallest pressure w' is obtained, as will be shown in what follows. Let us assume that, when the pump is at rest, there exists in the air chamber an average pressure $w_0 = (b + h_2)\gamma$, the chamber then containing the volume W_0 of air. If now the pump be set in motion, the air chamber will receive a volume Q of water per minute, this supply, for the sake of simplicity, being in the following calculation assumed to be furnished at a uniform rate, so that the water will flow with a velocity $v_0 = \frac{Q}{60f}$ in the supply pipe leading to the air chamber. This assumption is not exact on account of the variable delivery of the pump, but is sufficiently so for the purpose of the following investigation. The water delivered by the pump first enters the air chamber, compressing the air and increasing the pressure from w_0 to a greater value w. In consequence of this increase the air in the chamber imparts to the water $fl_2\gamma$ in the delivery pipe an acceleration

$$\frac{\partial v}{\partial t} = g\frac{w - w_0}{l_2\gamma} \quad . \quad . \qquad (4)$$

proportional to the excess of pressure $w - w_0$; in consequence hereof, the water in this pipe is started from its state of rest and gradually acquires the velocity v_0 corresponding to the normal running of the pump. Now if t represents the time that has elapsed from the beginning of the motion to any assumed instant, and v the velocity at this instant, then during the following element of time the quantity of water $fv_0\partial t$ will pass from the pump into the air chamber, and the quantity $fv\partial t$ will leave the air chamber and pass into the delivery pipe. Consequently the volume of air in the chamber will be diminished by the amount

$$\partial q = f(v_0 - v)\partial t \quad . \qquad (5).$$

The whole diminution of volume during the time t is therefore

$$q = fv_0 t - f\int v\partial t.$$

Now, according to *Mariotte's* law, we have

$$\frac{W_0}{W_0 - q} = \frac{w}{w_0},$$

and from this

$$q = W_0 \frac{w - w_0}{w};$$

differentiating, we obtain

$$\partial q = W_0 \frac{w_0}{w^2} \partial w.$$

This equation in combination with (5) gives

$$f(v_0 - v)\partial t = W_0 \frac{w_0}{w^2} \partial w,$$

and multiplying this by (4) we obtain

$$f(v_0 - v)\partial v = W_0 g \frac{w - w_0}{l_2 \gamma} \frac{w_0}{w^2} \partial w.$$

Now when v has reached the value v_0, the pressure w of the air in the chamber will have increased from its initial value w_0 to the required greatest pressure w'', hence, integrating as follows,

$$f\int_0^{v_0} (v_0 - v)\partial v = W_0 w_0 \frac{g}{l_2 \gamma} \int_{w_0}^{w''} \frac{w - w_0}{w^2} \partial w,$$

we obtain

$$f l_2 \gamma \frac{v_0^2}{2g} = W_0 w_0 \left(\text{hyp log} \frac{w''}{w_0} + \frac{w_0}{w''} - 1 \right) \quad . \qquad (6),$$

which equation will enable us to calculate the greatest pressure w'' of the air in the chamber at the start while the pump is acquiring its normal speed.

This value w'' is greater than the initial value $w_0 = b + h_2$, and it also exceeds the average pressure $w = b + h_2 + \phi_2$ that exists when the pump is running normally; it will therefore cause a further acceleration of the water in the delivery pipe, where the velocity accordingly will exceed the average value v_0. As a consequence, water again leaves the air chamber, the pressure falls below w'', and after some fluctuation it assumes an average value w, whereupon the variations of pressure are confined within the limits w_1 and w_2 prescribed by the periodical variations in the delivery of the pump.

On the other hand, when the pump is suddenly stopped, the water in the delivery pipe continues its motion by virtue of its living force, and by thus drawing water from the chamber causes the air space in the latter to increase from W to W′, this action continuing until the living force $fl_2\gamma\frac{v_0^2}{2g}$ of the water is consumed by the work performed by the retarding force. This work of retardation is evidently that required for expanding the air from the volume W_0 and pressure w_0 to the volume W' and pressure w'. This work, for an increment ∂W of the volume W, is expressed by

$$(w_0 - w)\partial W = \left(w_0 - \frac{W_0}{W}w_0\right)\partial W;$$

consequently when the whole living force of the water has been spent we have

$$fl_2\gamma\frac{v_0^2}{2g} = \int_{W_0}^{W'}\left(w_0 - \frac{W_0}{W}w_0\right)\partial W$$

$$= W'w_0 - W_0w_0 - W_0w_0 \text{ hyp log } \frac{W'}{W_0}$$

$$= W_0w_0\left(\frac{W'}{W_0} - 1 + \text{hyp log } \frac{W_0}{W'}\right),$$

or, since $\frac{W'}{W_0} = \frac{w_0}{w'}$,

$$fl_2\gamma\frac{v_0^2}{2g} = W_0w_0\left(\frac{w_0}{w'} + \text{hyp log } \frac{w'}{w_0} - 1\right) \qquad (7).$$

By means of this equation we can calculate the minimum pressure w' that will obtain in the air chamber when the pump is suddenly stopped, while the water in the delivery pipe continues its motion by virtue of its living force. Moreover, it is clear that, after the water has entirely lost its velocity, the pressure w' of the air in the chamber being now smaller than the average pressure w_0, the water will rush back and thus assume an oscillating motion which is only limited by the internal resistances.

The results found above are represented by the diagram (Fig. 77), where, on the horizontal base line AJ, the ordinate JH is made equal to the delivery head h_2, and HL to the height b of the water barometer. Then the sum $JL = AB = h_2 + b$ will

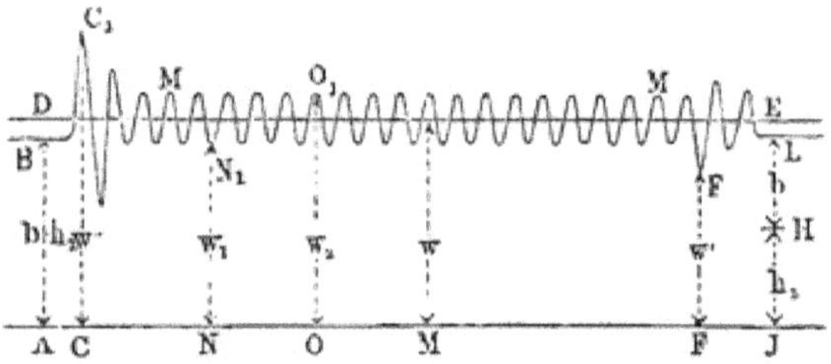

Fig. 77.

represent the pressure w_0 of the air in the chamber when the pump is at rest. At the beginning of motion the pressure will first rise to the greatest value $CC' = w''$ and then sink below its initial value, the oscillations gradually diminishing until the pump attains its normal condition of running, which is represented by the wavy line MM. During this state of motion the pressure in the chamber oscillates periodically between the smaller pressure $NN_1 = w_1$ and the greater $OO_1 = w_2$; the average pressure w is represented by the horizontal line DE, at a distance from the base of $JE = b + h_2 + \phi_2$. When the pump is suddenly stopped the pressure sinks to $FF_1 = w'$, and then, after some undulations of the water, it assumes the value $JL = w_0 = b + h_2$, which obtains when the pump is at rest.

It is evident that the fluctuations between $NN_1 = w_1$ and $OO_1 = w_2$ furnish a measure of the uniformity of the motion of the water, and in fountains are a measure of the fluctuations in the height of the stream. The greatest pressure $CC_1 = w''$ during the starting period forms the basis for determining the stresses to which the separate machine parts are subjected. Finally, the smallest pressure $FF_1 = w'$ occurring while the pump is being stopped determines the minimum volume W' to be given to the air chamber in order that no air may escape when the pressure falls to the lowest limit.

As the quantity of air absorbed by a volume of water is directly proportional to the pressure of the former, the reason is evident why the air gradually escapes from the delivery chambers unless care be taken to make good the loss occasioned by its absorption. On the other hand, air chambers on the suction pipes will never lack the necessary quantity of air, since the water, being saturated with air under atmospheric pressure, gives up a portion of it when subjected to the smaller pressure in the air chamber. In order to keep the chamber on the delivery pipe filled with air it is customary to place a small air-cock on the suction pipe, through which a small quantity of air may be sucked (or "snifted") in by the piston and allowed to rise to the chamber, or, after the latter has been filled, to pass off with the water. By this means a certain elasticity is also given to the water, the air contained in the latter serving to moderate any shocks that may arise.

Finally, as regards the *total* volume of the air chamber, *i.e.* the volume down to the place where the latter is attached to the delivery pipe, it is evident that in fire engines and other pumps which have no delivery head h_2 when at rest, the pressure in the air chamber for the state of rest can only be equal to that of the atmosphere. Not until the engine is at work will a delivery head h_2 exist, being then equal to the height to which, theoretically, the water can be thrown; consequently if, for the average pressure $w = b + h_2 + \phi_2$, we have determined in accordance with the above the volume W of the air chamber, the *total* volume must, according to *Mariotte's* law, be equal to

$$W_0 = W \frac{b + h_2 + \phi_2}{b}$$

On the other hand, in pumps which raise the water through a delivery pipe to a height h_2, the air, for the condition of rest, always has a pressure which is less than $b + h_2$, and it is therefore here sufficient to take the total volume of the air chamber equal to the greatest volume W′ occupied by the air after the pump has been suddenly stopped. It will then be impossible for the air to escape, and therefore, when the pump is again started, there will be a sufficient supply on hand to prevent the maximum pressure from exceeding the calculated value of w''.

Example.—A double-acting force pump has a piston of 0·4 m. [15·75 ins.] diameter, making 12 double strokes per minute, the length of stroke being 1 m. [39·37 ins.]; it forces the water through a delivery pipe having a diameter of 0·25 m. [9·84 ins.] and a length of 100 m. [328 ft.] into a reservoir located 20 m. [65·6 ft.] above the pump. Required, the pressures in the air chamber. In the first place, we have for the pressure w_0 of the air when the pump is at rest

$$w_0 = (b + h_2)\gamma = (10{\cdot}336 + 20)1000 = 30336 \text{ kg. per sq. metre}$$

[43·15 lbs. per sq. in.] Assuming the actual discharge to be 0·85 of the theoretical, we have

$$Q = 0{\cdot}85 \times F \times 2r \times 2n = 0{\cdot}85 \times 0{\cdot}1257 \times 1 \times 24 = 2{\cdot}562 \text{ cub. m. [90·48 cub. ft.]},$$

and consequently the average velocity of the water in the delivery pipe, whose sectional area $f = \frac{\pi}{4} 0{\cdot}25^2 = 0{\cdot}0491$ sq. m. [76·1 sq. ins.],

$$v_0 = \frac{Q}{60 f} = \frac{2{\cdot}562}{2{\cdot}946} = 0{\cdot}870 \text{ m. [2·85 ft.]};$$

from this we find the head, due to the resistance in the pipe, to be

$$\phi = \left(0{\cdot}01439 + \frac{0{\cdot}00947}{\sqrt{0{\cdot}870}}\right) \frac{100}{0{\cdot}25} \frac{0{\cdot}870^2}{2 \times 9{\cdot}81} = 0{\cdot}376 \text{ m. [14·8 ins.]},$$

and consequently the average pressure of the air in the chamber during the motion is

$$w = (b + h_2 + \phi)\gamma = 30712 \text{ kg. [43·68 lbs. per sq. in.]}$$

Now, if we assume the volume of air corresponding to this average pressure to be

$$W = 1{\cdot}6 \times F \times 2r = 1{\cdot}6 \times 0{\cdot}1257 = 0{\cdot}2 \text{ cub. m. [7·063 cub. ft.]},$$

and the fluctuating quantity of water in the double-acting pump to be

$$\nu V = 0{\cdot}21 \times 0{\cdot}1257 = 0{\cdot}0264 \text{ cub. m. [0·932 cub. ft.]},$$

we find the smaller pressure from

$$w_1 = -\frac{\nu V}{2W}w = \sqrt{w^2 + \left(\frac{\nu V}{2W}w\right)^2} = \left[-\frac{\nu V}{2W} + \sqrt{1+\left(\frac{\nu V}{2W}\right)^2}\right]w.$$

If we substitute in this formula the values of ν, V, W, and w, we obtain

$$w_1 = \left[-\frac{0 \cdot 0264}{0 \cdot 4} + \sqrt{1+\left(\frac{0 \cdot 0264}{0 \cdot 4}\right)^2}\right]w = (-0 \cdot 066 + \sqrt{1 \cdot 004356})w$$
$$= 0 \cdot 936w = 28746 \text{ kg. [40·89 lbs. per sq. in.]},$$

and in like manner we obtain for the greater pressure

$$w_2 = \left[+\frac{\nu V}{2W} + \sqrt{1+\left(\frac{\nu V}{2W}\right)^2}\right]w = 1 \cdot 068w = 32800 \text{ kg. [46·65 lbs. per sq. in.]}$$

The pressure of the air therefore fluctuates during the normal running of the pump some $6\frac{1}{2}$ per cent above and below the average pressure w. At the same time the volume of air in the chamber varies between

$$W_1 = \frac{1}{0 \cdot 936}W = \frac{0 \cdot 2}{0 \cdot 936} = 0 \cdot 214 \text{ cub. m. [7·56 cub. ft.]}$$

and

$$W_2 = \frac{0 \cdot 2}{1 \cdot 068} = 0 \cdot 187 \text{ cub. m. [6·60 cub. ft.]},$$

or in all about 27 litres [0·95 cub. ft.]

In order to determine the greatest pressure w'' that occurs during the starting of a pump we make use of equation (6)

$$fl_2\gamma\frac{v_0^2}{2g} = W_0 w_0\left(\text{hyp log}\frac{w''}{w_0} + \frac{w_0}{w''} = 1\right),$$

in which we can substitute $W_0 w_0 = Ww = 0 \cdot 2 \times 30712$. This equation can also be written as follows:

$$\text{hyp log}\frac{w''}{w_0} + \frac{w_0}{w''} = \frac{fl_2\gamma}{W_0 w_0}\,\frac{v_0^2}{2g} + 1 = \frac{0 \cdot 0491 \times 100 \times 1000}{0 \cdot 2 \times 30712}\,\frac{0 \cdot 870^2}{2 \times 9 \cdot 81} + 1$$
$$= \frac{4910}{6142}0 \cdot 0386 + 1 = 1 \cdot 031,$$

and it is satisfied by $\frac{w''}{w_0} = 1 \cdot 3$, which gives for the greatest pressure in the air chamber when the pump is started

$$w'' = 1 \cdot 3 \times w_0 = 1 \cdot 3 \times 30336 = 39437 \text{ kg. [56·09 lbs. per sq. in.]}$$

It is this maximum pressure which must form the basis in determining the dimensions of the parts of the pump. In case no air chamber were employed, the initial piston pressure, and consequently the stresses in all parts of the pump, would be much greater. The resistance experienced by the piston would then be equal to the weight of a water column having the height

$$h_2+\frac{Fl_2\gamma}{fg}\frac{u^2}{r},$$

in which the velocity of the crank pin is

$$u=\frac{2\pi rn}{60}=\frac{3\cdot14\times1\times12}{60}=0\cdot628 \text{ m. [2·06 ft. per sec.]}$$

This would make the height of the equivalent water column equal to

$$20+\frac{0\cdot4^2}{0\cdot25^2}\frac{100}{9\cdot81}\frac{0\cdot628^2}{0\cdot5}=20+\frac{256\times0\cdot394}{4\cdot905}=40\cdot56 \text{ m. [133 ft.]},$$

while the maximum excess of pressure in the air chamber only corresponds to a water column of

$$\frac{w''}{\gamma}-b=39\cdot437-10\cdot336=29\cdot101 \text{ m. [95·45 ft.]}$$

The smallest pressure in the air chamber when the pump is suddenly stopped may in like manner be found from equation (7)

$$\text{hyp log}\frac{w'}{w_0}+\frac{w_0}{w'}=1+\frac{fl_2\gamma}{W_0w_0}\frac{v_0^2}{2g}=1\cdot031.$$

which equation is satisfied by $\frac{w'}{w_0}=0\cdot79$. Therefore the smallest pressure in the air chamber is

$$w'=0\cdot79w_0=0\cdot79\times30336=23965 \text{ kg. [34·09 lbs. per sq. in.]},$$

and in order that the chamber may contain the assumed volume $W=0\cdot2$ cub. m. [7·06 cub. ft.] of air for the pressure w, its total volume must be

$$W'=W\frac{w}{w'}=0\cdot2\frac{30712}{23965}=0\cdot257 \text{ cub. m. [9·08 cub. ft.]}$$

§ 30. **Wasteful Resistances.**—The power required theoretically for driving a pump has already been computed above; we will now take the *wasteful resistances* into account with a view to determine the power actually absorbed.

In pumps with *hollow, valved pistons* the theoretical force necessary to *raise the latter* (see § 14) is

$$P=Fh\gamma;$$

as, however, the packing is pressed by the water against the inner surface of the pump barrel with a force represented by the head $b + h_2 - (b - h_1) = h_1 + h_2 = h$, it is evident that, as in water-pressure and steam engines, the force needed to overcome the piston friction will be

$$W = 4\phi \frac{b}{d} F h \gamma,$$

and consequently that necessary to lift the piston, taking this friction into account, will be expressed by

$$P = \left(1 + 4\phi \frac{b}{d}\right) F h \gamma,$$

where the coefficient of friction is $\phi = 0 \cdot 25$, b is the breadth of the piston packing in contact with the barrel, and d the diameter of the pump barrel (see Weisbach's *Mechanics*, vol. ii., also the volume on *Hoisting Machinery*, § 19).

The hydraulic resistances are almost exactly like those of water-pressure engines (see *Hydraulics and Steam Engines*).

Let ζ_0 be the coefficient of resistance for the entrance of the water into the suction pipe, d_1 the diameter of this pipe, v_s the velocity of entrance into the suction pipe, and v_1 the velocity of the ascending piston; then the head due to the hydraulic resistance at the entrance is

$$z_0 = \zeta_0 \frac{v_s^2}{2g} = \zeta_0 \left(\frac{d}{d_1}\right)^4 \frac{v_1^2}{2g}.$$

With a cylindrical opening not rounded we have $\zeta_0 = 0 \cdot 505$, while for a smooth and well-rounded opening $\zeta_0 = 0 \cdot 100$ (see Weisbach's *Mechanics*, vol. i.)

The head due to the friction of the water in the suction pipe is given by

$$z_1 = \zeta_1 \frac{l_1}{d_1} \frac{v_s^2}{2g} = \zeta_1 \frac{l_1}{d_1} \left(\frac{d}{d_1}\right)^4 \frac{v_1^2}{2g},$$

where l_1 is the length of the suction pipe, and $\zeta_1 = 0 \cdot 024$ the corresponding coefficient (see the volume just cited).

If ζ_m is the coefficient of resistance for the passage of the water through the suction valve, the corresponding head due to this resistance is

$$z_2 = \zeta_m \frac{v_1^2}{2g}.$$

Theoretically, the coefficient of resistance ζ_m can be determined from the coefficient of contraction a, the areas F of the piston and F_2 of the valve opening, according to the well-known formula

$$\zeta_m = \left(\frac{F}{aF_2} - 1\right)^2.$$

Assuming $a = 0\cdot6$ and $\frac{F}{F_2} = \frac{5}{2}$, we get

$$\zeta_m = 10 \text{ and } z_2 = 10\frac{v_1^2}{2g},$$

which agrees very well with the experimental results obtained by the author. The friction of the water in the pump barrel is further given by

$$z_3 = \zeta\frac{l}{d}\frac{v_1^2}{2g},$$

where l is the length of the barrel; the same formula is applicable to the friction in the delivery pipe, and accordingly if ζ_2 is the coefficient of friction, v_r the velocity, l_2 the length, and d_2 the diameter, we have

$$z_4 = \zeta_2\frac{l_2}{d_2}\frac{v_r^2}{2g} = \zeta_2\frac{l_2}{d_2}\left(\frac{d}{d_2}\right)^4\frac{v_1^2}{2g}.$$

Finally, for generating the velocity v_r of the water in the delivery pipe is required the head

$$z_5 = \frac{v_r^2}{2g} = \left(\frac{d}{d_2}\right)^4\frac{v_1^2}{2g}.$$

The sum of all these resistances constitutes the total hydraulic resistance:

$$W = (z_0 + z_1 + z_2 + z_3 + z_4 + z_5)F\gamma = \left[\zeta_m + \zeta\frac{l}{d} + \left(\zeta_0 + \zeta_1\frac{l_1}{d_1}\right)\left(\frac{d}{d_1}\right)^4 + \left(1 + \zeta_2\frac{l_2}{d_2}\right)\left(\frac{d}{d_2}\right)^4\right]\frac{v_1^2}{2g}F\gamma,$$

and therefore the total force needed to lift the valved piston is

$$P_1 = \left\{\left(1 + 4\phi\frac{b}{d}\right)h + \left[\zeta_m + \zeta\frac{l}{d} + \left(\zeta_0 + \zeta_1\frac{l_1}{d_1}\right)\left(\frac{d}{d_1}\right)^4 + \left(1 + \zeta_2\frac{l_2}{d_2}\right)\left(\frac{d}{d_2}\right)^4\right]\frac{v_1^2}{2g}\right\}F\gamma = \left[\left(1 + 4\phi\frac{b}{d}\right)h + \kappa_1\frac{v_1^2}{2g}\right]F\gamma,$$

where

$$\kappa_1 = \zeta_m + \zeta\frac{l}{d} + \left(\zeta_0 + \zeta_1\frac{l_1}{d_1}\right)\left(\frac{d}{d_1}\right)^4 + \left(1 + \zeta_2\frac{l_2}{d_2}\right)\left(\frac{d}{d_2}\right)^4$$

represents the sum of all the *hydraulic coefficients* of *resistance*.

During the *down stroke of the piston* the suction valve is closed and the piston valve open, consequently the water presses against the top and bottom of the piston with the same force $F(b + h_2)\gamma$, and therefore the pump load proper is equal to zero. If the piston packing is perfectly elastic the piston friction is also equal to zero, for the water then flows up past the outer surface of the piston, pushing the packing away from the sides of the cylinder. The only friction which the piston has to overcome on the down stroke is due to generating the velocity v_n of the water through the piston valve. The quantity which flows through the valve opening F_n during the down stroke is equal to that displaced by the piston head which has the area $F - F_n$; consequently we have

$$F_n v_n = (F - F_n) v_2,$$

and therefore

$$v_n = \frac{F - F_n}{F_n} v_2,$$

where v_2 is the velocity of the descending piston. The corresponding head due to this velocity is found, after multiplying F_n by a coefficient of contraction a_n determined experimentally, to be

$$h_n = \frac{v_n^2}{2g} = \left(\frac{F - a_n F_n}{a_n F_n}\right)^2 \frac{v_2^2}{2g},$$

and therefore the force needed to push down the piston is

$$P_2 = F h_n \gamma = \left(\frac{F - a_n F_n}{a_n F_n}\right)^2 \frac{v_2^2}{2g} F\gamma,$$

or

$$P_2 = \kappa_2 \frac{v_2^2}{2g} F\gamma,$$

where κ_2 represents the coefficient $\left(\dfrac{F - a_n F_n}{a_n F_n}\right)^2$.

In order to reduce this force as much as possible the valve

openings must be made as large as practicable. Moreover, the force needed to lift the piston is increased by the weight G of the latter and its rod, while that required to produce the downward motion is diminished by the same amount. The buoyant action of the water has the reverse effect. If V is the volume of the piston, and that part of the piston rod which, on an average, remains immersed during a double stroke, then the diminution of the lifting force, as well as the increase of the force pushing the piston downwards, due to the buoyant effort, is $V\gamma$. Accordingly neither the weight of the piston nor the buoyant effort exerted by the water on the latter involves any increase in the expenditure of work.

The *work absorbed* during a double stroke of the piston can now be determined by means of the well-known expression

$$A = P_1 s + P_2 s = (P_1 + P_2)s$$
$$= \left[\left(1 + 4\phi\frac{b}{d}\right)h + \kappa_1\frac{v_1^2}{2g} + \kappa_2\frac{v_2^2}{2g}\right]Fs\gamma$$
$$= \left[\left(1 + 4\phi\frac{b}{d}\right)h + \kappa_1\frac{v_1^2}{2g} + \kappa_2\frac{v_2^2}{2g}\right]F\gamma,$$

where s is the piston stroke, and V the theoretical discharge (Fs) per stroke. Now, if the pump makes n double strokes per minute, the work expended per second must be

$$L = \left[\left(1 + 4\phi\frac{b}{d}\right)h + \kappa_1\frac{v_1^2}{2g} + \kappa_2\frac{v_2^2}{2g}\right]\frac{n}{60}V\gamma,$$

or if the theoretical discharge per second be placed equal to $\frac{nV}{60} = Q_0$,

$$L = \left[\left(1 + 4\phi\frac{b}{d}\right)h + \kappa_1\frac{v_1^2}{2g} + \kappa_2\frac{v_2^2}{2g}\right]Q_0\gamma.$$

Letting v represent the *average piston speed* during a double stroke, we get

$$L = \left[\left(1 + 4\phi\frac{b}{d}\right)h + \kappa_1\frac{v_1^2}{2g} + \kappa_2\frac{v_2^2}{2g}\right]\frac{Fv}{2}\gamma.$$

The duration of a double stroke is $t = \frac{2ns}{60v}$, the duration of the up stroke is $t_1 = \frac{ns}{60v_1}$, and that of the down stroke $t_2 = \frac{ns}{60v_2}$, consequently,

$$\frac{2ns}{60v} = \frac{ns}{60v_1} + \frac{ns}{60v_2}, \text{ i.e. } \frac{2}{v} = \frac{1}{v_1} + \frac{1}{v_2},$$

and therefore the average piston speed is $v = \frac{2v_1v_2}{v_1 + v_2}$; when the values v_1 and v_2 differ but slightly we may employ the approximation

$$v = \frac{v_1 + v_2}{2}.$$

Placing the *actual discharge* $Q = \mu Q_0 = 0{\cdot}85 Q_0$, we have

$$Q_0 = \frac{Q}{\mu} = \frac{Q}{0{\cdot}85} = 1{\cdot}18Q,$$

and therefore the *expenditure of work* expressed in terms of actual discharge is

$$L = \left[\left(1 + 4\phi\frac{b}{d}\right)h + \kappa_1\frac{v_1^2}{2g} + \kappa_2\frac{v_2^2}{2g}\right]\frac{Q\gamma}{\mu}$$

$$= 1{\cdot}18\left[\left(1 + 4\phi\frac{b}{d}\right)h + \kappa_1\frac{v_1^2}{2g} + \kappa_2\frac{v_2^2}{2g}\right]Q\gamma.$$

If greater exactness is desired we must (according to Weisbach's *Mechanics*, vol. ii.) substitute for v^2 not the square of the average piston speed but the average of the square of the piston speed, hence for the up stroke

$$v_1^2 = 1{\cdot}645\left(\frac{s}{t_1}\right)^2,$$

and for the down stroke

$$v_2^2 = 1{\cdot}645\left(\frac{s}{t_2}\right)^2.$$

Finally, for the *efficiency* of a pump we obtain

$$\eta = \frac{Qh\gamma}{L} = \frac{\mu h}{\left(1 + 4\phi\frac{b}{d}\right)h + \kappa_1\frac{v_1^2}{2g} + \kappa_2\frac{v_2^2}{2g}}$$

$$= \frac{\mu}{1 + 4\phi\frac{b}{d} + \kappa_1\frac{v_1^2}{2gh} + \kappa_2\frac{v_2^2}{2gh}}.$$

In well-designed pumps, working under favourable conditions, we can assume $\eta = 0\cdot80$, in pumps of average perfection $\eta = 0\cdot75$, and in ordinary pumps $\eta = 0\cdot70$, and sometimes 0·65 only.

EXAMPLE.—In a *suction and lift pump* the diameter of piston is $d = 0\cdot3$ m. [0·98 ft.], the stroke $s = 1$ m. [3·28 ft.], the diameter of the suction pipe $d_1 = 0\cdot15$ m. [0·49 ft.], and its length $l_1 = 8$ m. [26·25 ft.], the diameter of the delivery pipe $d_2 = 0\cdot3$ m. [0·98 ft.], and the length of the latter together with the pump barrel $l_2 = 4$ m. [13·12 ft.]; further, the width of the piston packing is $b = 50$ mm. [1·97 ins.], and the two axes of the elliptic hole through the piston are $2a_1 = 0\cdot20$ m. [0·66 ft.] and $2b_1 = 0\cdot15$ m. [0·49 ft.], the orifice being rounded at each end. If the up stroke of the piston requires six and the down stroke four seconds, what is the power required to drive the pump?

The area of the piston is

$$F = \frac{\pi d^2}{4} = \frac{\pi}{4} 0\cdot09 = 0\cdot0707 \text{ sq. m. } [0\cdot761 \text{ sq. ft.}],$$

and that of the hole passing through it

$$F_n = \pi a_1 b_1 = \frac{\pi}{4} 0\cdot2 \times 0\cdot15 = \frac{F}{3} = 0\cdot0236 \text{ sq. m. } [0\cdot254 \text{ sq. ft.}]:$$

further, the average velocity of the plunger on its up stroke is

$$v_1 = \frac{s}{t_1} = \tfrac{1}{6} = 0\cdot167 \text{ m. } [0\cdot548 \text{ ft.}],$$

and on its down stroke

$$v_2 = \frac{s}{t_2} = \tfrac{1}{4} = 0\cdot25 \text{ m. } [0\cdot82 \text{ ft.}],$$

and consequently the average of the square of the velocity is, for the ascent and descent respectively,

$$v_1^2 = 1\cdot645\left(\frac{s}{t_1}\right)^2 = 1\cdot645 \times \tfrac{1}{36} = 0\cdot0457 \; [0\cdot4917],$$

$$v_2^2 = 1\cdot645\left(\frac{s}{t_2}\right)^2 = 1\cdot645 \times \tfrac{1}{16} = 0\cdot1028 \; [1\cdot1061].$$

The net load of the pump is

$$Fh\gamma = F(h_1 + h_2)\gamma = 0\cdot0707(8+4)1000 = 848\cdot4 \text{ kg. } [1870\cdot5 \text{ lbs.}],$$

whereas the load, including the friction at the piston, is

$$\left(1 + 4\phi\frac{b}{d}\right)Fh\gamma = (1 + 4 \times \tfrac{1}{4} \times \tfrac{1}{6})848\cdot4 = \tfrac{7}{6} \times 848\cdot4 = 990 \text{ kg. } [2182 \text{ lbs.}]$$

Taking the coefficient of resistance at the entrance to the

suction pipe equal to $\zeta_0 = 0\cdot5$, and for the passage through the suction valve equal to $\zeta_m = 16$, also assuming the coefficients of friction for the motion of the water in the suction pipe to be $\zeta_1 = 0\cdot026$, and in the pump barrel and delivery pipe $\zeta = \zeta_2 = 0\cdot038$, we obtain the total head corresponding to the resistances on the up stroke

$$\kappa_1 \frac{v_1^2}{2g} = \left[\zeta_m + \zeta\frac{l}{d} + \left(\zeta_0 + \zeta_1\frac{l_1}{d_1}\right)\left(\frac{d}{d_1}\right)^4 + \left(1 + \zeta_2\frac{l_2}{d_2}\right)\left(\frac{d}{d_2}\right)^4\right]\frac{v_1^2}{2g},$$

or, since $d_2 = d$,

$$\kappa_1 \frac{v_1^2}{2g} = \left[1 + \zeta_m + \zeta\frac{l + l_2}{d} + \left(\zeta_0 + \zeta_1\frac{l_1}{d_1}\right)\left(\frac{d}{d_1}\right)^4\right]\frac{v_1^2}{2g}$$

$$= \left[1 + 16 + 0\cdot038\frac{4}{0\cdot3} + \left(0\cdot5 + 0\cdot026\frac{8}{0\cdot15}\right)\left(\frac{0\cdot3}{0\cdot15}\right)^4\right]0\cdot051 \times 0\cdot0457$$

$$= (17\cdot5 + 30\cdot18)0\cdot00233 = 0\cdot111 \text{ m. } [0\cdot364 \text{ ft.}],$$

and hence the corresponding increase in the load on the piston

$$\kappa_1 \frac{v_1^2}{2g} F\gamma = 111 \times 0\cdot0707 = 7\cdot8 \text{ kg. } [17\cdot2 \text{ lbs.}],$$

which is very slight in comparison with the friction at the piston packing. The total force required for lifting the piston is now found to be

$$P_1 = 990 + 7\cdot8 = \backsim 1000 \text{ kg. } (2205 \text{ lbs.}]$$

Assuming a coefficient of contraction $a_n = \frac{2}{3}$, the effort needed to push down the piston will be

$$P_2 = \left(\frac{F - a_n F_n}{a_n F_n}\right)^2 \frac{v_1^2}{2g} F\gamma = \left(\frac{1 - \frac{2}{3} \times \frac{1}{2}}{\frac{2}{3} \times \frac{1}{2}}\right)^2 \times 0\cdot051 \times 0\cdot1028 F\gamma$$

$$= 0\cdot064 F\gamma = 4\cdot5 \text{ kg. } [9\cdot92 \text{ lbs.}]$$

We now obtain for the work expended per double stroke

$$A = (P_1 + P_2)s = (1000 + 4\cdot5)1 = 1004\cdot5 \text{ m. kg. } [7266 \text{ ft. lbs.}],$$

and consequently per second

$$L = \frac{A}{t} = \frac{A}{6 + 4} = 100\cdot5 \text{ m. kg. } [726\cdot6 \text{ ft. lbs.}]$$

Taking the volume of water raised per double stroke equal to

$$V = \mu F s = 0\cdot85 \times 0\cdot0707 \times 1 = 0\cdot060 \text{ cub. m. } [2\cdot12 \text{ cub. ft.}],$$

we have as the effect delivered per second

$$Qh\gamma = \frac{n}{60} Vh\gamma = \frac{1}{10} \times 60 \times 12 = 72 \text{ m. kg. } [522\cdot11 \text{ ft. lbs.}],$$

and consequently for the efficiency of the pump

$$\eta = \frac{72}{100\cdot5} = 0\cdot716.$$

This small value of η is essentially to be attributed to the friction at the piston packing, which, however, as a rule falls short of the value assumed above. (For further information on the resistance due to the piston packing see the volume on *Hoisting Machinery*, § 19.)

§ 31. **Wasteful Resistances** (*continued*).—In discussing the pumps provided with *solid pistons* or *plungers*, we must distinguish the case where the open end of the barrel is at the top, as in Figs. 34 and 35, from that in which the open end points downward, as represented by Figs. 36 and 37. In the former case water is sucked in on the up stroke and forced up the delivery pipe during the down stroke, while in the second arrangement the reverse takes place. In the following discussion we will assume a pump of the former kind (Fig. 78).

Let h_1 be the average height of suction estimated from low water-level to the average position of the piston, let d denote the diameter of the piston, b the width of the packing, v_1^2 the average of the square of the piston velocity, κ_1 the sum of all the coefficients of the hydraulic resistances, and F the area of the piston, then the *force needed to lift the piston* will be

$$P_1 = \left[\left(1 + 4\phi\frac{b}{d}\right)h_1 + \kappa_1\frac{v_1^2}{2g}\right]F\gamma .$$

On the other hand, let h_2 be the average height of delivery estimated from the average piston position to the point of discharge, v_2^2 the average of the square of the velocity of the descending piston, and κ_2 the sum of the coefficients of the hydraulic resistances on the down stroke of the piston, then the *force needed to push down the piston* will be

$$P_2 = \left[\left(1 + 4\phi\frac{b}{d}\right)h_2 + \kappa_2\frac{v_2^2}{2}\right]F\gamma .$$

Now, if we multiply each of these forces by the piston stroke s and add the products we shall find the *work required* per double stroke to be

$$A = P_1 s + P_2 s = \left[\left(1 + 4\phi\frac{b}{d}\right)(h_1 + h_2) + \kappa_1\frac{v_1^2}{2g} + \kappa_2\frac{v_2^2}{2g}\right]Fs\gamma$$

$$= \left[\left(1 + 4\phi\frac{b}{d}\right)h + \kappa_1\frac{v_1^2}{2g} + \kappa_2\frac{v_2^2}{2g}\right]V\gamma ,$$

where $h = h_1 + h_2$ is the total lift and $V = Fs$ the volume swept through by the piston in one stroke.

This formula agrees with that found for pumps with valved pistons, but differs from the latter in that the coefficients κ_1 and κ_2 have somewhat different values.

Let l represent the length of the pump barrel, l_1 the length and d_1 the diameter of the suction pipe, l_2 and d_2 the length and diameter respectively of the delivery pipe, ζ, ζ_1, ζ_2 the coefficients of friction of the water in the pump barrel and

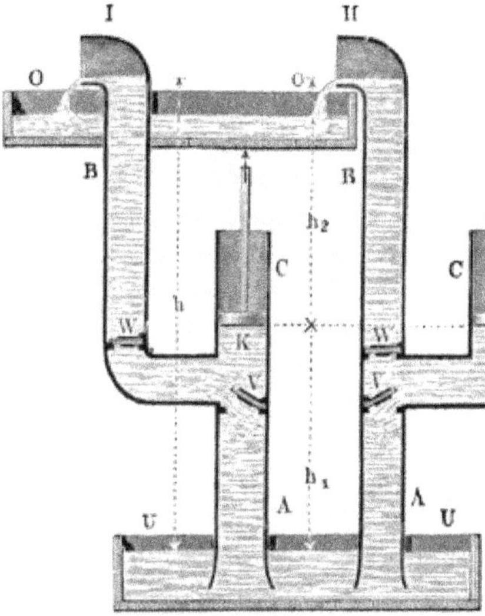

Fig. 78.

the two pipes, ζ_0 the coefficient of resistance to the entrance of water into the suction pipe, ζ_m and ζ_n the coefficients of resistance for the passage of the water through the valves, and ζ_{k_1}, ζ_{k_2} the coefficients of resistance due to changes in cross-section and direction of the communicating pipe; we accordingly have

$$\kappa_1 = \zeta\frac{l}{d} + \left(\zeta_0 + \zeta_1\frac{l_1}{d_1} + \zeta_m + \zeta_{k_1}\right)\left(\frac{d}{d_1}\right)^4,$$

and

$$\kappa_2 = \zeta\frac{l}{d} + \left(1 + \zeta_2\frac{l_2}{d_2} + \zeta_n + \zeta_{k_2}\right)\left(\frac{d}{d_2}\right)^4.$$

As a rule the term $\zeta\frac{l}{d}$ can be neglected owing to its small value. The coefficients of resistance $\zeta_m\left(\frac{d}{d_1}\right)^4$ and $\zeta_n\left(\frac{d}{d_2}\right)^4$ for the passage through the valves, on the other hand, depend on the areas F_m and F_n of the valve orifices, on the coefficients of contraction a_m and a_n, and on the sectional areas F_3 and F_4 of the valve chambers. If v_3 and v_4 represent the velocities of the water in these chambers, then the heads due to the resistances in the valve passages are

$$h_m = \left(\frac{F_3}{a_m F_m} - 1\right)^2 \frac{v_3^2}{2g} = \left(\frac{F_3}{a_m F_m} - 1\right)^2 \left(\frac{F}{F_3}\right)^2 \frac{v_1^2}{2g},$$

and

$$h_n = \left(\frac{F_4}{a_n F_n} - 1\right)^2 \frac{v_4^2}{2g} = \left(\frac{F_4}{a_n F_n} - 1\right)^2 \left(\frac{F}{F_4}\right)^2 \frac{v_2^2}{2g},$$

hence

$$\zeta_m\left(\frac{d}{d_1}\right)^4 = \left(\frac{F_3}{a_m F_m} - 1\right)^2 \left(\frac{F}{F_3}\right)^2,$$

and

$$\zeta_n\left(\frac{d}{d_2}\right)^4 = \left(\frac{F_4}{a_n F_n} - 1\right)^2 \left(\frac{F}{F_4}\right)^2.$$

The work required per second to drive the pump is, again,

$$L = \left[\left(1 + 4\phi\frac{b}{d}\right)h + \kappa_1\frac{v_1^2}{2g} + \kappa_2\frac{v_2^2}{2g}\right]\frac{nFs}{60}\gamma$$
$$= \left[\left(1 + 4\phi\frac{b}{d}\right)h + \kappa_1\frac{v_1^2}{2g} + \kappa_2\frac{v_2^2}{2g}\right]Q_0\gamma.$$

In pumps with *inverted cylinders* the force needed to lift the piston is equal to that determined above for the down stroke, and the force necessary to depress the piston the same as that absorbed in lifting it in the case just discussed. Consequently the above formulæ for the expenditure of work are applicable to both types of pumps.

Likewise for the *double-acting* pumps and *combinations of two single-acting pumps*, of the kinds illustrated in Figs. 38 and 39, the forces needed and the power absorbed can be found from the above formulæ, the only difference being that we must here substitute $\frac{2nFs}{60} = Fv$ for Q_0, and consequently obtain double the above value for L.

EXAMPLE.—A single-acting suction and force pump has a piston stroke $s = 0{\cdot}75$ m. [29·53 ins.], and is to lift 300 litres [10·6 cub. ft.] of water per minute to a height of 20 m. [65·6 ft.] What dimensions must be given to the pump, and what force must be exerted to drive it? With an average piston speed $v = 0{\cdot}2$ m. [0·656 ft.] per second, the number of double-strokes is $n = \frac{60v}{2s} = \frac{60 \times 0{\cdot}2}{2 \times 0{\cdot}75} = 8$, and since the actual discharge per second is $Q = \frac{300}{60} = 5$ litres [0·177 cub. ft.], and the theoretical discharge $Q_0 = \frac{5}{0{\cdot}85} = 5{\cdot}88$ litres [0·208 cub. ft.], the requisite area of the piston will be

$$F = \frac{2Q_0}{v} = \frac{2 \times 0{\cdot}00588}{0{\cdot}2} = 0{\cdot}0588 \text{ sq. m. [91·14 sq. ins.]},$$

which corresponds to a diameter of

$$d = \sqrt{\frac{4 \times 0{\cdot}0588}{3{\cdot}14}} = 0{\cdot}274 \text{ m. [10·79 ins.]}$$

As suitable diameters of the suction, delivery, and communicating pipes, we may now take $d_1 = d_2 = 0{\cdot}140$ m. [5·51 ins.], and for the valve chambers $d_3 = 0{\cdot}22$ m. [8·66 ins.] If the length of the suction pipe is $l_1 = 8$ m. [26·24 ft.], that of the delivery pipe $l_2 = 12$ m. [39·36 ft.], and we assume $4\phi\frac{b}{d} = 4 \times \frac{1}{3} \times \frac{1}{10} = \frac{2}{15}$, then, neglecting the hydraulic resistances, we find for the force needed to lift the piston

$$P_1 = \left(1 + 4\phi\frac{b}{d}\right)Fh_1\gamma = (1 + \tfrac{2}{15}) \times 58{\cdot}8 \times 8 = 533 \text{ kg. [1175 lbs.]},$$

and for depressing it

$$P_2 = \left(1 + 4\phi\frac{b}{d}\right)Fh_2\gamma = \frac{h_2}{h_1}P_1 = \frac{12}{8}P_1 = 800 \text{ kg. [1764 lbs.]}$$

As the water moves in the pump barrel with a velocity

$$\left(\frac{d}{d_1}\right)^2 v = \left(\frac{d}{d_2}\right)^2 v = \left(\frac{274}{140}\right)^2 v = 0{\cdot}765 \text{ m. [2·51 ft.] per sec.},$$

the corresponding coefficient of resistance will be $\zeta_1 = \zeta_2 = 0{\cdot}025$, and the head representing the frictional resistance in the suction pipe

$$\zeta\frac{l_1}{d_1}\left(\frac{d}{d_1}\right)^4 \frac{v^2}{2g} = 0{\cdot}025\,\frac{8}{0{\cdot}14}\left(\frac{274}{140}\right)^4 0{\cdot}051 \times 0{\cdot}2^2 = 0{\cdot}043 \text{ m. [0·14 ft.]},$$

while for the delivery pipe it will be

$$\frac{12}{8}0{\cdot}043=0{\cdot}065 \text{ m. [0·21 ft.]}$$

Moreover, if the diameter of the two valve orifices is $d_m = d_n = 0{\cdot}140$ m. [5·5 ins.], and the coefficient of contraction be $\alpha_m = \alpha_n = 0{\cdot}6$, the heads representing the resistances in these passages will be

$$h_m = h_n = \left(\frac{F_3}{\alpha_m F_m} - 1\right)^2 \left(\frac{F}{F_3}\right)^2 \frac{v^2}{2g} = \left[\frac{1}{\alpha_m}\left(\frac{d_3}{d_m}\right)^2 - 1\right]^2 \left(\frac{d}{d_3}\right)^4 \frac{v^2}{2g}$$

$$= \left[\frac{1}{0{\cdot}6}\left(\frac{220}{140}\right)^2 - 1\right]^2 \left(\frac{274}{220}\right)^4 0{\cdot}051 \times 0{\cdot}2^2 = 0{\cdot}047 \text{ m. [0·154 ft.]}$$

Now, if we assume the coefficient of resistance for entrance of the water into the suction pipe to be $\zeta_0 = 0{\cdot}25$, and the coefficient of resistance for the passage through the communicating pipes between the suction and delivery pipes and the pump barrel to be $\zeta_{k1} = \zeta_{k2} = 1{\cdot}5$, then the corresponding heads measuring these resistances will be

$$\zeta_0\left(\frac{d}{d_1}\right)^4 \frac{v^2}{2g} = 0{\cdot}25 \times 0{\cdot}051 \times 0{\cdot}765^2 = 0{\cdot}007 \text{ m. [0·023 ft.]}$$

and

$$\zeta_{k1}\left(\frac{d}{d_1}\right)^4 \frac{v^2}{2g} = \zeta_{k2}\left(\frac{d}{d_2}\right)^4 \frac{v^2}{2g} = 1{\cdot}5 \times 0{\cdot}051 \times 0{\cdot}765^2 = 0{\cdot}042 \text{ m. [0·138 ft.]},$$

and consequently the total hydraulic resistances

$$\kappa_1 \frac{v_1^2}{2g} = \left(\zeta_0 + \zeta_1 \frac{l_1}{d_1} + \zeta_m + \zeta_{k1}\right)\left(\frac{d}{d_1}\right)^4 \frac{v^2}{2g} = 0{\cdot}007 + 0{\cdot}043$$
$$+ 0{\cdot}047 + 0{\cdot}042 = 0{\cdot}139 \text{ m. [0·456 ft.]}$$

and

$$\kappa_2 \frac{v_1^2}{2g} = \left(1 + \zeta_2 \frac{l_2}{d_2} + \zeta_n + \zeta_{k2}\right)\left(\frac{d}{d_2}\right)^4 \frac{v_2}{2g} = 0{\cdot}029 + 0{\cdot}065$$
$$+ 0{\cdot}047 + 0{\cdot}042 = 0{\cdot}183 \text{ m. [0·6 ft.]}$$

If the piston velocity were three times as great, that is, equal to 0·6 m. [1·97 ft.] per second, the heads due to these hydraulic resistances would be increased nine-fold, and if in addition the suction and delivery pipes, valve openings, etc., had only three-fourths of the diameters assumed above, the resistances would become

$$9 \times \left(\frac{4}{3}\right)^4 = 28{\cdot}5$$

times as large, and consequently we should have

$$\kappa_1 \frac{v_1^2}{2g} = 0{\cdot}139 \times 28{\cdot}5 = 3{\cdot}96 \text{ m. [12·99 ft.]}$$

and

$$\kappa_2 \frac{v_2^2}{2g} = 0{\cdot}183 \times 28{\cdot}5 = 5{\cdot}21 \text{ m. [17·09 ft.]}$$

If in place of the square of the average velocity we use the average of the square of the velocity, we should have

$$\kappa_1\frac{v_1^2}{2g}=1\cdot645\times0\cdot139=0\cdot229 \text{ m. } [0\cdot75 \text{ ft.}],$$

and

$$\kappa_2\frac{v_2^2}{2g}=1\cdot645\times0\cdot183=0\cdot301 \text{ m. } [0\cdot99 \text{ ft.}]$$

The force needed to lift the piston under the first supposition is

$$P_1=\left[\left(1+4\phi\frac{b}{d}\right)h_1+\kappa_1\frac{v_1^2}{2g}\right]F\gamma=533+0\cdot229\times58\cdot8=546\cdot5 \text{ kg. } [1205 \text{ lbs.}],$$

and that necessary to push down the piston is

$$P_2=\left[\left(1+4\phi\frac{b}{d}\right)h_2+\kappa_2\frac{v_2^2}{2g}\right]F\gamma=800+0\cdot301\times58\cdot8=817\cdot7 \text{ kg. } [1803 \text{ lbs.}]$$

Hence we have for the work absorbed during a double stroke

$$A=(P_1+P_2)s=(546\cdot5+817\cdot7)0\cdot75=1023 \text{ m. kg. } [7399 \text{ ft. lbs.}],$$

and therefore the expenditure of work per second is

$$L=\frac{n}{60}A=\frac{8}{60}1023=136\cdot4 \text{ m. kg. } [986\cdot6 \text{ ft. lbs.}]$$

Theoretically the work necessary is

$$Qh\gamma=5\times20=100 \text{ m. kg. } [723\cdot3 \text{ ft. lbs.}],$$

and consequently we have for the efficiency of the pump

$$\eta=\frac{100}{136\cdot4}=0\cdot733.$$

Under the second supposition, that the piston velocity is three times as great, we should have

$$L=\frac{8}{60}0\cdot75[533+800+(3\cdot96+5\cdot21)1\cdot645\times58\cdot8]$$
$$=0\cdot1(533+800+886)=222 \text{ m. kg. } [1606 \text{ ft. lbs.}],$$

which would correspond to an efficiency of

$$\eta=\frac{100}{222}=0\cdot450.$$

CHAPTER IV

TYPES OF RECIPROCATING PUMPS

§ 32. **Hand Pumps.**—Small pumps are frequently operated by hand, either directly by means of a handle attached to the piston rod or by the aid of a lever or crank. The first mentioned arrangement has but limited application, since in this case the load must not exceed the force corresponding to the direct application of muscular power, or about 12 kilos [26 lbs.] With the lever pumps the case is different, since the lever arm of the effort may be made from three to six times the length of that of the load, thus admitting of a correspondingly increased load, and besides several operators may be more readily engaged at the pump at the same time. Denoting by s the path travelled by the point of application of the effort, by a the lever arm of the effort, and by b that of the load, the corresponding stroke of the piston will be

$$s_1 = \frac{b}{a} s,$$

and consequently for a path of the effort of $s = 0{\cdot}9$ m. [3 ft.], which corresponds to the sweep of a man's arm, we have

$$s_1 = \frac{b}{a} 0{\cdot}9 \text{ m.} \left[s_1 = \frac{b}{a} 3 \text{ ft.} \right]$$

For the ratio $\frac{b}{a} = \frac{1}{3}$ we should have a stroke of $s_1 = 0{\cdot}3$ m. [1 ft.], and for $\frac{b}{a} = \frac{1}{6}$ the stroke would be only 0·15 m. [6 ins.] The lift of lever pumps is therefore always small, mostly between the limits 0·15 and 0·3 m. [6 ins. to 1 ft.] As muscular effort can be employed to better advantage in pressing down than in pulling up a lever, it is customary to arrange the pump lever in such a manner that it will require essentially a downward pressure or pull. In suction pumps it is therefore as a rule made with two arms, while force pumps usually have a one-armed lever.

The simplest form of a suction and lift pump is that shown in Fig. 79. It is used for disposing of smaller accumulations of water in building excavations, and consists of four planed boards which are joined tightly and thus form the prismatic pump barrel A; the latter is closed at the lower end by a hollow block B provided with a leather clack C forming a very simple suction valve. The piston K, which is moved by means of a rod S having a handle H, is likewise made of a square block of wood having a hole through the centre and provided with a leather clack D, the tight fit at the sides of the barrel being obtained by strips of leather nailed on the sides of the block. Impurities are kept out by the perforated boards E at the lower end; the discharge spout is also formed in a simple manner by boards G.

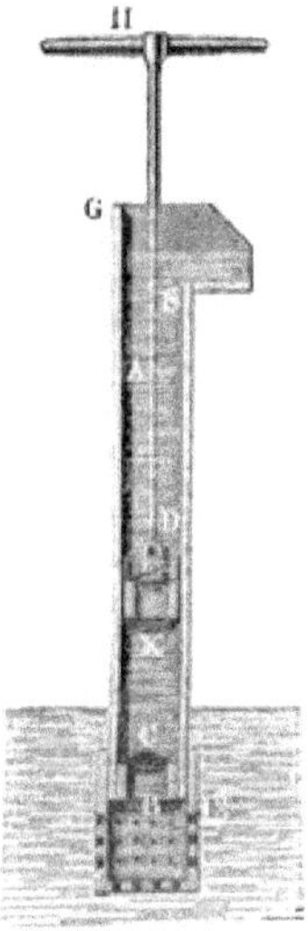

Fig. 79.

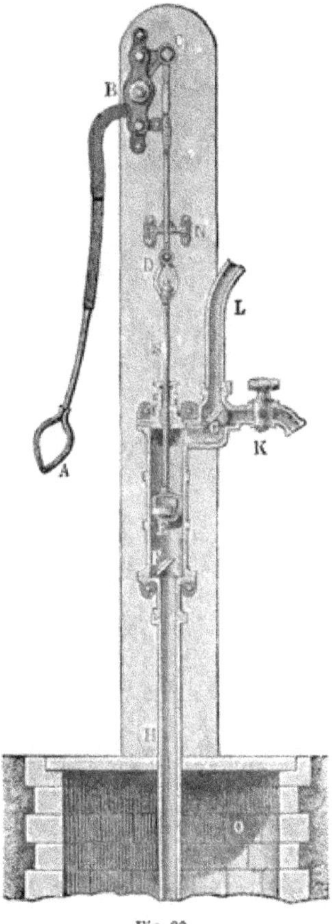

Fig. 80.

In Fig. 80 is illustrated a well-designed and very common

type of suction and lift pump. Here the piston E is of brass and packed by leather; the piston rod S is guided by a stuffing box and is reciprocated by means of a bell-crank lever ABC having its fulcrum at B and provided with a handle at A. The end C is connected to the piston rod by the forked rod CD, which passes through a stuffing box held by a bracket at N. Water is discharged either through the cock K or, when this is closed, through the pipe L leading to a more elevated point. A delivery valve G serves to retain the water in the delivery

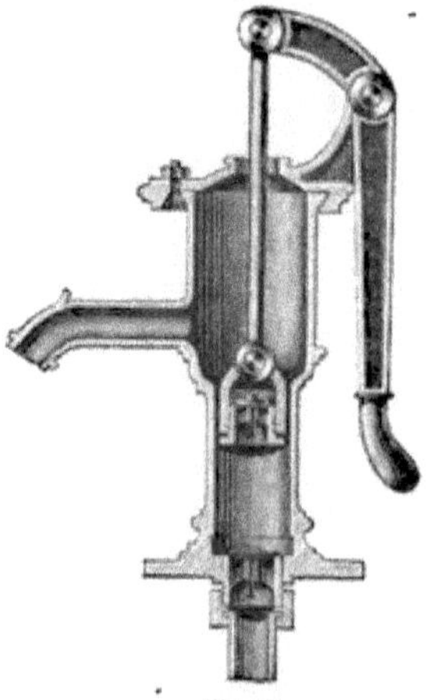

Fig. 81.

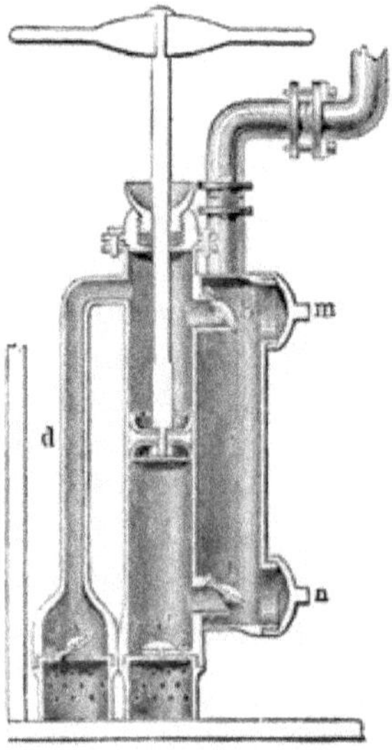

Fig. 82.

pipe when the pump remains at rest for some time; this valve is frequently omitted, since it is not needed for the working of the pump. The handle A usually serves as a counterweight, being raised to a certain height by the operator during the down stroke of the unloaded piston and assisting the upward movement of the latter. It is evident that for very deep wells it is necessary to place the pump at a sufficient depth to make suction possible.

Norton's so-called *tube wells* are also nothing but single-acting lever pumps, the peculiarity of which consists in that they require no special well to be built, the suction pipe being simply driven into the ground. For this purpose it is pointed at the lower end and, for the admission of water, it is provided with a number of small perforations near this end. The pump, coupled to the upper end of the pipe, is shown in Fig. 81.

In Fig. 82 is seen the arrangement of a double-acting hand pump. It is readily understood that of the valves either a and b or a' and b' simultaneously open or shut.

It is a matter of special practical importance in all pumps that the valves should be *easily accessible*, so that they may be readily examined whenever occasion demands. As a model construction with reference to this point may be considered the so-called *California pump* first introduced by *Hansbrow*, an

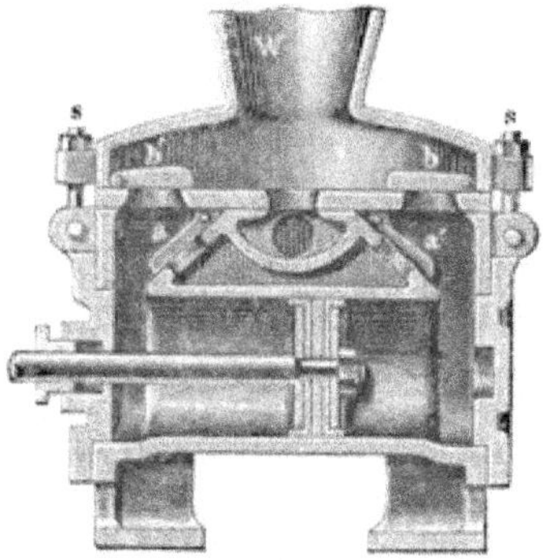

Fig. 83.

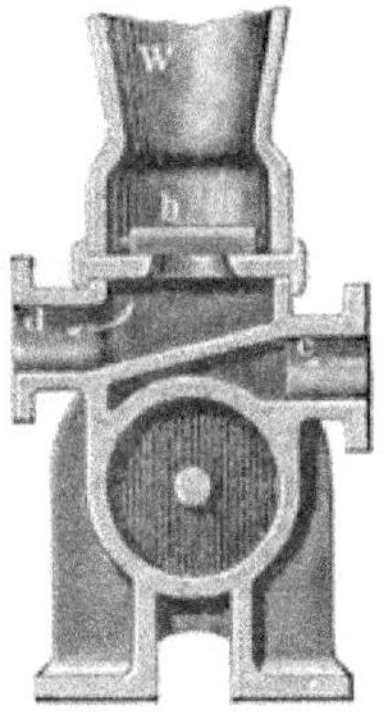

Fig. 84.

improved design of which, as made by *Werner*,[1] is shown in Figs. 83 and 84.

Also in this pump the valves a and b or a' and b' always open or shut at the same time when the piston moves back and forth horizontally. Water enters through the pipe c and is discharged through d, and all the valves are made accessible by unscrewing the eye bolts s and removing the air chamber W.

§ 33. **Fire Engines** (and *garden pumps*) are portable pumps which discharge the water, not into pipes, but by throwing it up in streams. For the purpose of obtaining a uniform jet of water, they are always provided with an air chamber, and for easy transportation they are commonly mounted on wheels. In factories *stationary* fire engines driven by water or steam power are usually employed, while in cities and towns portable *steam fire engines* are nowadays largely in use.

[1] See *Zeitschr. deutsch. Ing.* 1870, p. 196.

A fire engine consists essentially of one or two pump cylinders made of brass and provided with the requisite suction and delivery valves, the water being forced from the latter through one or two short pipes into the air chamber of copper.

Fig. 85.

Water is conducted to the suction valve through a suction or supply pipe, and discharged from the air chamber through an eduction or discharge pipe to which is coupled a *hose* with *nozzle.* In hand engines the piston is reciprocated simply by a

Fig. 86.

lever, while in steam fire engines the pump piston is usually attached to the piston rod of the steam engine and thus driven directly by the latter. The pump is generally placed in a box-shaped reservoir or cistern from which it takes its water, and which is supplied either by buckets or from the suction pipe, or in some cases by a special feed pump.

The general arrangement of a two-cylinder hand fire engine is shown in Figs. 85 to 87. Fig. 85 represents a half elevation and a half longitudinal section, Fig. 86 the plan, and Fig. 87 a vertical cross-section through the air chamber. At B, B the cylinders are seen, at A the suction pipe, at C, C the communicating pipes, at R the air chamber, and at D the discharge pipe. The suction pipe has two inlets E and F, the former from the cistern and the latter from without. When the water is to be taken from the cistern, the inlet F is closed by a cap, and E is provided with a strainer for the exclusion of impurities. On the other hand, when the supply is to be derived from some other source, E is closed and a hose is screwed on to F and carried to the supply reservoir, the strainer then being attached to the end of the hose. Below the air chamber is a cylindrical casting which is divided into three chambers by vertical partitions; two of these chambers communicate with the pump barrels and are covered by sector-shaped valves, while the third, which connects with the discharge pipe, is open at the top. A board GG placed on the top of the cistern along its entire length, and firmly secured to the latter and to the woodwork below it by eight bolts, serves as support not only for the pump lever bracket K but also for the two guides LL which prevent side motion of the lever, and for the spring buffers which limit the motion of the latter.

Fig. 87.

Many other constructions of fire engines have been brought into practical use. For instance, the plungers have been replaced by valved lifting pistons, and the lever arrangement for their working by a crank mechanism.

In Fig. 88 is illustrated the pumping arrangement of a fire engine with valved piston, designed by *Levesque*. The air chamber W is here placed directly above the pump barrel, and the piston rod is surrounded by the lower end of the outlet pipe and guided by a stuffing box. A complete fire engine of this kind consists of two such pumping apparatus, and is operated by a double crank. In order to obtain a uniform motion of the crank shaft, the latter is provided with two fly wheels which, during the transportation of the machine, serve as waggon wheels.

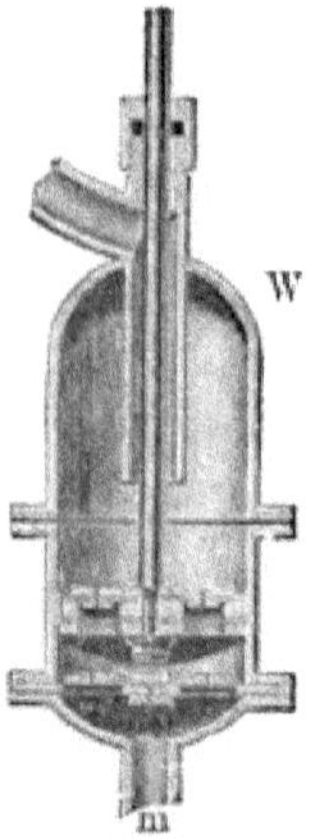

Fig. 88.

Letestu's fire engine also belongs to this class. The characteristic features of this apparatus are a valved piston of peculiar construction, and that the water enters the cylinder from the top and is forced downward into the air chamber. As, first, regards *Letestu's* piston, it consists of a perforated funnel (Fig. 89, I.), made of iron plate, and a leather funnel (Fig. 89, II.), covering the former on the inside and projecting some 10 mm. [$\frac{3}{8}$ in.] below it, thus serving at the same time as valve and packing. The leather cone is not sewn up, consisting simply of a sector with the radial edges somewhat bevelled and lapping over each other. The pumping mechanism is shown in Fig. 90, which represents a section through the rear end of the engine. When the piston is moving upward, the water entering the cylinder C from the cistern W above it forces the leather covering away from the iron funnel K, and flows through the perforations in the latter. On the other hand, during the down stroke the water below the piston forces the leather against the iron cone, and upon lifting the delivery valve V enters the air chamber above it. In ABC is seen one half of the pump lever, which oscillates about C and is provided with wooden handles or *brakes* at A.

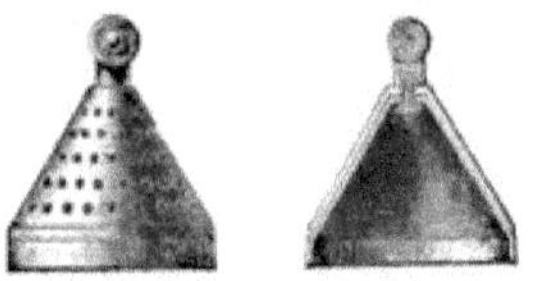

Fig. 89.

Another departure in fire-engine construction consists in the use of a *diaphragm pump*, AKA, Fig. 91.[1] In place of a

[1] Diaphragm pumps of various types are nowadays frequently employed, also on vessels and railways, in draining excavations for buildings, and for other service involving the pumping of water containing sand, gravel, or gritty matter, these impurities not affecting the action of the diaphragm. — TRANSLATOR'S REMARK.

piston a leather (or rubber) cone AA is here employed, which is clamped to the pump casting by means of an iron ring, and at

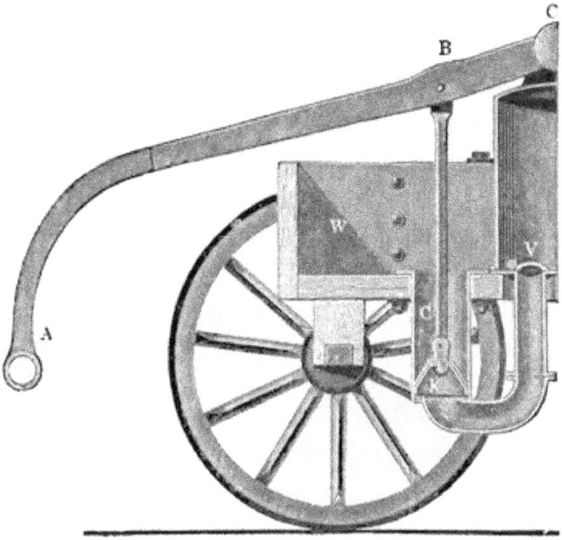

Fig. 90.

the upper end is fastened to the piston rod KL. The further details of the arrangement are evident from the figure. When

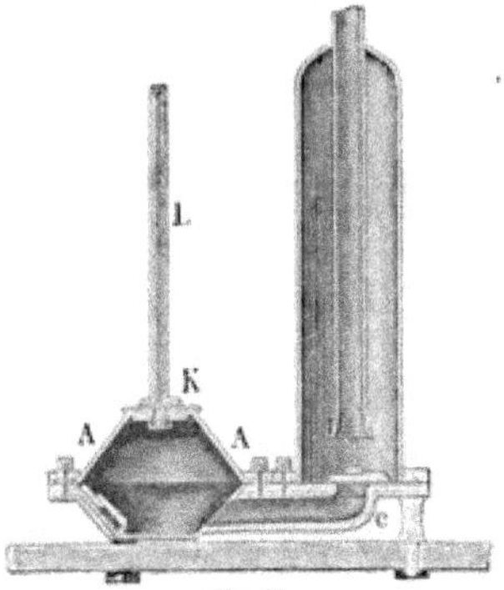

Fig. 91.

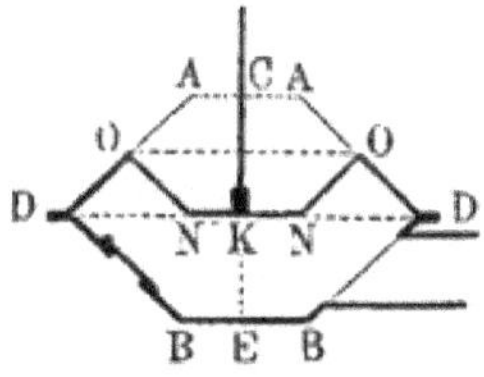

Fig. 92.

the diaphragm AA (Fig. 92) is depressed to the bottom BB of the barrel, the whole space enclosed between the two conical frusta AD and DB is emptied. Denoting by r_1 and r_2 the radii of the upper and lower bases AA and DD, and by h the height CK = EK of each frustum, consequently the stroke

$s = 2h$, then the theoretical quantity of water delivered by the pump per minute will be

$$V = \tfrac{2}{3}\pi h(r_1^2 + r_1r_2 + r_2^2).$$

As a rule the stroke s is smaller, however, and were it to be made equal to CK $= h$ only, we should have for the corresponding volume AONA

$$V = \tfrac{1}{3}\pi h\left[r_1^2 + r_1\frac{r_1 + r_2}{2} + \left(\frac{r_1 + r_2}{2}\right)^2\right]$$
$$= \frac{\pi h}{6}\left(\tfrac{7}{2}r_1^2 + 2r_1r_2 + \frac{r_2^2}{2}\right).$$

Horizontal pump cylinders have also been employed in fire engines, especially for feed pumps. As examples of this type may be mentioned the designs made by *Etter*, *Kronauer*, and others (see Kronauer's *Zeitschrift für Technologie*, vol. i.)

I.

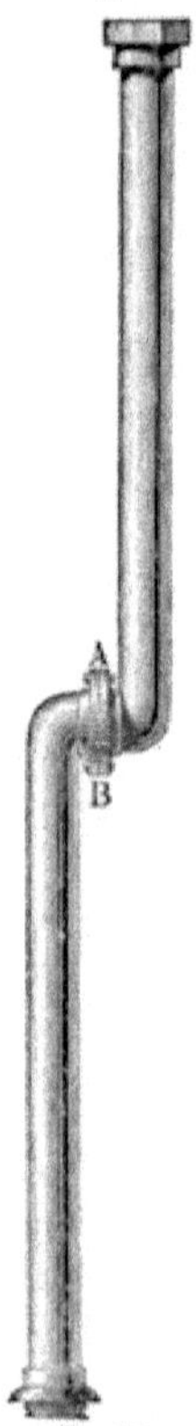

II.

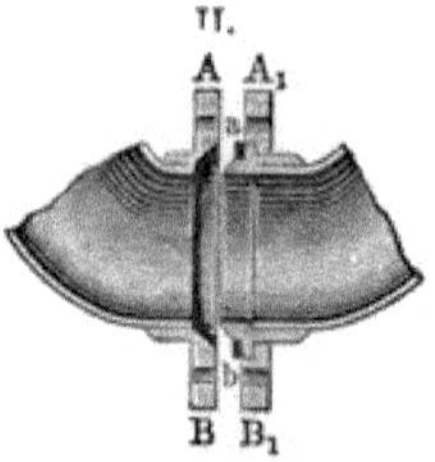

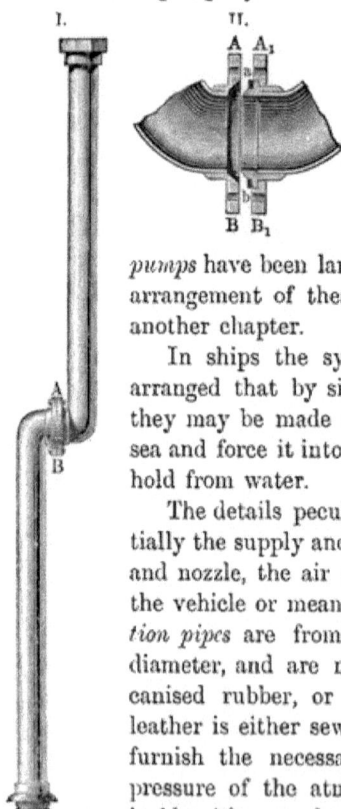

Fig. 93 (I. and II.)

Also the so-called *rotary pumps* have been largely used in fire engines. The arrangement of these pumps will be described in another chapter.

In ships the system of pipes is generally so arranged that by simply turning a four-way cock they may be made either to draw water from the sea and force it into the ship or to free the ship's hold from water.

The details peculiar to fire engines are essentially the supply and outlet or discharge pipes, hose and nozzle, the air chamber, the pump lever, and the vehicle or means of transportation. The *suction pipes* are from 5 to 8 mm. [2 to 3 ins.] in diameter, and are made either of leather or vulcanised rubber, or of copper. Suction hose of leather is either sewn or riveted, and, in order to furnish the necessary resistance to the external pressure of the atmosphere, has a lining on the inside either made of wire 4 to 5 mm. [$\frac{3}{16}$ in.] thick, and wound in a spiral, or of copper rings

3 to 5 cm. [$1\frac{1}{4}$ to 2 ins.] wide and 6 to 15 mm. [$\frac{1}{4}$ to $\frac{5}{8}$ in.] apart. Suction pipes of copper are provided at the middle with a joint as at AB (Fig. 93) made tight by a leather ring *ab*, the object of this joint being to admit of a change in the direction of the pipe when necessary. The inlet of the pipe should be provided with a strainer.

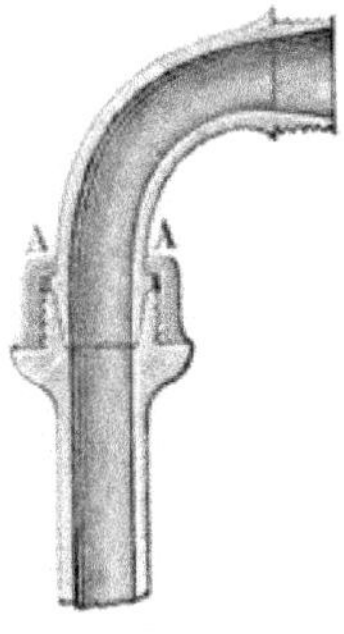

Fig. 94.

The *air chamber* is either cast of brass or made of sheet copper or brass, and is usually cylindrical with a segment-shaped top. Its capacity should be at least eight times that of the pump cylinder, and the thickness of metal is to be calculated in the same manner as for steam boilers. The *communicating pipes* are given a diameter equal to half that of the cylinder, and lead to the bottom of the air chamber, while the *outlet* or *stand pipe* is either located directly above the bottom or carried through the top of the chamber. Closely above the cistern an elbow is coupled to the stand pipe by means of a joint A (Fig. 94), and to this elbow the hose with its nozzle is screwed. In large engines a swivelling pipe placed directly on the top of the air chamber is instead made use of. Such a pipe has several joints, as shown in Fig. 95, also two elbows at the top, and a branch pipe and cocks at the central portion. A suitable joint is shown in Fig. 96. The split coupling AB connecting the pipe ends C and D is screwed on to the flange of D, and grasps the flange of the other pipe C; after swinging the pipe round, it is clamped in the desired position by means of the screw S.

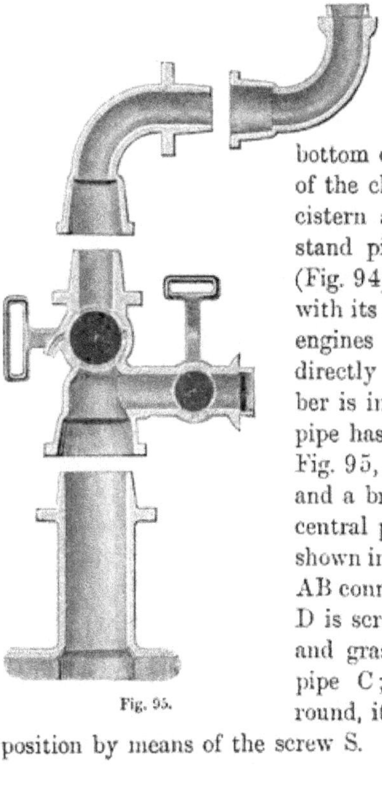
Fig. 95.

The hose for conducting the water from the fire engine to distant points is either of leather or of hemp; it has a diameter of from 3 to 5 cm. [1¼ to 2 ins.], and is made in sections from

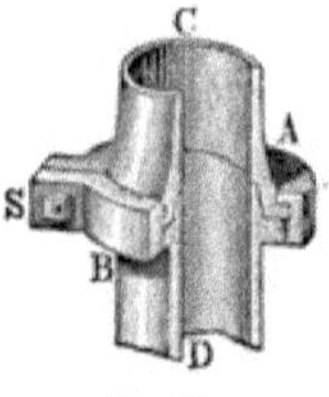

Fig. 96.

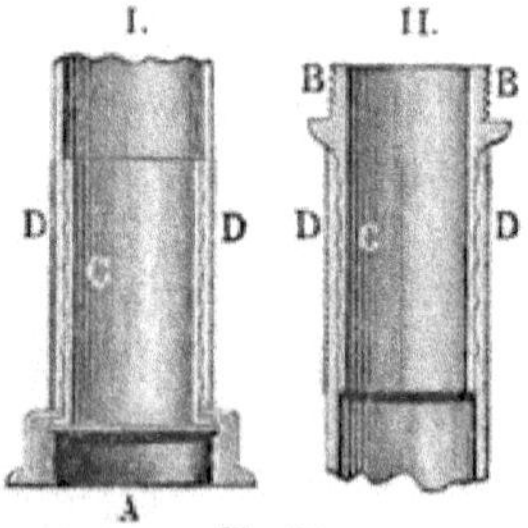

Fig. 97.

6 to 10 metres [20 to 33 ft.] in length. Leathern hose is either sewn with waxed thread or riveted by means of copper rivets about 1 mm. [0·04 in.] in diameter, while hempen hose is woven without a seam. The manner of joining the sections by screw couplings is evident from Fig. 97, I. and II. Both the nut A (Fig. I.) and the nipple BB (Fig. II.) form the terminations of short brass pipes C, over which the hose ends DD are slipped, being secured by cord or wire.

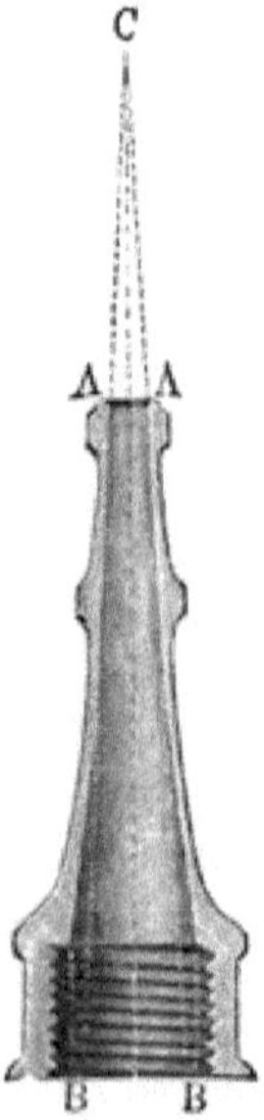

Fig. 98.

The short pipe through which the water is finally discharged is usually given an inside diameter of from 2·5 to 4 cm. [1 to 1½ in.] only, and a length of at least 0·3 m. [1 ft.]; at one end it is provided with a nut for coupling to the stand pipe or hose, and at the other it is threaded for reception of the *nozzle*. The latter is a conical brass pipe AB (Fig. 98) 15 to 20 cm. [6 to 8 ins.] long, having at the inlet end a diameter of 2·5 to 4 cm. [1 to 1½ ins.], while at the outlet it is contracted to 10 or 15 mm. [0·4 to 0·6 in.] (the opening being frequently made adjustable). The sides may suitably be made to converge at an angle BCB of 5 degrees. According to numerous experiments made by the author, the coefficient of resistance of such a nozzle has been found not to exceed 0·03, that is, the reduction in the height of stream due to friction of the water in the nozzle will be only 3 per cent.

The *lever* for operating the pistons of large hand fire-engines is composed of two bars rigidly connected by means of cross stays and located on each side of the swivel pipe and the platform for an attendant. The ends of the lever receive the wooden handle bars or *brakes*, which are from 5 to 7 cm. [2 to 2¾ ins.] in diameter, and about 3 metres [10 ft.] long, thus providing room for ten operators at the most. Evidently the lever is to be given sufficient length and such curvature that the framework of the waggon will offer no hindrance to the firemen.

To obtain easy access to the valves for cleaning, which is a matter of great importance in all pumps, and especially in fire engines, various plans have been adopted. A common construction is to make the valve seat conical on the outside and grind it to an accurate fit in a taper hole, as in cocks; by this method the seat, together with its valve, may be easily removed, but at the same time the valve is apt to become rather small or the seat casting inconveniently large. This objection has been overcome in a satisfactory manner by the arrangement introduced in the fire engines built by *J. Beduwe* of Aachen, and illustrated in Figs. 99 to 101. Here the four valve clacks, two for suction *s* and two for delivery *d*, are hinged on a peculiarly-shaped piece C (Fig. 99), which can be slid in and out like a drawer through a side opening below the air chamber W (Fig. 100). This clack slide is made to fit tightly against the wall of the air chamber V by means of a wedge-shaped piece K, held in place by a screw *s*, and easily removable after unscrewing the nut or hand wheel M. The further details of the construction are readily understood from the vertical section in Fig. 100 and the plan in Fig. 101, where AA are the cylinders, S is the suction, and D the discharge pipe.

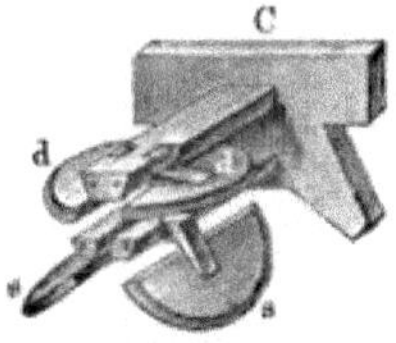

Fig. 99.

The complete arrangement of a hand fire engine mounted on wheels is shown in Fig. 102. The cistern R, internally lined with sheet-metal, is supported by the hind wheels A and the fore wheels B, and, besides containing the whole pumping apparatus, receives the bearings C for the lever D. Further,

G is the communicating pipe, CM the stand pipe terminating in the nozzle M, etc.

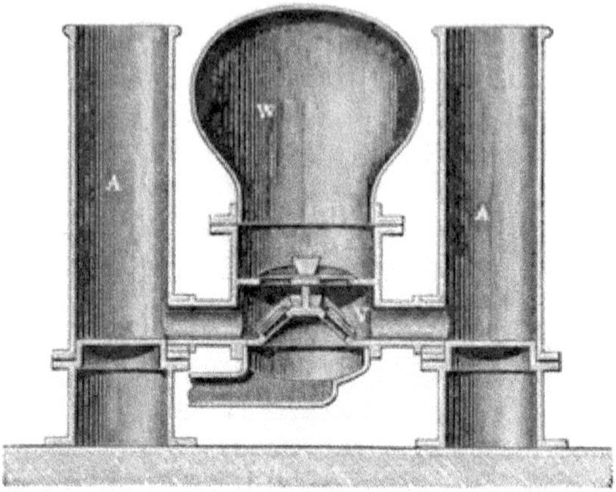

Fig. 100.

As, owing to lack of space, the number of firemen that can be engaged at the brakes is in most cases limited to about twenty,

Fig. 101.

it is obvious that for the usual height of stream of from 20 to 30 metres [65 to 100 ft.] the size of plungers and the volume of water that can be thrown are also limited. The plungers are rarely made over 0·18 m. [7 ins.] in diameter, usually only 0·15 m. [6 ins.], the corresponding volume thrown being about 0·3 cub. m. [10·6 cub. ft.] per minute. With a view to throwing greater volumes, 1 to 1·5 cub. m. [35 to 53 cub. ft.] per minute to heights ranging from 40 to 50 metres [130 to 160 ft.], *steam fire engines* have been introduced, first in America and England and afterward all over the continent of Europe. In order to enable a steam fire engine to be worked to advantage, it is evidently necessary that an adequate supply of water should be available, and as this can be obtained only with

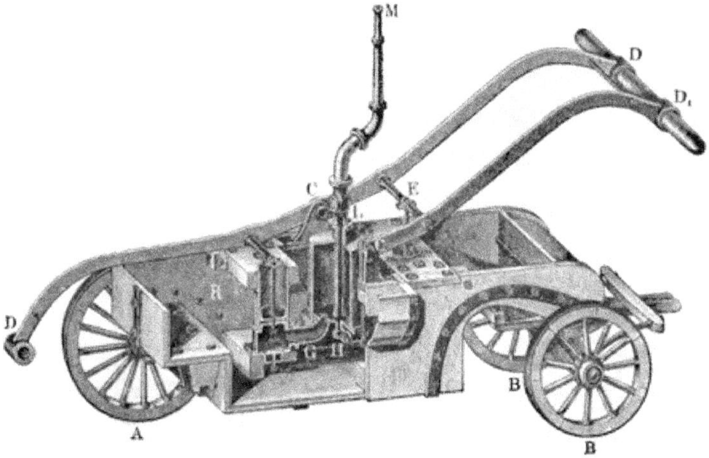

Fig. 102.

difficulty by hand supply from buckets or hand engines, the use of steam fire engines is principally limited to cities or towns, where the water may be taken from hydrants. In some cases they are also utilised as *supply engines* furnishing water through long lines of hose from a river or pond to a greater number of hand engines.

A steam fire engine must combine lightness of construction with capacity to generate steam of sufficiently high pressure in the shortest time possible, and therefore the heating surface of the boiler must be large and the water space small. By the use of fuel that ignites easily and generates intense heat it has been made possible to raise steam of 8 or 10 atmospheres pressure in less than 10 minutes. Horses are always employed for the transportation of steam fire engines, experiments to use steam power for this purpose, as in road locomotives, having proved of little consequence.

In Fig. 103 is shown a section of a steam fire engine built by *Egestorff* of Hanover.[1] The vertical boiler A has a cylindrical fire box B, and contains 199 tubes of 35 mm. [$1\frac{3}{8}$ ins.] diameter, the total surface exposed to the fire or gases being about 28 sq. metres [300 sq. ft.] The steam pump D is

Fig. 103.

located on the top of the cylindrical suction air chamber C, from which the water is sucked through the pipe *c* and forced into the delivery air chamber W. Steam is furnished to the engine through the pipe *a*, and exhausted through *d* into the smoke stack S with a view to increasing the draught. The pump is double acting, and has four valves V, the arrangement of which is shown in the section (Fig. 104). The steam piston is directly connected to the pump plunger by

[1] *Mittheil. des Hannov. Gewerbe-Vereins*, 1864, and Rühlmann's *Allgem. Maschinenlehre*, vol. iv.

means of the piston rod, which, by the aid of a slider crank F drives an auxiliary shaft provided with a fly wheel R in the manner explained below in the article on steam pumps. The diameter of the steam piston is 0·215 m. [8·46 ins.], that of the pump plunger 0·178 m. [7 ins.], and the stroke of both is 0·228 m. [9 ins.]; the steam pressure is about 7 kg. per sq. cm. [100 pounds above the atmosphere]. At a maximum speed of 161 revolutions of the fly wheel, or about $1\frac{1}{4}$ metre [4·1 ft.] piston velocity, the engine threw 1·5 cub. m. [53 cub. ft.] of water per minute to a height of 47·5 m. [156 ft.], the thickness of stream being 30 mm. [1·18 in.]

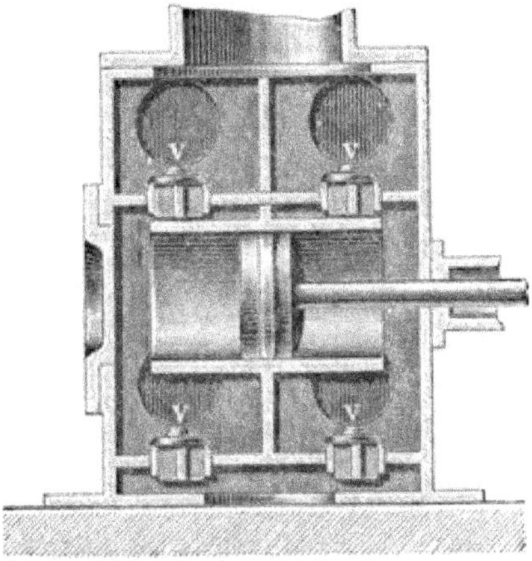

Fig. 104.

Modern portable steam fire engines are frequently made of a capacity of 5 cub. m. [176 cub. ft.] per minute, and are able to throw a 50 mm. [2-inch] stream to a height of 60 metres or more. The general outline of their design often differs from that shown in the cut in that the pumps are placed vertically and together, with the air chambers grouped more closely about the boiler. Minor steam fire engines are sometimes mounted on two wheels only, and arranged to be drawn by a single horse; a notable fire engine of this type has been designed by *Granell* of Stockholm.—TRANSLATOR'S REMARK.

§ 34. **Calculations for Fire Engines.**—Among the data given for the design of a fire engine is the desired height of stream h, and hence the velocity of efflux at the nozzle is determined. This height ranges in hand engines from 15 to 30 metres [50 to 100 ft.], while in steam fire engines, as mentioned, it may reach 60 metres [176 ft.] or more. In case no resistance were offered by the air to the motion of the water the velocity of efflux w from the nozzle would correspond to a height of stream $h_0 = \frac{w^2}{2g}$, but in consequence of the existing resistance the *actual* height will be only

$$h = \tfrac{3}{4}h_0 = \tfrac{3}{4}\frac{w^2}{2g} = 0{\cdot}038w^2 \text{ m. } [0{\cdot}012w^2 \text{ ft.}],$$

as has been proved by experiments. In order to obtain an actual height of stream h it is therefore necessary to generate a velocity of efflux

$$w = \sqrt{\frac{4}{3}2gh} = 5{\cdot}11\sqrt{h} \text{ m. } [9{\cdot}26\sqrt{h} \text{ ft.}],$$

that is, the pressure to be exerted on the water must be equal to that produced by a water column having a height $\frac{w^2}{2g} = \frac{4}{3}h$. In other words, the machine is to be proportioned as a pump which is required to force the water to a height of $\frac{4}{3}h$.

To determine the force needed for this purpose, let Q denote the water discharged per second, and let F designate the area, and d the diameter of each piston or plunger. The resistance in the suction pipe may be neglected, since the cylinders take their water directly from the cistern. Now, assuming the common arrangement of two single-acting pump cylinders with plungers reciprocated by a lever, and denoting further by d_1 the diameter of the discharge pipe or hose, by l_1 the length of the latter, by d_2 the diameter of the delivery valve, and by d_m that of the nozzle, then (according to the rules explained in Weisbach's *Mechanics*, vol. i.) we have for the head corresponding to the hydraulic resistances of the water on its way from the delivery valve to the nozzle

$$\left[1 + \zeta + \zeta_1\frac{l_1}{d_1}\left(\frac{d_m}{d_1}\right)^4 + \zeta_2\left(\frac{d_m}{d_2}\right)^4\right]\frac{w^2}{2g} = \kappa\frac{w^2}{2g},$$

where $\zeta = 0{\cdot}05$ is the coefficient of resistance at the entrance to the fireman's pipe, ζ_1 that due to the friction in the hose, and ζ_2 the coefficient for the delivery valve. Assuming as suitable values the diameter of the hose $d_1 = 0{\cdot}05$ metre [0·164 ft.], the length $l_1 = 20$ m. [65·6 ft.], the diameter of the nozzle orifice $d_m = 0{\cdot}016$ m. [0·052 ft.], $\zeta_1 = 0{\cdot}03$ or about $\frac{1}{3}$ greater than is usual for pipes, and placing

$$\zeta_2\left(\frac{d_m}{d_2}\right)^4 = \tfrac{1}{15} = 0{\cdot}067\,,$$

we obtain

$$\kappa = 1 + 0{\cdot}05 + 0{\cdot}12 + 0{\cdot}067 = 1{\cdot}24\,,$$

while for the case that no hose is used, the water being thrown directly from the stand pipe, we can place

$$\kappa_1 = 1 + 0{\cdot}05 + 0{\cdot}067 = 1{\cdot}12.$$

The pressure which must be exerted on the water by the plunger in its descent is now determined to be

$$P_0 = F\gamma\kappa\frac{w^2}{2g} = \frac{4}{3}F\kappa h\gamma,$$

and hence taking into account the friction at the plunger we obtain for the force which must be communicated to the latter

$$P = \left(1 + 4\phi\frac{b}{d}\right)P_0 = \frac{4}{3}\left(1 + 4\phi\frac{b}{d}\right)F\kappa h\gamma.$$

Assuming in this formula

$$\left(1 + 4\phi\frac{b}{d}\right) = 1{\cdot}15,$$

and introducing the values found above $\kappa = 1{\cdot}24$ and $\kappa_1 = 1{\cdot}12$, we accordingly have

$P = 1{\cdot}90Fh\gamma$, when the water is forced through hose, and
$P = 1{\cdot}72Fh\gamma$, when it escapes directly from the stand pipe.

In hand fire engines the available force P is, as a rule, given by the maximum number of men in each working shift, and hence for a given height of stream h, the volume thrown Q and the size of the pump are already determined. As the men engaged at the engine remain at work only a short time, we can assume the effect produced by them much greater, empirically about three times greater, than when a working day of eight hours is taken as the basis (6 m. kg. or 43·4 ft. lbs.) If we therefore fix the effect of each workman per second at

18 m. kg. [130·2 ft. lbs.], and suppose the brakes to move at a velocity of 1·6 m. [$5\frac{1}{4}$ ft.], it is obvious that, since each man exerts a downward pressure only or through an average distance of 0·8 m. [2·62 ft.], the force derived from each individual will be

$$K = \frac{18}{0 \cdot 8} = 22 \cdot 5 \text{ kg. } [49 \cdot 6 \text{ lbs.}]$$

If in all $2z$ men are at work, or z men on each side of the fire engine, and if further a denotes the distance of the brakes, and b that of the pumps from the fulcrum of the lever, then, neglecting the trifling journal friction, we have

$$P = zK\frac{a}{b} = \frac{4}{3}\,1 \cdot 15Fh\kappa\gamma,$$

and with $K = 22 \cdot 5$ kg. [49·6 lbs.], when pumping through hose,

$$z\frac{a}{b}\,0 \cdot 085Fh\gamma = 85Fh \left[z\frac{a}{b} = 0 \cdot 038Fh\gamma = 2 \cdot 39Fh\right],$$

or

$$F = \frac{za}{85bh} \text{ sq. m. } \left[F = \frac{za}{2 \cdot 39bh} \text{ sq. ft.}\right],$$

while, for the case that the water escapes directly from the stand pipe, we obtain

$$z\frac{a}{b} = 0 \cdot 076Fh\gamma = 76Fh \left[z\frac{a}{b} = 0 \cdot 035Fh\gamma = 2 \cdot 16Fh\right]$$

and

$$F = \frac{za}{76bh} \text{ sq. m. } \left[F = \frac{za}{2 \cdot 16bh} \text{ sq. ft.}\right]$$

The area of plunger must be taken twice as great when there is only *one* single-acting pump cylinder, and only half as great when two double-acting pump cylinders are used.

From the velocity $v = 1 \cdot 6$ m. [$5\frac{1}{4}$ ft.] of the brakes follows the average velocity of the plungers $v_0 = \frac{b}{a}v$, and hence, placing the actual volume Q delivered equal to 85 per cent of the theoretical, we obtain for the quantity thrown by *both* pumps per second

$$Q = 0 \cdot 85F\frac{b}{a}v = 1 \cdot 36\frac{b}{a}F \text{ cub. m. } \left[Q = 4 \cdot 46\frac{b}{a}F \text{ cub. ft.}\right]$$

After this value has been computed, we can determine the diameter of the nozzle from

$$\frac{d_m^2\pi}{4}w = Q, \text{ giving } d_m = 1{\cdot}13\sqrt{\frac{Q}{w}}.$$

Denoting by s the distance passed through by each brake per stroke, which distance in larger engines varies from 1 to 1·2 metre [3·28 to 3·94 ft.], the number of single strokes n of *each* plunger will be given by

$$n = \frac{60v}{s} = \frac{96}{s}\text{ m.} \left[= \frac{315}{s}\text{ ft.}\right],$$

or, in other words, it may be assumed at from 80 to 90. As two plungers are used, n cylinder volumes are also forced into the air chamber, and hence we have the volume $Q = 0{\cdot}85nFs_0$, where $s_0 = \frac{b}{a}s$ is the stroke of plunger, which varies from 0·16 to 0·25 metre [6 to 10 ins.]

The size of the air chamber is determined according to the rules laid down in § 29, from the fluctuations in volume of the water per stroke, and the variations allowed in the height of stream. In the latter paragraph we found for the pressures w_1 and w_2 in the air chamber, which pressures are to be assumed proportional to the corresponding heights of stream, the relation

$$\frac{w_1}{w_2} = \frac{W_2}{W_2 + \nu V},$$

derived from *Mariotte's* law, when νV is the fluctuating volume of the water, and W_2 the minimum volume of the air in the chamber. Giving this equation the form

$$\frac{w_2 - w_1}{w_2} = \frac{\nu V}{W_2 + \nu V},$$

and denoting by δ the value $\frac{w_2 - w_1}{w_2}$, which may serve as a measure of the variations in the height of stream, we can determine the volume W_2 of the air in the chamber at the moment when it contains the most water from

$$W_2 = \nu V \frac{1 - \delta}{\delta},$$

where the fluctuating volume νV is dependent on the movement of the plungers. Assuming the movement of the latter, as produced by the brakes, to be subjected to the same variations as characterise the crank motion, we shall have for νV the value found in § 29 for a double-acting pump, or $0{\cdot}210V = 0{\cdot}210F\frac{b}{a}s$, and consequently obtain for W_2 the expression

$$W_2 = 0{\cdot}210\ V\frac{1-\delta}{\delta}.$$

For instance, if the value

$$W_2 = 5V = 5F\frac{b}{a}s$$

were chosen, the inequality in the height of stream would be

$$\delta = \frac{0{\cdot}210}{5 + 0{\cdot}210} = 0{\cdot}0403,$$

or about 4 per cent.

As already explained, W_2 does not signify the total volume of the air chamber, but the volume which the original quantity of air of atmospheric pressure assumes when compressed so as to exert a pressure $\frac{w_2}{2g} = h_0 = \frac{4}{3}h$. If therefore b is the height of a water column corresponding to the pressure of the atmosphere, the total volume W of the air chamber down to the inlet of the discharge pipe must be made equal to

$$W = \frac{b+h}{b}W_2.$$

Remark.—For ordinary fire engines with two single-acting pump cylinders, and worked by a crew of from eight to thirty-two men, the diameter of plungers is from 0·12 to 0·18 metres [4¾ to 7 ins.], the volume of water thrown per minute to a height of 25 to 30 metres [80 to 100 ft.] being 0·30 to 0·60 cub. m. [10·6 to 21·2 cub. ft.], with a nozzle ranging from 12 to 20 mm. [0·47 to 0·79 ins.] in diameter. The number of *single* strokes of each plunger per minute varies from 80 to 120, being usually about 90.

EXAMPLE.—If by a single-acting, double-cylinder fire engine with a crew of $2z = 16$ men and a lever ratio of $\frac{a}{b} = 5$, water is to be thrown from the stand pipe to a height of 30 metres [98·4 ft.], the area of each plunger will be

$$F = \frac{za}{76bh} = \frac{8 \times 5}{76 \times 30} = 0{\cdot}0175 \text{ sq. m. } [0{\cdot}1883 \text{ sq. ft.}],$$

which corresponds to a diameter $d = 0{\cdot}150$ m. [5·9 ins.]

Assuming the distance travelled by each brake on the down stroke to be 1 metre [3·28 ft.], the stroke of plunger will be $s\frac{b}{a} = 0{\cdot}2$ m. [7·9 ins.] and the capacity of each pump cylinder

$$V = Fs\frac{b}{a} = 0{\cdot}0175 \times 0{\cdot}2 = 0{\cdot}0035 \text{ cub. m. } [0{\cdot}1236 \text{ cub. ft.}]$$

The number of single strokes of each plunger per minute is

$$n = \frac{60 \times 1{\cdot}6}{1} = 96,$$

and the actual volume of water thrown per second will be

$$Q = 0{\cdot}85F\frac{b}{a}v = 1{\cdot}36 \times \tfrac{1}{5} \times 0{\cdot}0175 = 0{\cdot}00476 \text{ cub. m.} = 4{\cdot}76 \text{ litres } [0{\cdot}1681 \text{ cub. ft.}]$$

Hence we now obtain for the diameter of the nozzle

$$d_m = 1{\cdot}13\sqrt{\frac{Q}{w}} = 1{\cdot}13\sqrt{\frac{0{\cdot}00476}{\sqrt{2g\frac{4}{3}30}}} = 0{\cdot}0147 \text{ metres} = \text{about 15 mm. } [0{\cdot}59 \text{ ins.}]$$

If the volume W_2 of the air in the chamber is to be equal to four times the capacity of each pump cylinder, or if

$$W_2 = 4Fs\frac{b}{a} = 4 \times 0{\cdot}0035 = 0{\cdot}014 \text{ cub. m.},$$

we shall have for the total volume of the air chamber

$$W = \frac{b + h_0}{b}W_2 = \frac{10{\cdot}34 + \frac{4}{3} \times 30}{10{\cdot}34}0{\cdot}014 = 0{\cdot}068 \text{ cub. m.} = 68 \text{ litres } [2{\cdot}4 \text{ cub. ft.}],$$

which volume may be furnished by a cylinder 0·38 metre [15 ins.] in diameter and 0·6 metre [2 ft.] high.

More detailed information on fire engines may be obtained from *Frick's* work: *Die Feuerspritze, Anleitung zu deren Bau, Berechnung, Behandlung und Prüfung*, Braunschweig, Fr. Vieweg und Sohn.

§ 35. **Driving Gear for Pumps.**—Mine pumps deriving their power from a *water-wheel* are usually driven by *cranks.* If an ordinary *vertical* water-wheel is used, no gears are needed, since the number of revolutions (four to eight per minute) of such a wheel corresponds to the usual number of double strokes of the pump. On the other hand, when *turbines* are employed, the number of revolutions is so great that one or more pairs of gears are always required in order to give the crank shaft the desired speed.

The pumps may be driven with or without the aid of *levers*. In the latter case the pump rod is suspended directly from the crank, which is either in one piece with the water-wheel shaft, or on a shaft coupled to the latter, while in the former case it is suspended from a lever which is reciprocated by the crank by means of a special connecting rod. When one pair of gears is employed, the smaller is placed on the water-wheel or turbine shaft, and the larger on the crank shaft; when two pairs of gears are made use of, a third shaft is interposed between the turbine and crank shafts, receiving the power from the former by means of a large gear, and provided with a smaller gear for driving the crank shaft. If, for instance, the turbine makes sixty revolutions per minute, and only five are required for the crank shaft, that is, for a velocity ratio of $\psi = \frac{5}{60} = \frac{1}{12}$, two pairs of gears may be employed, having the ratios $\psi_1 = \frac{1}{3}$ and $\psi_2 = \frac{1}{4}$. Owing to these necessary reductions in speed, turbines are less suited for driving mine pumps than the slow-running vertical water-wheels.

Smaller pumps that are intermittently in use, and then only for a short period, are occasionally driven from a *vertical shaft* rotated by *horse power*. The arrangement is then practically the same as when a water-wheel is the source of power, only, a pair of gears is always needed on account of the slow speed of the vertical shaft, which, with the usual length of sweep, makes only two revolutions per minute.

Pumps used for feeding boilers and for various other purposes in mills, factories, etc., are often driven from a horizontal line of shafting. Such pumps are for this purpose usually provided with a fast and a loose belt pulley mounted on a shaft, which, by a crank, either directly or indirectly through gears, connects with the piston rod. Various types of power pumps of this class are met with in practice. A novel form of power pump is the *electric pump*, which is at the present time coming into the field, especially for pumping in mines. The power is derived from an electric motor which, in some instances, by means of double inclined tooth gearing and a crank shaft with cranks at angles to each other, drives three pump pistons working in a corresponding number of cylinders. The multi-cylinder type is adopted in order to make the variations in resistance to the very regular driving force as slight as possible. In this manner the maximum variation, as stated by some builders, may be brought down to about 5 or 6 per cent of the mean resistance.—TRANSLATOR'S REMARK.

Finally, pumps are frequently operated by *windmills.* The simplest arrangement is then obtained by making a bend in the windmill shaft and attaching the pump rod to the crank thus formed. As, however, by this method a very large number of revolutions is obtained, and hence a very short stroke must be

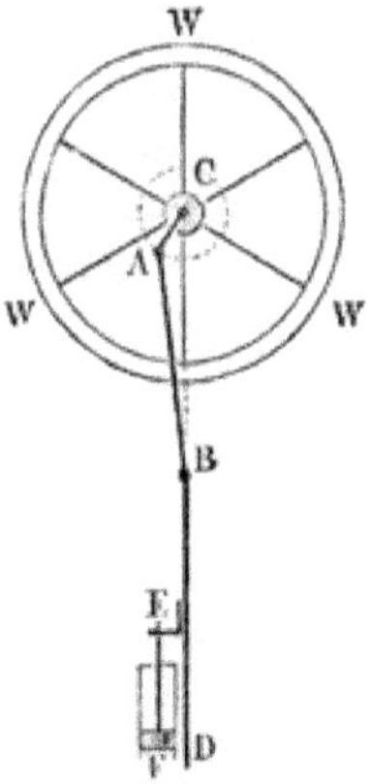

Fig. 105.

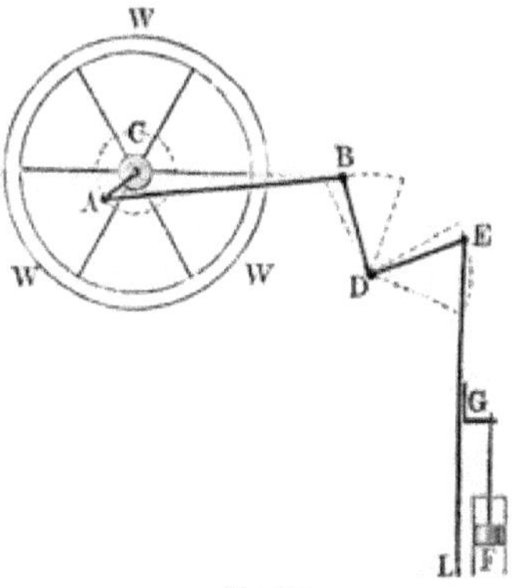

Fig. 106.

given to the pump, which materially reduces the useful effect of the latter, the better plan is to make use of a pair of gears reducing the motion of the crank shaft to one revolution for every three revolutions of the windmill shaft. Since the wind-wheel must always be pointed against the wind, and therefore must be made movable about the vertical axis of the mill building, it is necessary, when no gears are employed, to attach the pump rod to the connecting rod by means of a swivel joint, while in windmills with gears an upright gear shaft must be introduced in the centre of the building in order that the proper contact of the gears shall not be interfered with as the direction of the wind-wheel shaft changes.

The sketches shown in Figs. 105 to 110 indicate some of the arrangements above referred to.

Fig. 105 represents a simple method of driving a pump by a water-wheel W; CA is the crank, AB the connecting rod, BD the pump rod, and EF the piston rod connected to BD by means of the arm E. Figs. 106 and 107 show two lever

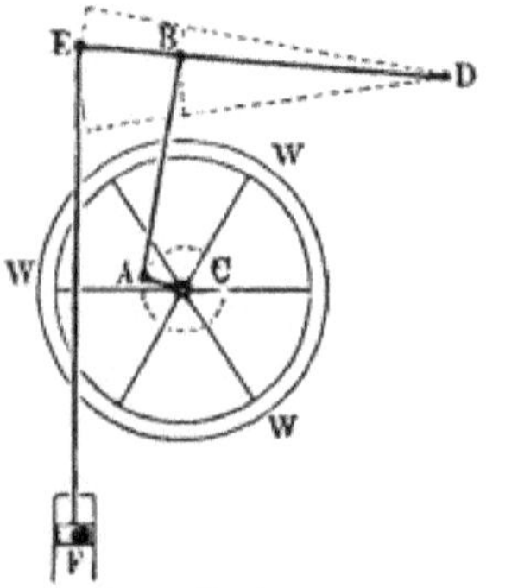

Fig. 107.

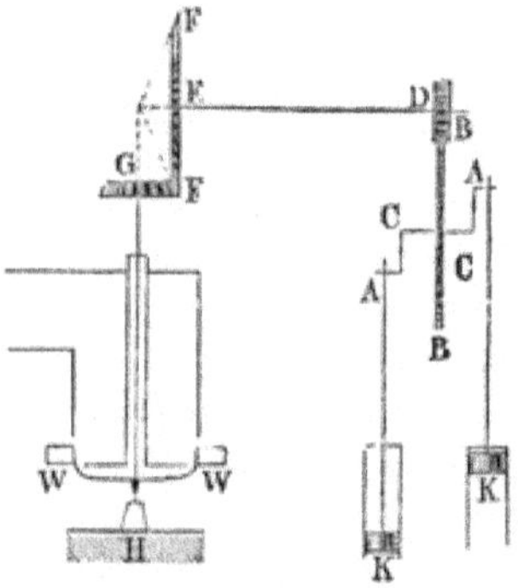

Fig. 108.

motions; in the former a bell crank or so-called *bob* BDE is used, while in the latter a straight lever DBE is acted on by the connecting rod AB with the short arm DB, and receives the load at the end of the long arm DE. In Fig. 108 the pump is driven by a turbine through the medium of two pairs

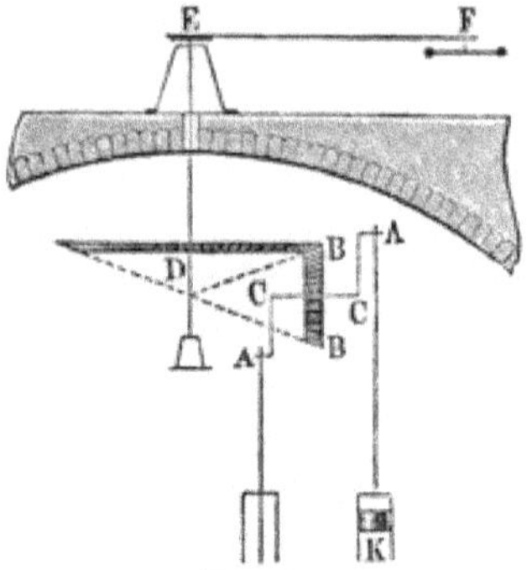

Fig. 109.

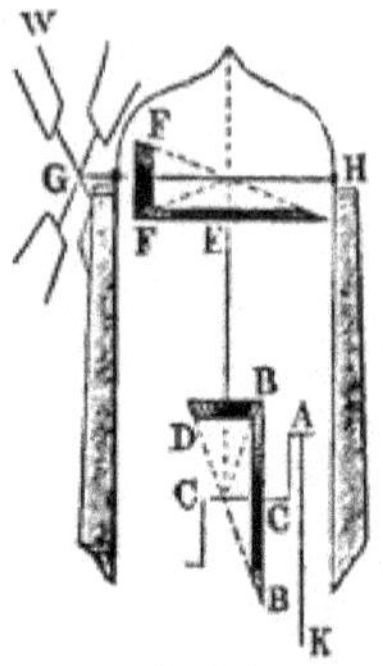

Fig. 110

of gears. W is the turbine wheel, GH the shaft belonging to the latter, DE the gear shaft, CC the crank shaft with the cranks CA, which reciprocate the pump rods AK. Fig. 109 illustrates a pump driven by horse power, DE representing the upright shaft with the sweep EF and the driving gear DB engaging the gear B on the crank shaft C. Finally, in Fig. 110 is shown a windmill pump with two pairs of gears. By means of the bevels F and E the wind-wheel shaft GH drives the vertical shaft DE, which in turn rotates the crank shaft C by the aid of the bevels D and B.

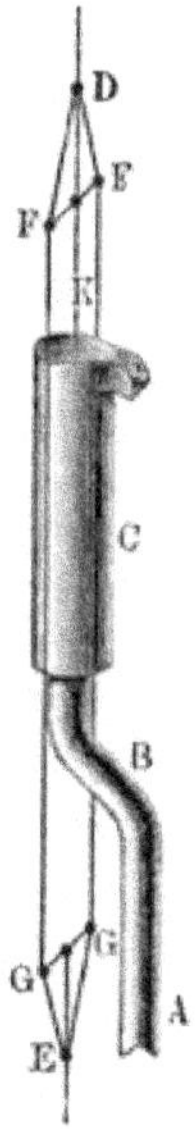

Fig. 111.

The piston rods of *mine pumps* are attached to long *rods* or systems of rods or beams working up and down in the so-called *engine shaft*. The axes of these rods as well as those of the pump cylinders will therefore fall in the line of direction of this shaft, being vertical when the shaft is vertical, and inclined when the latter inclines to the horizon; the connection to the piston rods is usually one-sided, rarely central, being accomplished by means of cross bars. Evidently a central connection would be mechanically more perfect, as it would bring a uniform tensile stress on the main pump rod, which is otherwise subjected to a bending action; but, on the other hand, such an arrangement would involve greater complication, since the rods would have to be made forked, and bends must be introduced in the suction and in some cases also in the delivery pipe. One method of applying a central connection is shown in Fig. 111, where AB is the suction pipe with a bend at B, C the pump barrel, K the piston rod attached to the arm FF, and FFGG the forked portion of the pump rod.

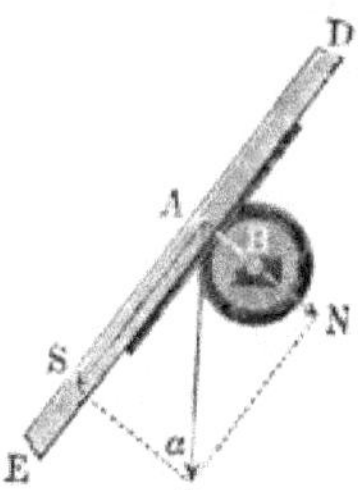

Fig. 112.

Both inclined and vertical pump rods with eccentric connections require to be guided by rollers. When the rod is inclined at an angle a to the horizon, as DE, Fig. 112, the roller B sustains the component $N = G \cos a$ of its weight G. If ϕ is the coefficient of friction, r the radius of the roller, and ρ the radius of its journal, then the journal friction reduced to the axis of the rod will be

$$F = \phi \frac{\rho}{r} N = \phi \frac{\rho}{r} G \cos a,$$

or, if the total weight G_1 of the rollers is taken into account, approximately

$$F = \phi \frac{\rho}{r} (G + G_1) \cos a.$$

Combining this resistance with the remaining lateral force $S = G \sin a$, we obtain—

(1) the force required for pulling up the rod alone,

$$P = S + F = G \sin a + \phi \frac{\rho}{r} (G + G_1) \cos a,$$

and (2) the force which assists the rod in its descent,

$$P_1 = S - F = G \sin a - \phi \frac{\rho}{r} (G + G_1) \cos a.$$

When the rod is vertical and the connection is eccentric, guide rollers, D and E, Fig 113, are introduced to keep the rod from bending. For this case, if Q is the load of the pump, a the eccentricity or the distance AB from the axis of the piston rod to that of the pump rod, and l the distance apart DE of the two rollers D and E, then each of the latter will be acted on by a force

$$N = \pm \frac{Qa}{l},$$

since the couple $(-Q, Q)$ of moment Qa must be in equilibrium with the couple $(-N, N)$ of moment Nl.

If there are several pumps attached to *the same* side of the rod, we must substitute for Qa the sum $Q_1a_1 + Q_2a_2 + \ldots$, where $Q_1, Q_2 \ldots$ represent the respective loads of the pumps and $a_1, a_2 \ldots$ the corresponding lever arms; we then obtain

$$N = \pm \frac{Q_1a_1 + Q_2a_2 + \ldots}{l}.$$

On the other hand, when Q_1 and Q_2 act on *opposite* sides of the rod, we have

$$N = \pm \frac{Q_1a_1 - Q_2a_2 + \ldots}{l}.$$

In the latter instance, if $Q_1a_1 - Q_2a_2 + \ldots$ were equal to zero, we should also have $N = 0$, that is, guide rollers would be unnecessary.

If, when the pump rod is inclined, the load is attached to the upper or lower side of it, the pressure on the rollers will be dependent both on the load on the pumps and the weight of the rod. Supposing the centre of gravity S of the rod DE (Fig. 114) to be located midway between the two rollers, then, owing to its weight, the rod will in D and E exert a normal

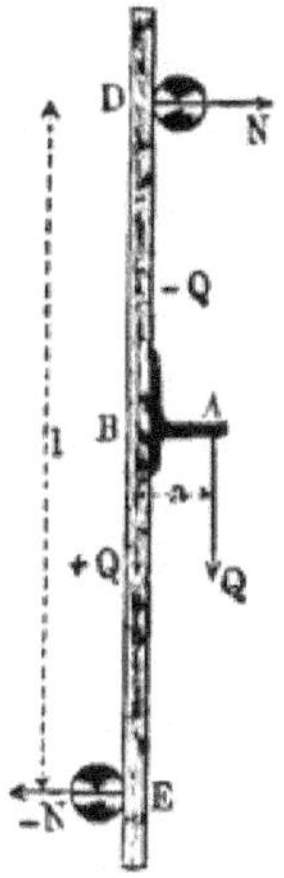

Fig. 113.

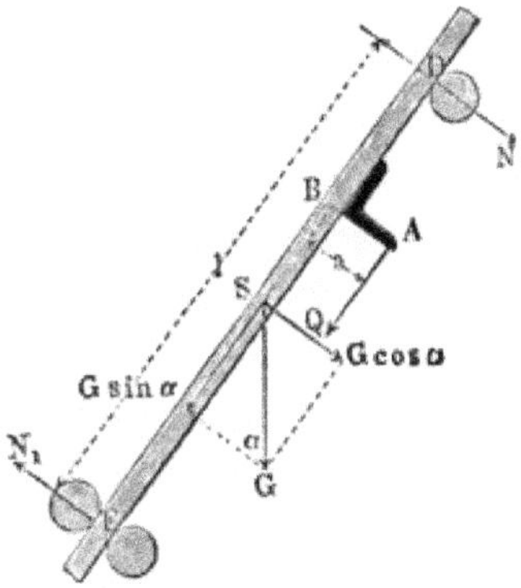

Fig. 114.

downward pressure of $\frac{1}{2}G \cos \alpha$, while, in consequence of the load Q on the pump, it will press downward with a force $\frac{Qa}{l}$ at D, and upward with the same force at E. The total pressures on the rollers will therefore be, at D,

$$N = \frac{Qa}{l} + \frac{1}{2}G \cos \alpha,$$

and at E

$$N_1 = \frac{Qa}{l} - \frac{1}{2}G \cos \alpha.$$

The journal friction of the rollers due to both of these pressures, and reduced to the axis of the pump rod, will be

$$F = \phi \frac{\rho}{r} (N + N_1),$$

which becomes either

$$=\phi\frac{\rho}{r}\frac{2Qa}{l}$$

or

$$=\phi\frac{\rho}{r}G\cos a,$$

according as $\frac{Qa}{l}$ is greater or smaller than $\frac{1}{2}G \cos a$.

When the pumps in a mine are operated by one pump rod only, the latter must be balanced. This can be done either by means of counterweights or by a hydraulic balancing arrangement (see Weisbach's *Mechanics*, vol. iii. 1, chap. 9), or by using both suction and force pumps and making the load due to the forcing action equal to that due to suction together with the weight of the rod.

Two rods are frequently employed, so connected to the driving motor or to each other that one ascends while the other descends, the two being therefore always in balance. For further details of the latter mode of connection we refer to the volume of Weisbach's *Mechanics* just cited.

§ 36. **Mine Pumps operated by a Water-Wheel.**—The general features of a pumping arrangement of this kind are illustrated in Fig. 115. The motive power is taken from an overshot water-wheel RCO, the driving water W being introduced to the wheel through the sluice gate S. The water-wheel shaft C has two cranks, one at each end, and set at 180° angle to each other. By means of connecting rods and bell cranks or bobs these cranks cause two equally loaded pump rods to move alternately up and down. In the figure only one crank CA is shown, together with its connecting rod AB, bob BDE, and pump rod EF. Of the suction pumps attached to the latter only the upper one KL is completely visible, while the next lower pump H_1, which supplies water to that above it, is but partly shown. The arms G and G_1, to which the piston rods are attached, are secured to opposite sides of the pump rod, and consequently the couples due to the eccentric connection neutralise each other, so that the bending stress in the rod is very slight. The lifted water is discharged at L, and flows away through the adit or tunnel M, while the driving water after being utilised is carried off by the day

S
W
R
A
B
C
D
E
P
O
G
L
M
G1
K
H1
F

level P. Usually in a plant of this kind there is a second fall between the day level and the tunnel; this fall is available for other purposes, the water after having been used being also carried away by the tunnel.

In making the calculations for a pumping arrangement of above description, it is evidently necessary, besides computing the load on the pistons in accordance with the rules already given, to take into consideration the friction at the bearings and pins of the bell cranks as well as of the driving cranks. The efficiency η of the whole arrangement is a product $\eta_1 \eta_2 \eta_3$ of the efficiencies η_1 of the water-wheel, η_2 of the cranks, pump bobs, connecting rods, and pump rods, and η_3 of the pumps; when these three efficiencies are known, the efficiency of the whole machine will therefore be known, and we can determine the ratio of the net load on the pumps to the driving force required.

Denoting by Q the quantity of driving water used per second, by h the fall, and letting Q_1, Q_2 . . . represent the quantities to be raised per second by the pumps to the heights h_1, h_2 . . ., we have the following general formula—

$$\eta Q h \gamma = Q_1 h_1 \gamma + Q_2 h_2 \gamma + \ldots,$$

or simpler,

$$\eta Q h = Q_1 h_1 + Q_2 h_2 + \ldots$$

From the given volumes Q_1, Q_2 . . . and the corresponding lifts h_1, h_2 . . . we can now compute the necessary water power, and from the known fall we can determine the driving water required—

$$Q = \frac{Q_1 h_1 + Q_2 h_2 + \ldots}{\eta h}.$$

If s is the stroke of the pistons, n the number of double strokes per minute, μ the coefficient of discharge (see § 25), and F_1, F_2 . . . the areas of the pistons, then we have

$$Q_1 = \mu \frac{n}{60} F_1 s, \quad Q_2 = \mu \frac{n}{60} F_2 s, \text{ etc.};$$

and consequently

$$Q = \frac{\mu n s}{60 \eta h}(F_1 h + F_2 h_2 + \ldots).$$

The number of revolutions of the water-wheel is the same as the number of double strokes n made by the pumps; consequently, if a is the radius of the wheel, its circumferential velocity will be

$$v = \frac{\pi n a}{30}.$$

From the well-known formula $Q = \epsilon d c v$, where d is the width of shrouding, c thè width of the wheel, and ϵ the coefficient of fill, the wheel must have a width

$$c = \frac{Q}{\epsilon d v} = \frac{30Q}{\epsilon \pi n d a} = \frac{9{\cdot}55Q}{\epsilon n d a},$$

in order to receive the requisite volume of driving water Q.

EXAMPLE.—A pumping plant driven by a water-wheel is to be designed for a mine; 0·1 cub. m. [3·53 cub. ft.] of water per minute are to be raised from the bottom to a height of 48 metres [157·48 ft.], then from the latter level 0·15 cub. m. [5·29 cub. ft.] to a height of 60 metres [196·85 ft.], and finally from this point 0·25 cub. m. [8·83 cub. ft.] are to be lifted per minute to the tunnel 80 metres [262·47 ft.] above it. The available water power has a fall of 10 metres [32·81 ft.]

Assuming the efficiency of the whole arrangement to be $\eta = 0{\cdot}50$, we obtain in the first place the driving water required per second,

$$Q = \frac{1}{60}\,\frac{0{\cdot}1 \times 48 + 0{\cdot}15 \times 60 + 0{\cdot}25 \times 80}{0{\cdot}5 \times 10} = 0{\cdot}113 \text{ cub. m. [3·99 cub. ft.]}$$

Giving the wheel a radius $a = 4{\cdot}5$ m. [14·76 ft.], and a speed of five revolutions per minute, we have for the circumferential velocity

$$v = \frac{3{\cdot}14 \times 4{\cdot}5 \times 5}{30} = 2{\cdot}355 \text{ m. [7·72 ft.]},$$

and thus for a radial depth of the shrouding $d = 0{\cdot}3$ m. [0·98 ft.], and a coefficient of fill $\epsilon = \frac{1}{4}$, we obtain the width of the rim

$$c = \frac{Q}{\epsilon d v} = \frac{0{\cdot}113}{\frac{1}{4} \times 0{\cdot}3 \times 2{\cdot}355} = 0{\cdot}64 \text{ m. [2·1 ft.]}$$

If we desire to overcome the respective delivery heights by using bucket pumps having lifts not exceeding 8 metres [26·25 ft.], we shall require for the lowest section 6 pumps of 8 metres [26·25 ft.] lift, in the middle section 8 pumps of 7·5 metres [24·6 ft.] lift, and in the upper section 10 pumps of 8 metres [26·25 ft.] lift; of these respectively 3, 4, and 5 pumps will be attached to each rod. Giving to all the pumps a stroke of 1 m. [3·28 ft.], and assuming the coefficient of efflux to be $\mu = 0{\cdot}85$, we obtain for the piston areas the following values—

$$F_1 = \frac{0{\cdot}1}{0{\cdot}85 \times 5 \times 1} = 0{\cdot}0235 \text{ sq. m. } [0{\cdot}2528 \text{ sq. ft.}] \text{ for the lowest pumps,}$$

$$F_2 = \frac{0{\cdot}15}{0{\cdot}85 \times 5} = 0{\cdot}0353 \text{ sq. m. } [0{\cdot}3798 \text{ sq. ft.}] \text{ for the middle ones, and}$$

$$F_3 = \frac{0{\cdot}25}{0{\cdot}85 \times 5} = 0{\cdot}0588 \text{ sq. m. } [0{\cdot}6327 \text{ sq. ft.}] \text{ for the upper pumps.}$$

Hence the corresponding diameters will be

$$d_1 = 0{\cdot}173 \text{ m. } [6{\cdot}81 \text{ ins.}], \quad d_2 = 0{\cdot}212 \text{ m. } [8{\cdot}35 \text{ ins.}],$$

and

$$d_3 = 0{\cdot}274 \text{ m. } [10{\cdot}79 \text{ ins.}]$$

§ 37. **Water-Pressure Pumps.**—The *water-pressure engines* are specially adapted to serve as prime movers for pumping plants, inasmuch as they move in the same manner and at the same velocity as pumps. *Water-pressure pumps* are, therefore, always direct acting. The methods of connecting the pump rods to the piston rods of the water-pressure engines are illustrated in Figs. 116 to 120.

In the arrangement shown in Fig. 116 the pump piston L is attached to the same rod KL as the driving piston K, and while the latter ascends water is discharged by the pump into the delivery pipe BC, whereas during the descent, which is caused by the weight of the rod, water is sucked in through the suction pipe AB. The rod KL must be guided by a stuffing box both at the driving and at the pump cylinder. The arrangement indicated in Fig. 117 is quite different, the driving rod KL being carried upward and the pump rod DS being attached to one side of it. Owing to the eccentric application of the load a couple is formed, which necessitates guiding the joint of the two rods between rollers. The weight of the rods is balanced by a counterpoise attached to a lever ACB. A more perfect joint is that shown in Fig. 118, where the pump rod BS is united to the piston rod KL by means of a forked or two-sided connection CD, and therefore the application of the load is perfectly central. In order to balance

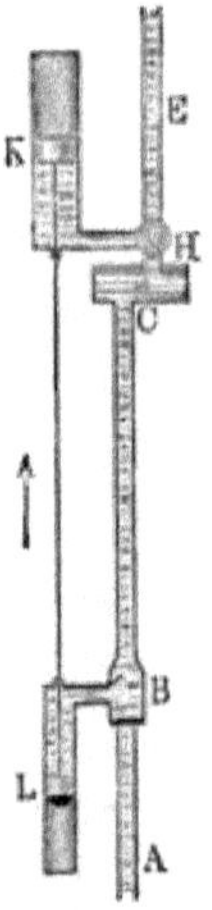

Fig. 116.

the weight of the rod in its descent, a so-called *hydraulic balance* HA is made use of, through which the driving water is discharged upward; the rod will be completely balanced by giving the water column in the hydraulic balance a height $z = \frac{G}{F\gamma}$.

When a water-pressure pump is to be employed for an inclined shaft, the driving cylinder may either, together with

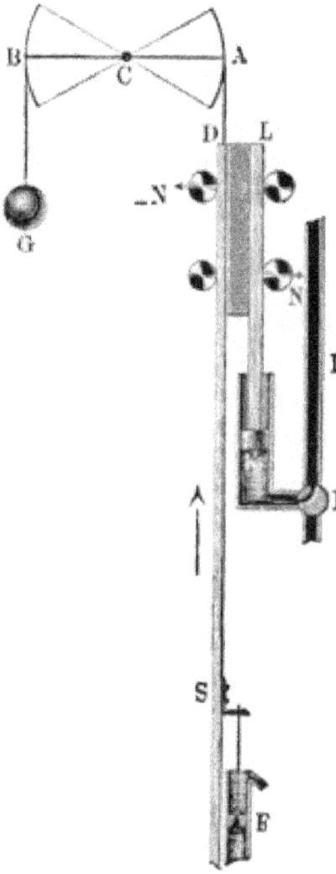

Fig. 117.

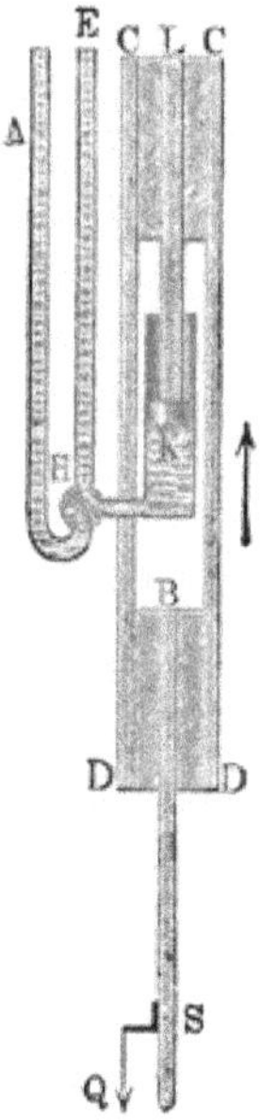

Fig. 118.

the pumps, be made to work in the direction of the shaft, or it may be placed vertically, the pump rods BS (Fig 119) being, in this case, by means of a bell crank or sweep ABC, attached to the piston rod KL, whose upper end is provided with a friction roller L travelling between vertical guides DE.

Some older constructions consist of two water-pressure engines having the same supply pipe EH (Fig. 120). Their pistons move alternately up and down, and the two engines are connected by means of an equal armed balance beam ACA_1 in such a manner that the rods keep each other in equilibrium, no further counterpoising being necessary. In the apparatus illustrated there are two force pumps LL_1 with a common suction pipe D, and a common delivery pipe S.

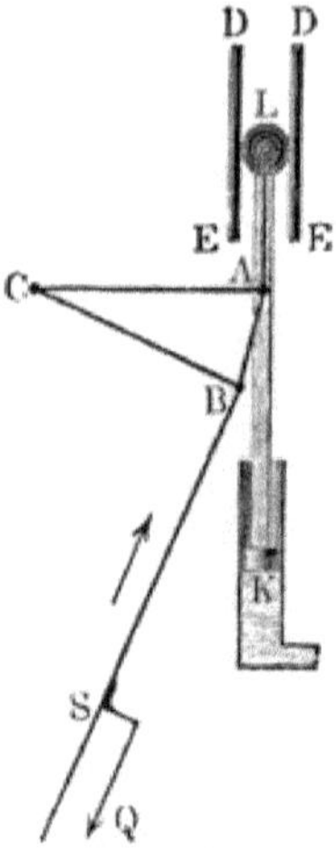

Fig. 119.

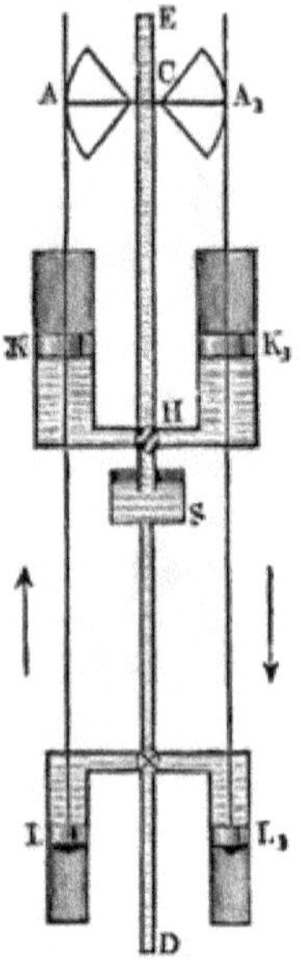

Fig. 120.

The general arrangement of a pumping plant operated by water-pressure engines is evident from Figs. 121 to 124, which represent such a plant designed by *Braunsdorf*, and in actual operation in a mine near Freiburg. In Figs. 121 and 122 the water-pressure engine is shown; EE is the supply pipe, D is a branch pipe leading to the valve cylinder and containing a stop valve, TT is the driving cylinder, CC a pipe which connects the latter and the valve cylinder, and AA is the discharge pipe with the necessary regulating valve. The pump rod PQ is attached to the piston rod K (as in Fig. 118) by means of a forked connection on each side of the driving cylinder, consisting of the coupling rods MN, etc. The main valve consists of the distribution piston S and the counter valve G; an auxiliary piston valve H is also used, attached to a rod suspended from a lever L, which is moved alternately up and down by lugs on the piston rod K. At *c* is a branch pipe leading from the supply pipe EE, and at *a* the discharge pipe for the water used in the valve cylinder. The cuts represent the driving piston on its down stroke, during which period both the main valve SG and the auxiliary piston valve H are in their highest positions. Towards the end of the down stroke of the driving piston the auxiliary valve H reaches its lowest position, and now the space above the counter piston G is brought in communication with the driving

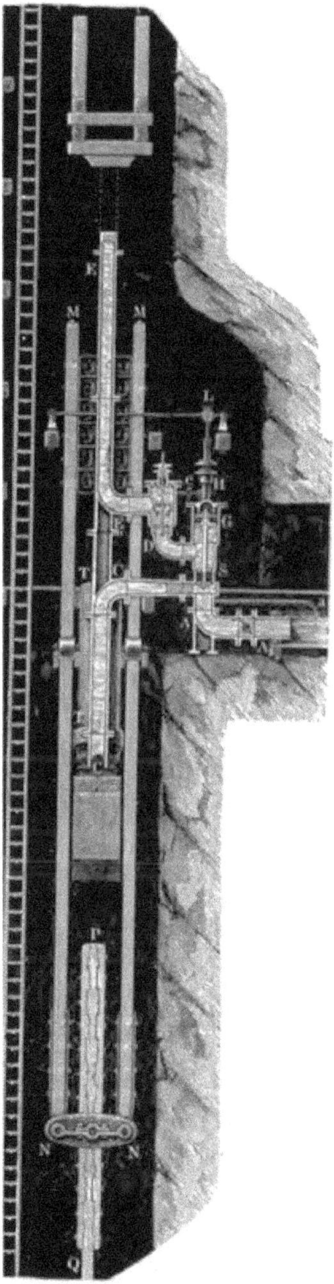

Fig. 121.

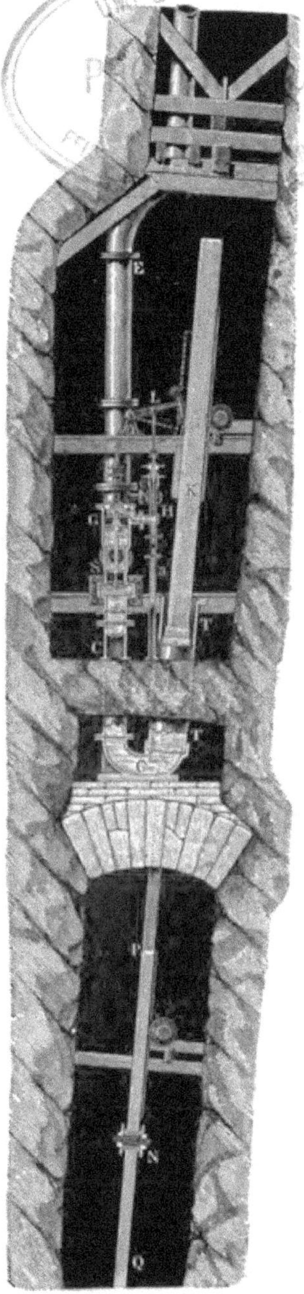

Fig. 122.

water through the pipe *c*, the double piston valve SG in consequence being forced downward. When the distribution valve S has arrived to its lowest position, the pipe CC, and accordingly the space below the driving piston, will be in communication with the driving water column EED, and the driving piston will therefore begin to ascend, etc.

Figs. 123 and 124 illustrate the arrangement of the *force pumps* driven by the above-described water engine. Also at this point the pump rod PQ forms a fork MN, which straddles the pump barrel C, and is united to the piston rod K by the scarf M. The suction and delivery pipe DS contains the suction valve V and delivery valve W, and communicates with the lower end of the barrel by means of a short pipe. During the down stroke of the pump piston K, which is shown in the figure, V is closed and W open, and consequently the water is forced upward through the delivery pipe WD; on the other hand, when the piston K ascends, V is opened and W closed, and a fresh supply of water is sucked in through the suction pipe SV from the cistern S, which obtains its water from the delivery pipe AB of the next lower pump.

The mechanical relations of a water-pressure pump of the common construction just described are to be calculated as follows.

The efficiency $\eta = \eta_1\eta_2$ of the whole machine is here a product of the efficiencies η_1 of the water-pressure engine and η_2 of the pumping apparatus. Hence, denoting by Q the volume of driving water, by h the fall for driving the water-pressure engine, and finally by Q_1, Q_2 . . . the volumes which are to be raised by the pumps to the heights h_1, h_2 . . . , we can write

$$\eta Q h = Q_1 h_1 + Q_2 h_2 + \ldots$$

Now, if s is the stroke of piston, μ the coefficient of discharge for the pumps, and n the number of double strokes per minute, we also have

$$Q = \frac{n}{60} F s, \quad Q_1 = \frac{n}{60} \mu F_1 s, \quad Q_2 = \frac{n}{60} \mu F_2 s, \text{ etc.,}$$

and consequently

$$\eta F h = \mu (F_1 h_1 + F_2 h_2 + \ldots),$$

from which we obtain the formula

$$F = \frac{\mu}{\eta} \frac{F_1 h_1 + F_2 h_2 + \ldots}{h}$$

for determination of the necessary area of driving piston. This

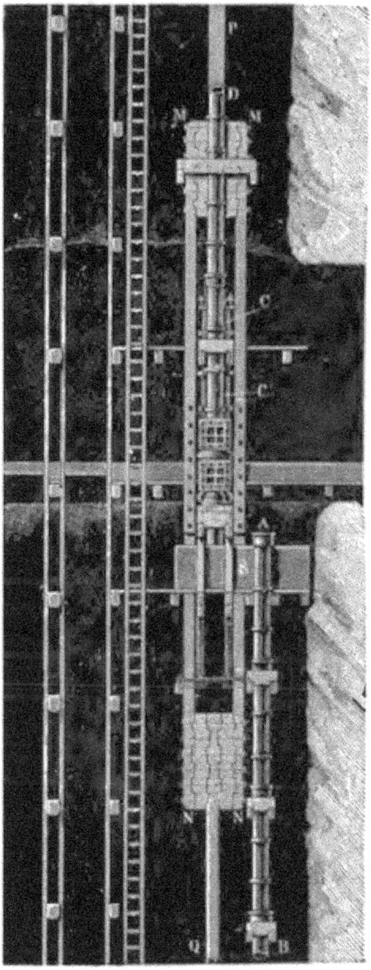

Fig. 123.

Fig. 124.

area is, as a rule, made larger in a new plant in order that sufficient power may be available in case it should be necessary to attach additional pumps; the surplus of power must evidently be disposed of by adjusting the regulating valve.

In the water-pressure pump above described the water is forced upward by the weight G of the descending rod, while the driving piston of the water engine merely sucks in water and lifts the rod.

If we place

$$h = y - z$$
$$h_1 = y_1 + z_1$$
$$h_2 = y_2 + z_2, \text{ etc.},$$

where y and z are the heads corresponding to the ascent and descent respectively of the driving piston, y_1, y_2 the heights of suction, and z_1, z_2 the delivery heights of the individual pumps, then we have

$$\eta(\mathrm{F}y\gamma - \mathrm{G}) = \mu(\mathrm{F}_1y_1 + \mathrm{F}_2y_1 + \ldots)\gamma \quad . \quad (1)$$

and

$$\eta(\mathrm{G} - \mathrm{F}z\gamma) = \mu(\mathrm{F}_1z_1 + \mathrm{F}_2z_2 + \ldots)\gamma; \quad (2)$$

hence we obtain by addition

$$\frac{\eta}{\mu}\mathrm{F}(y - z) = \mathrm{F}_1(y_1 + z_1) + \mathrm{F}_2(y_2 + z_2) + \ldots,$$

or, as above,

$$\frac{\eta}{\mu}\mathrm{F}h = \mathrm{F}_1h_1 + \mathrm{F}_2h_2 + \ldots$$

After the piston area F has been determined from this formula, we can now, from equations (1) and (2), compute either the necessary weight G of the rod or the head z of the hydraulic balance.

EXAMPLE.—If, as in the example, § 36, the volumes 0·10, 0·15, and 0·25 cub. m. [3·53, 5·29, and 8·83 cub. ft.] are to be raised per minute to the respective heights of 48, 60, and 80 metres [157·48, 196·85, and 262·47 ft.], for which purpose a fall of 75 metres [246·07 ft.] is available, the proportions of a water-pressure engine suitable for the conditions mentioned may be computed in the following manner.

Assuming again for the total efficiency of the plant $\eta = 0{\cdot}50$, we obtain, as the necessary volume of driving water per second,

$$\mathrm{Q} = \frac{1}{60}\,\frac{0{\cdot}1 \times 48 + 0{\cdot}15 \times 60 + 0{\cdot}25 \times 80}{0{\cdot}5 \times 75} = 0{\cdot}0150 \text{ cub. m. } [0{\cdot}5297 \text{ cub. ft.}]$$

Making use of three *force pumps* of the kinds shown in Figs. 123 and 124, and placing the stroke $s=2$ metres, the number of double strokes per minute at 4, and assuming the coefficient of discharge $\mu=0·85$, we have for the piston areas and diameters of the three pumps

$$F_1=\frac{0·1}{0·85\times4\times2}=0·0147 \text{ sq. m.}, d_1=0·137 \text{ m. } [F_1=0·158 \text{ sq. ft.}, d_1=5·39 \text{ ins.}]$$

$$F_2=\frac{0·15}{0·85\times4\times2}=0·0221 \text{ sq. m.}, d_2=0·168 \text{ m. } [F_2=0·238 \text{ sq. ft.}, d_2=6·61 \text{ ins.}]$$

$$F_3=\frac{0·25}{0·85\times4\times2}=0·0368 \text{ sq. m.}, d_3=0·217 \text{ m. } [F_3=0·396 \text{ sq. ft.}, d_3=8·54 \text{ ins.}]$$

For the driving piston we get

$$F=\frac{0·015\times60}{4\times2}=0·1125 \text{ sq. m.}, d=0·379 \text{ m. } [F=1·210 \text{ sq. ft.}, d=14·92 \text{ ins.}]$$

According to equation (1) the necessary pressure column is

$$y=\frac{\mu}{\eta}\frac{F_1y_1+F_2y_2+F_3y_3}{F}+\frac{G}{F\gamma}.$$

Assuming the weight of the trussed rod to be 9000 kilos [19,840 lbs.] and the height of suction of each pump to be 5 metres [16·4 ft.], we obtain

$$y=\frac{0·85}{0·5}\frac{0·0147+0·0221+0·0368}{0·1125}5+\frac{9000}{112·5}=5·56+80=85·56 \text{ m. } [280·7 \text{ ft.}]$$

As, however, the available fall is only 75 metres [246·07 ft.], the machine must be placed at a distance

$$z=85·56-75=10·56 \text{ m. } [44·63 \text{ ft.}]$$

below the level of the tunnel. By deducting the suction of 5 metres from the total heads of the pumps we obtain the heights 43 m., 55 m., 75 m. [141·08, 180·45, 246·07 ft.] to which the water is forced.

§ 38. **Water-Pressure Pumps** (*continued*).—On the pipe line between Berchtesgaden, Reichenhall, Traunstein, and Rosenheim, in Upper Bavaria, nine water-pressure pumps constructed by the celebrated *Reichenbach* are used for pumping brine over the mountain. In these machines three different systems may be distinguished, the most modern and preferable being that employed at Berchtesgaden and illustrated in Fig. 125. This machine differs from the ordinary water-pressure pumps essentially in that it has two driving pistons secured to the same piston rod; the smaller of the two is pushed upward

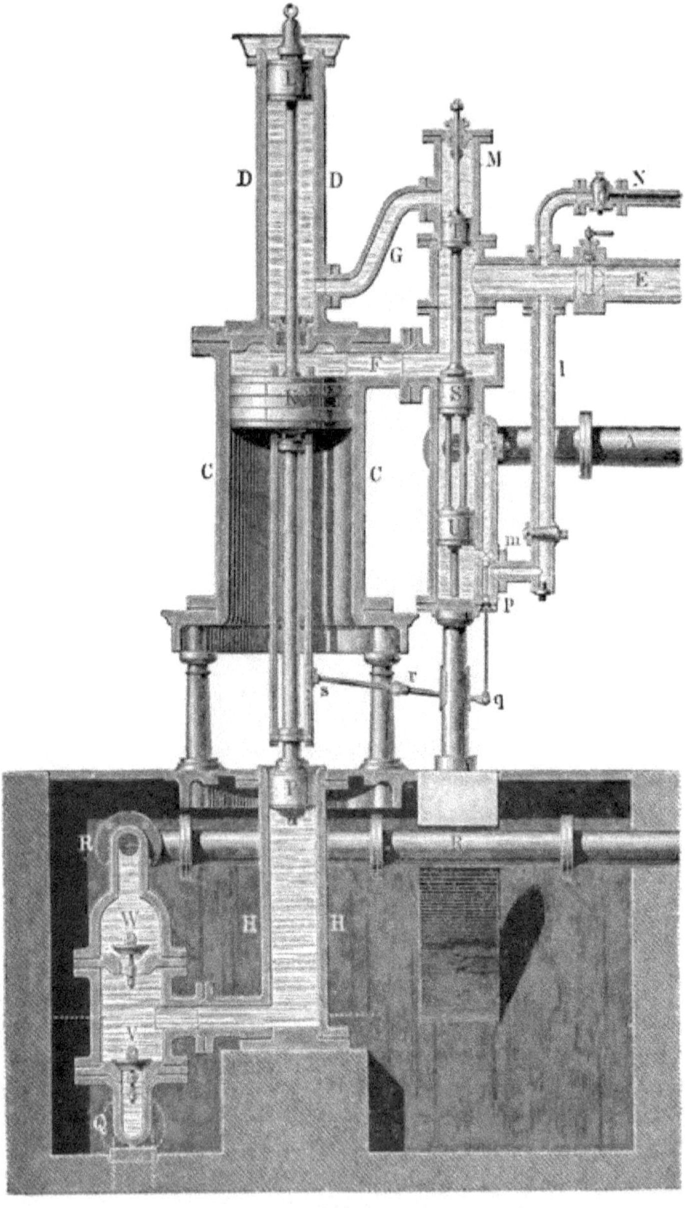

Fig. 125.

by the driving water and creates the suction necessary to cause the brine to rise, while the larger, which is moved downward by the driving water, forces the brine into the delivery pipe. The fall utilised for this machine is 116 metres [380·58 ft.], and the height to which it raises the brine is 378 metres [1240·18 ft.]

The water-pressure engine proper consists of the driving cylinders C and D, with the driving pistons K and L, the supply pipe E, the discharge pipe A, the valve rods with the three piston valves S, T, and U all enclosed in one valve cylinder, and an auxiliary valve composed of the two pistons *m* and *p* and operated by means of a lever *srq* from the driving piston rod KP. The pump is composed of the cylinder with the plunger P attached to the end of the driving piston rod, further, the valve chamber in which the suction valve V and delivery valve W are located, and the suction pipe connected at Q as well as thé delivery pipe attached at R. In the piston position shown in the figure the driving water is admitted through EF above the larger piston K, and forces the combination of pistons LKP downward, causing the water contained below the smaller piston L to flow back through the pipe G into the valve cylinder and to discharge through the orifice M; at the same time the brine contained in the pump below the plunger P is forced through the delivery valve W into the delivery pipe R. Towards the end of the down stroke the piston valve *mp* is pushed upward by the lever *srq*, thus exposing the lower side of the reversing piston U to the driving water entering from the small pipe *l*. As a result, the valve combination UST ascends, the main valve S shutting off the water from the driving piston K, while the piston valve T places the smaller driving cylinder DD in communication with the supply pipe E.

The pressure of the water on the small driving piston L now causes the trussed driving rod PKL to ascend, together with the pump plunger P, which sucks in the brine through the suction pipe at Q and suction valve V, while the water above the large piston K is discharged through FA. At the latter portion of this up stroke the double piston valve *mp* is pulled down by the lever *srq*, thus removing the pressure from the piston U; as the pressure on the larger piston S is greater than that on the smaller T, the valve combination will now descend and a new stroke will commence.

Denoting by h the fall of the water engine, by h_1 the height of suction, h_2 the delivery height, and by ϵ the specific gravity of the brine; further, letting F designate the area of the large piston K, F_1 that of the smaller driving piston L, and F_2 that of the pump plunger P, then, neglecting all wasteful resistances, we can place

$$F_1 h = \epsilon F_2 h_1,$$

and

$$(F - F_1)h = \epsilon F_2 h_2;$$

we consequently have

$$\frac{F_1}{F - F_1} = \frac{h_1}{h_2},$$

and the ratio

$$\frac{F_1}{F} = \frac{h_1}{h_1 + h_2}.$$

A pipe located above the supply pipe, and provided with a cock N, serves to let in air when it is desired to draw off the water after the machine has been stopped. In other respects the water engine does not differ from the machines described in the volume on Hydraulics. The apparatus just described has a stroke of nearly 1 metre [3·28 ft.], the diameter of the large piston is $d = 0·738$ m. [2·42 ft.], and that of the smaller driving piston as well as of the pump plunger is $d_1 = 0·292$ m. [0·96 ft.]; consequently the driving water absorbed per stroke is

$$V = \frac{\pi(d^2 + d_1^2)s}{4} = 0·495 \text{ cub. m. [17·48 cub. ft.]}$$

Adding the water needed for operating the valves, or 0·015 cub. m. [0·53 cub. ft.], we obtain for the theoretical effect of the motor per double stroke

$$A = Vh\gamma = (0·495 + 0·015)116 \times 1000 = 59160 \text{ m. kg. [427917 ft. lbs.]}$$

The theoretical volume of brine raised per double stroke is

$$V_1 = \frac{\pi d_1^2}{4}s = 0·067 \text{ cub. m. [2·37 cub. ft.]},$$

and therefore, if the specific gravity of the brine is $\epsilon = 1{\cdot}20$, we have for the theoretical effect of the pump per double stroke

$$A_1 = V_1(h_1 + h_2)\gamma = 0{\cdot}067 \times 378 \times 1200 = 30391 \text{ m. kg. } [219828 \text{ ft. lbs.}],$$

which corresponds to an efficiency for the whole machine of

$$\eta = \frac{A_1}{A} = \frac{30391}{59160} = 0{\cdot}514.$$

Some very interesting water-pressure engines have been placed in the "Königin-Marien-Schachte" at Clausthal. They were designed by *Jordan*[1] after the manner of the two-cylinder steam engines, having two horizontal driving cylinders A (Fig. 126), the piston rods by means of the connecting rods B rotating a fly-wheel shaft C having two cranks D at right angles to each other. The pumps K are in a line with the driving cylinders, each piston rod k transmitting the motion of the driving pistons a directly to the pump plungers p, while the fly-wheel shaft C merely serves to make the motion more uniform, and to operate the valve pistons S by means of the eccentrics E. Both driving cylinders and pumps are double acting, thus producing great uniformity in the discharge of the water.

The column of driving water has a height of 368·4 metres [1208·7 ft.], and the water is raised by the pumps to a height of 228·87 metres [750·7 ft.], of which 4 metres [13·1 ft.] is overcome by suction. On account of the arrangement chosen the use of a pump rod is entirely avoided, inasmuch as the machines are located at M (Fig. 127), or about 225 metres [738·2 ft.] below the level of the adit or tunnel A, the pumps sucking in water from B and forcing it up to A, while the driving water is admitted at C and conducted to the machine through the pipe D; the latter water, after being utilised in the cylinders, rises together with the water delivered by the pumps through the discharge pipe E to the tunnel. Accordingly the driving column is given by the height between A

[1] See *Zeitschr. f. Berg-, Hütten- u. Salinenwesen*, Jahrg. 1878, pp. 233 and 240.

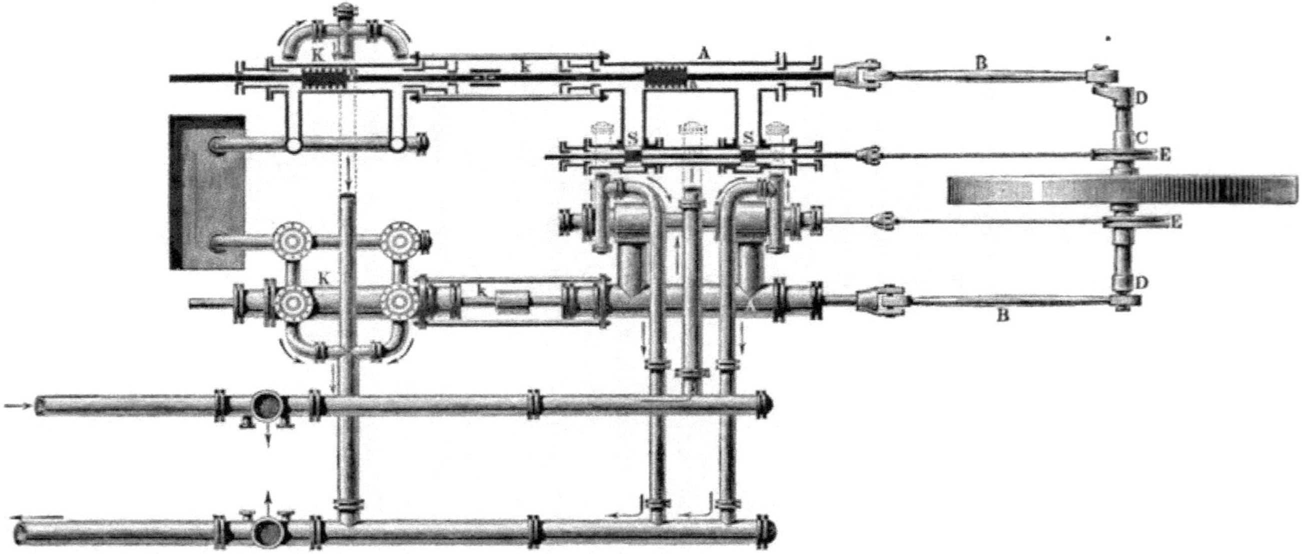

Fig. 126.

and C. As already mentioned, the necessity of using a pump rod is hereby avoided, an advantage in comparison with which the accompanying disadvantages are trifling; the latter are due to the facts that the driving water must pass through an extra length of 2 × 224·8 m. = 449·6 m. [1474·7 ft.] from A to M and again back to A, and that the driving cylinders are subjected to a correspondingly increased pressure. In modern practice the use of pump rods in mines is also avoided by the employment of steam pumps placed *underground*, as will be further referred to below.

The double-cylinder water-pressure pumps at Clausthal have given satisfactory results ever since they were started, being characterised by smoothness of running with no signs of shocks, such as are easily apt to occur in water-pressure engines, even when the regular speed of 12 revolutions per minute was increased to 16 revolutions. The driving pistons are 0·310 metres [11·81 ins.] in diameter, the pump cylinders 0·328 metres [12·92 ins.], and the stroke common to both is 0·625 metres [2·05 ft.], and consequently for 12 revolutions per minute the mean piston velocity is 0·25 m. [0·82 ft.], and the water raised per minute 1·878 cub. m. [66·32 cub. ft.]; all pipes are made of such proportions that for this effect the velocity of the water never exceeds 1 metre [3·28 ft.]

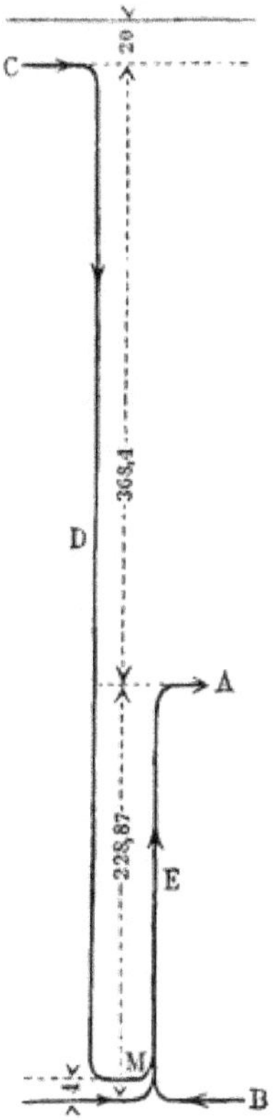

Fig. 127.

Special calculations and data regarding the exact results obtained may be found in the interesting articles by Hoppe in *Zeitschr. f. Berg-, Hütten- und Salinenwesen*, Jahrg. 1878 and 1879. According to these articles, the efficiency of the pumps increased with the velocity, amounting to 0·35 at 12 revolutions per minute. Nearly the same result was obtained by a calculation which took into account the various losses. By efficiency is here understood the ratio $\eta = \frac{Q_1 h_1}{Q h}$, where h is the fall of the driving water Q, and h_1 the height to which the volume Q_1 is pumped.

§ 39. **Steam Pumps.**—Analogous to the use of the terms steam hammer, steam pile-driver, etc., the name steam pump is applied to pumps in which the plunger is reciprocated *directly* by a steam piston and rod, in distinction from pumping engines where the motion is transmitted in a less direct manner, as by the use of walking beams, pump rods, etc. Smaller steam pumps are largely employed as boiler feed-pumps, also in procuring water for factories or public buildings, and for steam fire engines (§ 33). The modern pumping engines located in the shafts of mines are usually of the direct-acting type. In nearly all the cases cited the pumps are double-acting suction and force pumps, only the boiler feed-pumps, owing to their comparatively small capacity, being often made single-acting, with plungers of the kind shown in Fig. 64. Even in the latter instance, however, the steam cylinders are always double acting, only that the effective piston areas, by the use of a large piston rod, are made to differ materially in size, so as to conform to the great difference in the resistances offered to the plunger on forward and return stroke. With a view to obtaining a still more uniform motion a fly-wheel is sometimes employed, being placed on a revolving shaft which may also be used to operate the valve by means of an eccentric. The double-acting pumps are likewise frequently provided with a fly-wheel shaft, though in modern practice the latter is occasionally dispensed with.

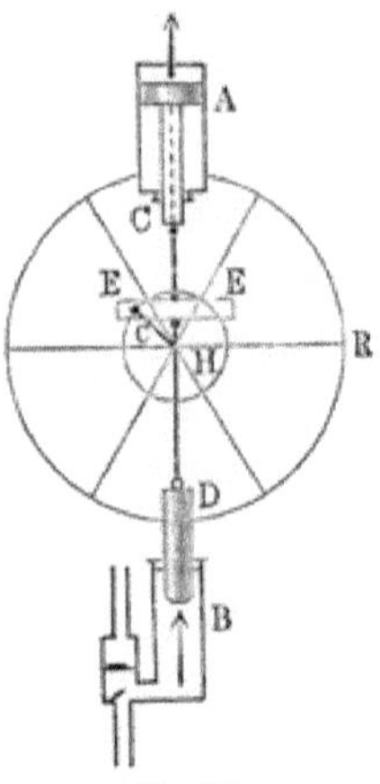

Fig. 128.

The arrangement of a single-acting steam pump can be seen in Fig. 128. Here the steam piston A is connected to the plunger D by a rod which at the middle is expanded into a slide E with a slot in which the block C receiving the crank pin of the auxiliary shaft H is guided. This movement, which has been termed the *slider crank* (see Weisbach's *Mechanics*, vol. iii. 1), offers greater resistance to motion than the ordinary crank, but where simplicity of construction is of prime importance, as is the case with small pumps, it may be used to advantage, more especially as the forces transmitted to the fly-wheel shaft, which are due only to the inequalities of pressure on the pistons, are there quite insignificant; for larger pumps this arrangement is not to be recommended.

As the height of delivery for a pump of the above description feeding a boiler against a pressure of n atmospheres when the water-level is at the height h_2 above the pump cylinder, we must introduce $h_2 + nb = h_2 + 10{\cdot}34n$ metres [$h_2 + 33{\cdot}9n$ ft.], and consequently, if friction is neglected, the resistances opposing the plunger on the up and down strokes are to each other as $h_1 : h_2 + nb$, when the height of suction is denoted by h_1. In order that steam pressure and pump resistance may correspond as nearly as possible, it would be necessary to proportion the surfaces exposed to the steam in the same ratio, *i.e.* we should make

$$\frac{h_1}{h_2 + nb} = \frac{F - f}{F} = \frac{D^2 - d^2}{D^2},$$

where F is the area and D the diameter of the steam piston A, while f is the sectional area and d the diameter of the piston rod C. But, owing to the insignificance of the resistance due to suction as compared with that due to the forcing action, disproportionately large piston rods would then be needed in boiler feed-pumps, and therefore the above rule is usually disregarded, smaller rods being used and the fly-wheel being depended on for uniformity of motion. Besides, it should be borne in mind that the pump must be capable of feeding the boiler also when the steam pressure has fallen considerably below the average, and also on this account the rod must not exceed a certain size if the annular surface of the piston is to be sufficient for producing the suction.

Such steam pumps are generally quick running (making 100 double strokes per minute and more), and therefore, according to § 26, they require an air chamber on the suction pipe, except when the height of suction is very slight. In boiler feed-pumps this height is usually but trifling, often being equal to zero or negative, in which case the water flows into the pump of its own accord.

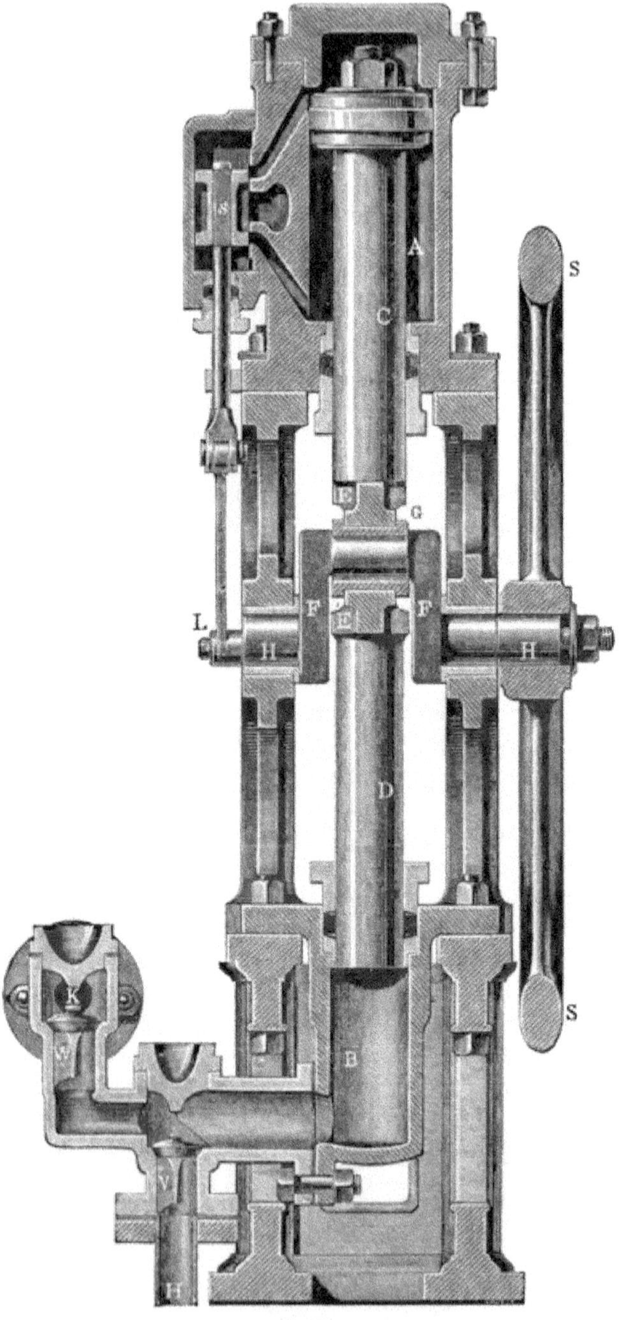

Fig. 129.

In Fig. 129 is shown a section of a vertical steam pump of this kind,[1] which can be easily understood from what has been said above. The plunger D is joined to the piston rod C by the slide E (sometimes termed "Scotch yoke"), which in its slot receives the block G for the crank pin of the fly-wheel shaft H, which is provided with an eccentric pin L for moving the valve S. In the machine referred to the diameter of steam piston is 105 mm. [4·13 ins.], that of the piston rod 65 mm. [2·56 ins.], and of the plunger 59 mm. [2·32 ins.], the stroke being 144 mm. [5·67 ins.]; the ratio of piston areas is therefore $\frac{F-f}{F} = 0{\cdot}62$, and the volume passed through by the plunger $\frac{\pi 0{\cdot}59^2}{4} 1{\cdot}44 = 0{\cdot}394$ litres [0·014 cub. ft.]

In the pump illustrated in Fig. 130, which was constructed by *Weise* and *Monski* in Halle, and is designed to be fastened to a wall, the slide is replaced by the forked connecting rod EF attached by means of the pin E and rotating the auxiliary crank shaft H with the fly-wheel R. The manner of operating the valve rod S by the eccentric L is evident from the figure, as is also the mode of attaching the suction pipe J and delivery pipe K to the valve chamber with the air chamber W.

Fig. 130.

The arrangement of a double-acting steam pump is obvious

[1] Wiebe's *Skizzenbuch*, Heft 1.

from the diagram shown in Fig. 131, where A is the steam cylinder and B the pump barrel provided with four valves. The piston rod CD between the two cylinders carries a cross bar E, from which the crank shaft H is rotated by means of the forked connecting rod EG. Also, in this case, this shaft is a mere auxiliary shaft which receives the fly-wheel R and the eccentric L for operating the slide valve S.

Such double-acting pumps have, of late, been largely constructed, without an auxiliary shaft, for use in the shafts of mines, being placed near the bottom only a few metres above the sump,[1] and made to force the water directly to the surface. For moving the steam distribution valve various methods may then be employed. Either a small double piston-valve is made use of, operated by leading steam through suitably arranged passages alternately to each side of it, and which moves the main valve,[2] or the valve is moved by an arm on the piston

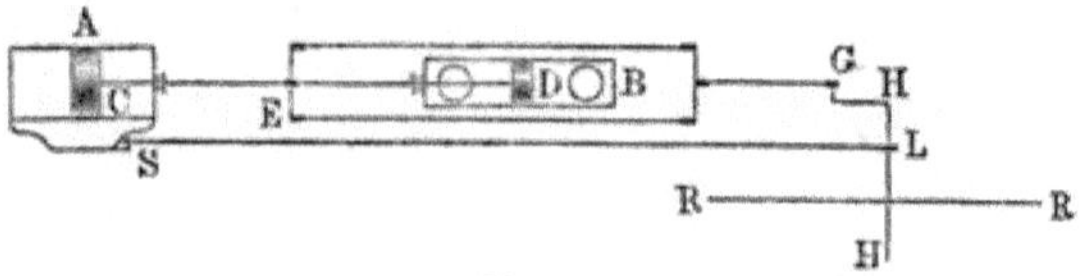

Fig. 131.

rod which strikes against collars (tappets) secured to the valve rod. The latter arrangement is used in the pump shown in Fig. 132, which was designed to overcome a difference of level amounting to 100 metres [328 ft.] at a mine in Silesia.[3] Here also the steam piston at A and the double-acting plunger are connected by a piston rod in common to both, and provided with an arm C which in its reciprocating motion strikes against the collars D and E alternately. By this means the valve rod G moves the valve contained in the

[1] In sinking or recovering a shaft vertical, direct-acting pumps are often used, which are kept suspended in the shaft, and are gradually lowered (see, for instance, *Deane's* sinking pump, illustrated in *Engineering*, London, May 1, 1885); large bodies of water may also be removed from old mines by means of horizontal pumps located on a platform floating on the surface and descending with it.—TRANSLATOR'S REMARK.

[2] See Maxwell's pumps, *Zeitschr. deutsch. Ing.* 1870, p. 196, also the pumps constructed by Tangye Brothers, and described in Rühlmann's *Allg. Maschinenlehre*, vol. iv. p. 693.

[3] See *Zeitschr. deutsch. Ing.* 1872, p. 545.

steam chest S in such a manner that the steam admitted at a_1 acts on the side of the piston exposed to it, and, after being utilised, escapes through a_2. The stroke of the pistons can be regulated by adjusting the tappets D and E. In order to prevent a flooding of the machine, the latter is placed at a height of 5·65 metres [18·53 ft.] above the sump, from which water

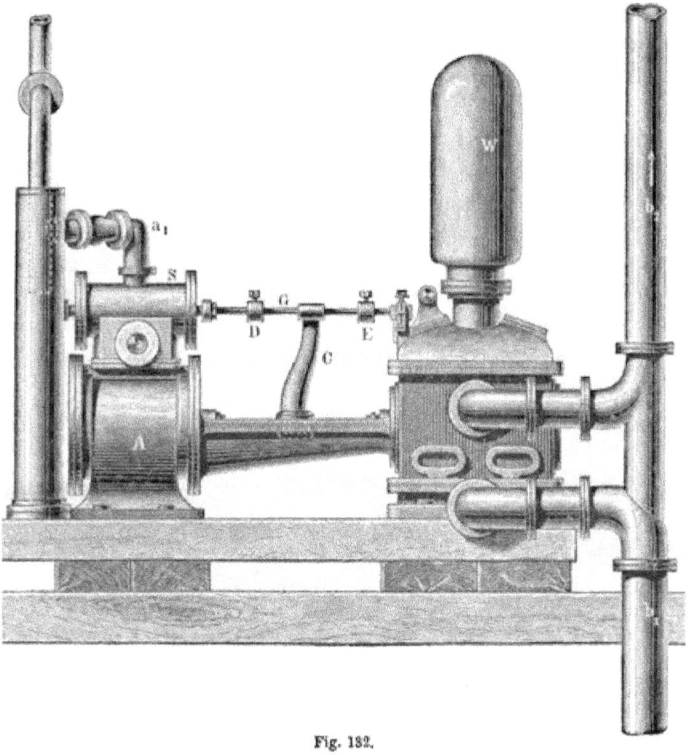

Fig. 132.

is sucked through the pipe b_1, and then forced through b_2 to the day level 100 metres [328 ft.] above.

The diameter of the pump cylinder is 0·275 m. [10·83 ins.], and that of the steam piston 0·550 m. [21·65 ins.], the stroke being 0·435 m. [17·13 ins.] The average speed is from 40 to 50 single strokes per minute, and, according to tests made, the actual capacity is from 0·89 to 0·91, or, on an average, 0·9 of the theoretical. The *duty* of the pump working under a boiler pressure of 62 lbs. [4·2 atm.] above the atmosphere was 1,306,995 m. kg. [9,453,495 ft. lbs.] per 1 cwt. of coal. Steam is conducted to the machine through pipes leading from the boilers above ground, and exhausted from the shaft through a special pipe, though the exhaust steam may be used for the ventilation of the air shaft. More recently a simple means of condensation has been introduced in pumps of this kind by conducting the exhaust steam from a_2 directly to the suction pipe b_1 of the pump, the condensing water being thus carried off together with the lifted water by the delivery pipe b_2. Besides the gain in power effected by the condensation, an advantage of this arrangement consists in that it is possible to dispense with the elaborate system of piping for discharge of the steam.

Such underground pumping engines, both with and without an auxiliary shaft, have gained extensive application in modern mining practice, owing to the fact that by their use the heavy pump rods which greatly encroach on the space in the shafts are entirely dispensed with, although a disadvantage results from the cooling of the steam caused by the great length of the steam pipes. To avoid this objection, it has been proposed to place the boilers also below ground, but this arrangement seems to have met with but little favour.

An arrangement for operating the main valve by means of an auxiliary piston, as used in the direct-acting steam pumps built by the *Knowles Steam Pump Works* at Warren, Massachusetts, is shown in Fig. 133. Here a roller on a tappet-arm T, attached to the piston rod A of the pump, near the end of each stroke depresses a rocking bar R which is fulcrumed on the central framework, and by means of a connecting link L and a short arm on the valve rod V gives a slight rotary motion to the auxiliary piston P. By this rotation small steam ports at the under-side of this piston are brought into proper relation to corresponding ports in the steam chest C. In the position shown in the cut, the space at the left of the auxiliary piston connects with the exhaust, while steam enters behind it at the right, causing it to move towards the left-hand side, and push the slide valve S with it over to the left, thereby reversing the motion of the pump piston B. When the auxiliary piston P has moved a short distance, it closes its steam port at the right and exhaust port at the left, and subsequently uncovers a port at the left through which steam enters and provides a cushion which prevents its further progress in this direction.

When the pump piston arrives near the opposite end of its stroke the roller on the tappet arm gives the auxiliary piston a rotary movement in the reverse direction, thereby bringing this piston into such a position that it will take steam at the left- and exhaust at the right-hand end through the corresponding ports in the steam chest, and enabling it to repeat the same action as on the preceding stroke.

Direct-acting pumps are frequently made of the "twin" or "duplex" type, *i.e.* in the form of two complete single steam pumps placed together side by side. In this case the steam valves are usually operated in substantially the same manner as in the example shown in Fig. 132, with the difference that the piston rod of each single pump operates, not its own slide valve, but that of the other

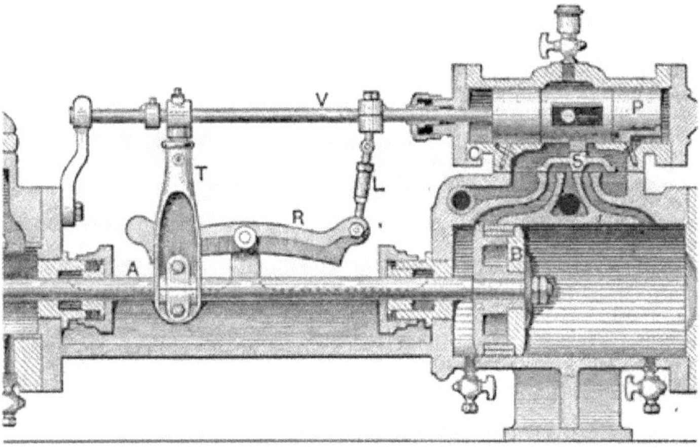

Fig. 133.

pump, by means of long vibrating arms. With this arrangement one pump takes up the motion when the other is about to leave it off, thus effecting a more uniform delivery than when a single pump is used. One half of a pump of this class, as built by *Henry R. Worthington*, New York, is shown in Fig. 134, where A is the steam piston, B the double-acting plunger working through a deep metallic packing ring, which can be easily taken out together with the plunger, C is the suction chamber and D the delivery chamber, with their respective rubber lift valves, and E is the plain slide valve, operated by a vibrating arm connecting with the piston rod of the other pump, and corresponding to the arm F shown, which moves the steam valve of the opposite engine. Pumps of this and similar types are often made to use steam expansively by "compounding" (compare § 41 and following), a high- and a low-pressure cylinder placed in line or "tandem" being used for each division of the pump. In another form of duplex pump, built by the *Hall Steam Pump Company* of New York, the engine piston of one division is made to operate the slide valve of the other division by letting in steam from the cylinder at the proper moment to an auxiliary piston; by this arrangement the use of valve rods, tappets, and vibrating arms are entirely dispensed with.

In double-acting mining pumps it is of importance that the plunger packing should be easily accessible for inspection and repairs; such pumps of the duplex class are therefore often made with four water cylinders placed in pairs, with a space between the ends, and the plungers that are visible between the latter working in exterior stuffing boxes (see *Engineering*, London, 1889, vol. i. p. 798, and *American Machinist*, New York, August 13, 1891). Duplex pumps for high-pressure hydraulic service are, for the same

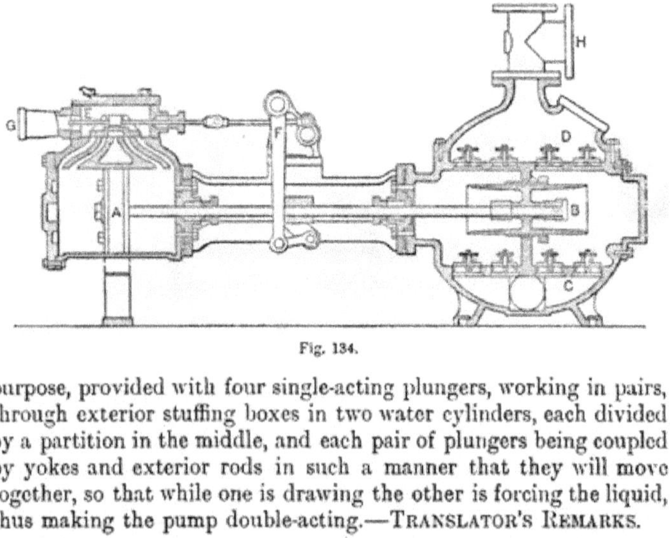

Fig. 134.

purpose, provided with four single-acting plungers, working in pairs, through exterior stuffing boxes in two water cylinders, each divided by a partition in the middle, and each pair of plungers being coupled by yokes and exterior rods in such a manner that they will move together, so that while one is drawing the other is forcing the liquid, thus making the pump double-acting.—TRANSLATOR'S REMARKS.

EXAMPLE.—A double-acting steam pump is to suck 1 cub. m. [35·316 cub. ft.] of water per minute from a well 6 metres [19·69 ft.] deep, and force it into a reservoir located 20 metres [65·62 ft.] above the pump. It is required to determine the proportions of the latter under the assumption that the average piston velocity of the latter is 0·6 m. [1·97 ft.]

Placing the actual capacity equal to 0·80 of the theoretical, we obtain, in accordance with the above requirements, the area F of the plunger from

$$0{\cdot}8F \times 60 \times 0{\cdot}6 = 1 \text{ to be } F = 0{\cdot}0347 \text{ sq. m. } [0{\cdot}3734 \text{ sq. ft.}],$$

which corresponds to a diameter of D = 0·210 m. [8·27 ins.] Letting the pump make 36 double strokes per minute, we obtain for the length of stroke

$$l = \frac{60 \times 0{\cdot}6}{2 \times 36} = 0{\cdot}5 \text{ m. } [1{\cdot}64 \text{ ft.}]$$

If the velocity of the water in the pipes is not to exceed 1 m. [3·28 ft.] the sectional area of the latter will be

$$f = 0{\cdot}6\text{F} = 0{\cdot}0208 \text{ sq. m. } [0{\cdot}2238 \text{ sq. ft.}],$$

and their diameter d = 0·163 m. [6·42 ins.]

For an absolute boiler pressure of 5 atmospheres [73·65 lbs.], and a pressure in the steam cylinder of $\frac{3}{4}$ of the former, or $\frac{15}{4} = 3{\cdot}75$ atmospheres [55·24 lbs.], and assuming the back pressure (no condenser being used) to be 1·15 atm. [16·94 lbs.]; further, placing the efficiency of the pump at 0·75, and that of the steam engine at $\frac{2}{3}$, then the area F_1 of the steam piston will be obtained from

$$\frac{2}{3}\text{F}_1(3{\cdot}75 - 1{\cdot}15)10{\cdot}334 \times 0{\cdot}6 = \frac{1000(6+20)}{0{\cdot}75 \times 60},$$

$$\left[\frac{2}{3}\text{F}_1(3{\cdot}75 - 1{\cdot}15)14{\cdot}73 \times 144 \times 1{\cdot}97 = \frac{62{\cdot}5(19{\cdot}69 + 65{\cdot}62)35{\cdot}316}{0{\cdot}75 \times 60}\right],$$

which gives F_1 = 0·0538 sq. m. [0·578 sq. ft.], corresponding to a diameter D_1 = 0·262 m. [10·31 ins.]

Example 2.—A single-acting steam pump is to feed 50 litres [1·7658 cub. ft.] of water per minute into a boiler from a well 5 metres [16·4 ft.] deep. What must be the area of plunger when the velocity of the latter is not to exceed 0·4 metres [1·31 ft.], and what size must be given to the steam piston if feeding is to be possible even when the boiler pressure has fallen to *one* atmosphere (14·73 lbs.) above the outside atmosphere?

Assuming the capacity to be 0·80 of the theoretical, we obtain the area of plunger from

$$\frac{0{\cdot}80\text{F} \times 0{\cdot}4}{2} = \frac{0{\cdot}050}{60}, \text{ which gives F} = 0{\cdot}0052 \text{ sq. m. } [0{\cdot}056 \text{ sq. ft.}],$$

corresponding to a diameter of D = 0·081 m. [3·19 ins.] Further, giving the pump a stroke s = 0·150 m. [5·91 ins.], we have for the number of revolutions of the fly-wheel per minute

$$n = \frac{60 \times 0{\cdot}4}{2 \times 0{\cdot}15} = 80.$$

The resistance of the water in pipes and valves, together with the friction of the pump plunger, may be estimated at 25 per cent of the useful effect, and consequently the work absorbed per double stroke of plunger will, for a boiler pressure of one atmosphere, or 10·334 m. [33·9 ft.] water column, be given by

$$A = 1{\cdot}25F(5 + 10{\cdot}334)1000 \times 0{\cdot}150 = 2875F = 14{\cdot}95 = \backsim 15 \text{ m. kg.}$$

$$\left[A = 1{\cdot}25F(16{\cdot}4 + 33{\cdot}9)62{\cdot}5 \times \frac{5{\cdot}91}{12} = 1935F = 108{\cdot}36 = \backsim 108{\cdot}5 \text{ ft. lbs.}\right]$$

Now assuming that the piston rod has the same sectional area as the plunger, or 0·0052 sq. m. [0·056 sq. ft.], and denoting the area of steam piston by F_1, the steam pressure behind the piston by p, and the back pressure by p_0, then the work done, theoretically, by the steam piston per double stroke will evidently be expressed by

$$[F_1 p - (F_1 - F)p_0 + (F_1 - F)p - F_1 p_0]s = (2F_1 - F)(p - p_0)s.$$

If, on account of the small size of the engine, we place the efficiency equal to 0·60 only, we shall have for the determination of F_1 the equation

$$0{\cdot}60(2F_1 - F)(p - p_0)0{\cdot}15 = A$$

$$\left[0{\cdot}60(2F_1 - F)(p - p_0)\frac{5{\cdot}91}{12} = A\right].$$

Placing here the back pressure $p_0 = 1{\cdot}1 \times 10334$ kg. [$p_0 = 1{\cdot}1 \times 33{\cdot}9 \times 62{\cdot}5$ lbs.], and p equal to 80 per cent of the boiler pressure, or 0·80 × 2 × 10334 kg. [0·80 × 2 × 33·9 × 62·5 lbs.], and after substituting the values found above, $F = 0{\cdot}0052$ [$F = 0{\cdot}056$], $A = 15$ [$A = 108{\cdot}5$], and $s = 0{\cdot}15$ $\left[s = \frac{5{\cdot}91}{12}\right]$, we obtain

$$0{\cdot}60(2F_1 - 0{\cdot}0052)(1{\cdot}6 - 1{\cdot}1)10334 \times 0{\cdot}15 = 15$$

$$\left[0{\cdot}60(2F_1 - 0{\cdot}056)(1{\cdot}6 - 1{\cdot}1)33{\cdot}9 \times 62{\cdot}5 \times \frac{5{\cdot}91}{12} = 108{\cdot}5\right],$$

or

$$930F_1 = 15 + 2{\cdot}418 = 17{\cdot}418,$$

which gives $F_1 = 0{\cdot}0187$ sq. m. [0·2012 sq. ft.], and thus a piston diameter of $D_1 = 0{\cdot}154$ m. [6·06 ins.]

It is evident that feeding can be done also for higher steam pressures. For instance, if the boiler pressure is five atmospheres in excess of the barometric pressure, the work required per double stroke will be given by

$$A = 1{\cdot}25 \times 0{\cdot}0052(5 + 5 \times 10{\cdot}334)1000 \times 0{\cdot}150 = 55{\cdot}3 \text{ m. kg.}$$

$$\left[A = 1{\cdot}25 \times 0{\cdot}056(16{\cdot}4 + 5 \times 33{\cdot}9)62{\cdot}5 \times \frac{5{\cdot}91}{12} = 400 \text{ ft. lbs.}\right],$$

of which the fraction

$$A_1 = \frac{5}{56{\cdot}67}A = 4{\cdot}7 \text{ m. kg. [34 ft. lbs.]}$$

is absorbed for the suction on the up stroke, while

$$A_2 = \frac{51{\cdot}67}{56{\cdot}67}A = 50{\cdot}6 \text{ m. kg. [366 ft. lbs.]}$$

is expended during the down stroke. The pressure p now required in the steam cylinder will be obtained from the equation

$$55{\cdot}3 = 0{\cdot}6(2F_1 - F)(p - 1{\cdot}1 \times 10334)0{\cdot}150 = 0{\cdot}09(2 \times 0{\cdot}0187 - 0{\cdot}0052)(p - 11367)$$

$$\left[400 = 0{\cdot}6(2 \times 0{\cdot}2012 - 0{\cdot}056)(p - 1{\cdot}1 \times 33{\cdot}9 \times 62{\cdot}5)\frac{5{\cdot}91}{12}\right],$$

from which we deduce

$$p = \frac{55{\cdot}3}{0{\cdot}0029} + 11367 = 30436 \text{ kg. } [p = 43{\cdot}3 \text{ lbs. per sq. in.}],$$

or less than three atmospheres. On the up stroke the steam performs an amount of work

$$A' = 0{\cdot}6[(F_1 - F)p - F_1 p_0]0{\cdot}150 =$$
$$0{\cdot}6[(0{\cdot}018 - 0{\cdot}052)30436 - 0{\cdot}0187 \times 11367]0{\cdot}15 = 17{\cdot}85 \text{ m. kg. } [129 \text{ ft. lbs.}],$$

and on the down stroke an amount

$$A'' = 0{\cdot}6[F_1 p - (F_1 - F)p_0]0{\cdot}15 =$$
$$0{\cdot}6[0{\cdot}0187 \times 30436 - (0{\cdot}0187 - 0{\cdot}0052)11367]0{\cdot}15 = 37{\cdot}45 \text{ m. kg. } [271 \text{ ft. lbs.}]$$

Consequently during each up stroke an amount of energy

$$A' - A_1 = 17{\cdot}85 - 4{\cdot}7 = 13{\cdot}15 \text{ m. kg. } [95{\cdot}11 \text{ ft. lbs.}]$$

is expended in accelerating the fly-wheel, and during each down stroke the same amount

$$A_2 - A'' = 50{\cdot}6 - 37{\cdot}45 = 13{\cdot}15 \text{ m. kg. } [95{\cdot}11 \text{ ft. lbs.}]$$

must be given out by the fly-wheel. From this condition the size of the wheel may easily be computed (see Weisbach's *Mechanics*, vol. iii. 1, chap. 9) for any maximum variation in speed desired.

§ 40. **Cornish Pumping Engines.**—The characteristic feature of all pumping arrangements for mines, except when the pump is located near the bottom of the shaft, consists in the presence of a pump rod or a system of rods set in motion by the steam piston either directly or indirectly by the use of a walking beam; the mode of attaching the piston rods of the individual pump cylinders has already been explained in § 35 and following.

The direct method of driving the pump rod is evidently the simpler and least expensive, besides allowing of greater speed than when a walking beam is used, but the latter cannot very well be dispensed with when it is desired to leave a clear space directly above the shaft. For this reason the older pumping engines were always constructed as beam engines.

A special advantage of the latter type lies in the possibility of giving the steam piston a greater velocity than the pump plungers by the use of a beam of unequal arms; a common proportioning, with this object in view, being to make the arm to which the pump rod is attached about $\frac{3}{4}$ or $\frac{4}{5}$ of that acted on by the steam piston.

In single-acting pumping engines the steam acts on one side of the piston only, lifting the heavy rod, while the weight of the latter is utilised on the return stroke to force water into the delivery pipe. As a consequence the steam pressure must evidently act on the lower side of the piston in direct-acting engines, while in beam engines the upper surface is exposed to its force. When the pump rods are very heavy, which is the case in deep shafts, their weight frequently exceeds the force required for overcoming the resistances in the pumps, and it is then necessary to balance the excess of weight by means of counterpoises (compare Weisbach's *Mechanics*, vol. iii. 1, chap. 9).

Beam engines are either of *Watt's* or of the *Cornish* type. The essential difference between the two is that, while the former works with the low pressure of 1·1 atmospheres (16·2 lbs.), the latter makes use of a pressure of 5 atmospheres (73·7 lbs.) Otherwise, both constructions are condensing and use the steam expansively, in the Cornish engine from eight to twelve times, while in Watt's the expansion is only threefold. The latter circumstance explains the vast difference in coal consumption of the two engines. The maximum duty of Watt's engine was 27,500,000 foot pounds [1] per 1 bushel of Newcastle coal à 42·638 kg. [94 lbs.], or 89,167 m. kg. per kilogram of coal,[2] whereas in the best Cornish engines an average duty of 60,000,000 foot pounds was obtained with the same quantity (94 lbs.) of coal, corresponding to a duty of 194,547 m. kg. per kilogram of coal, and an hourly coal consumption of only 1·39 kg. [3·06 lbs.] per horse-power. According to later accounts of Cornish engines of the ordinary proportions,—*i.e.* from 2·5 to 3 metres

[1] See Kley, *Die Woolf'schen Wasserhaltungsmaschinen*, where data are given in great detail regarding the duty of Cornish engines.

[2] One million English foot pounds per 1 bushel of coal is equal to 3242·45 m. kg. per 1 kg.

[8¼ to 10 ft.] stroke, 1·5 to 2·5 m. [5 to 8¼ ft.] cylinder diameter, and making from 6 to 10 strokes per minute,—the coal consumption may be placed at 2 kg. [4·4 lbs.] per horse-power per hour.

Two large Cornish pumping engines, built by the celebrated machine works in Seraing, are in operation at Bleiberg near Aix-la-Chapelle.[1] Each of these engines have steam cylinders of 2·67 m. [8¾ ft.] diameter and 3·66 m. [12 ft.] stroke, and pump cylinders of 1 m. [3·28 ft.] diameter and 2·86 m. [9·38 ft.] stroke. When running at seven double strokes per minute, and without expansion, each engine develops from 700 to 800 h.p., while with five expansions 234 h.p. are developed with an economy of only 1·45 kg. [3·19 lbs.] of coal per h.p. per hour, as compared with 4 or 5 kg. for the ordinary Belgian engines.

With reference to the condensing action in these single-acting engines, which are provided with *cataracts* for producing the periodical working of the valves (see Weisbach's *Mechanics*, vol. ii., on Steam Engines), it may here only be noted that the steam, after pushing the piston ahead, is not conducted directly to the condenser, but is allowed to enter to the other side of the piston by the opening of an equalising valve introduced for this purpose. The same pressure is thus obtained on both sides of the piston, which completes its return stroke under influence of the weight of the pump rod, in the meantime forcing the steam from one side to the other. If, now, at the proper moment the equalising valve is closed, the admission valve opened, and on the exhaust side the cylinder is placed in communication with the condenser, then the steam will again force the piston ahead with its total pressure in excess of that of the condenser. Double-acting engines, having communication between cylinder and condenser at the end of each stroke, were early abandoned on account of the necessity of applying excessively large counterweights to balance the great weight of the pump rod.

In Fig. 135 an elevation of a Cornish pumping engine is shown. At A is the steam cylinder, at B the piston rod, at CE the walking beam with its gudgeons D, and at F the pump

[1] Armengaud, *Publication industrielle*, vol. iv., and John Cockerill's *Porte-feuille*.

rod attached to the short arm of the beam. With the exception of the lowest one, all the pumps are mere force pumps. During the down stroke of the steam piston the resistance to

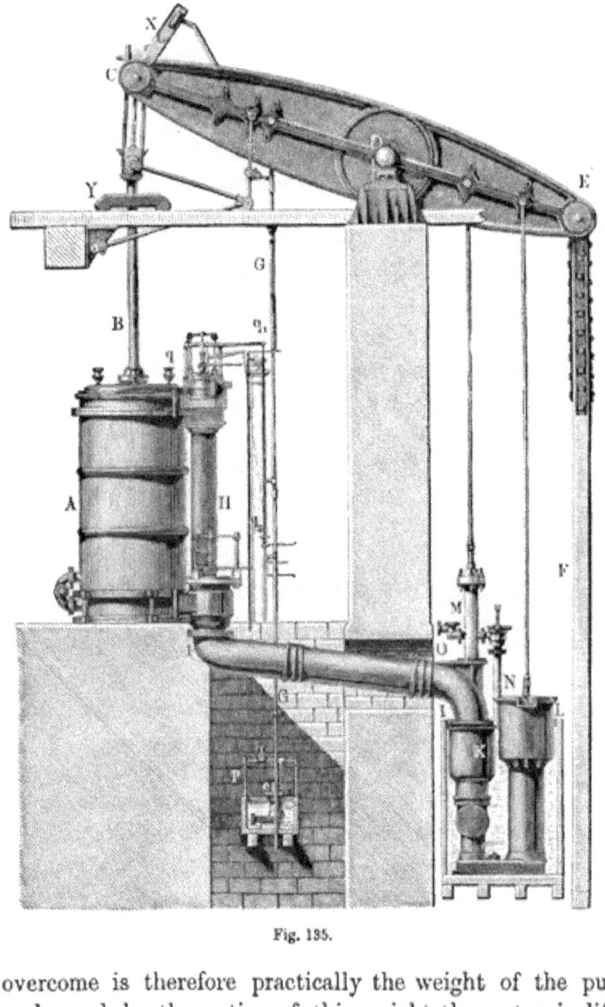

Fig. 135.

be overcome is therefore practically the weight of the pump rod only, and by the action of this weight the water is lifted on the next stroke, when, as already mentioned, the steam which was previously acting on the upper side of the piston is admitted to the lower side also. In order to limit the stroke in case of breakage of the pump rod, and prevent the piston from striking the cylinder heads, a strong lug or *catch pin* X is secured to the driving arm of the beam, and similar stops are attached at intervals to the pump rod. In case the rod should break, the catch pin would strike against the elastic block Y, and the disconnected portion of the rod would drop on heavy timber supports. From the walking beam are also suspended the valve rod G, the piston rods of the boiler feed-pump M and of the air and hot-water pump L.

We further distinguish at K the condenser, and at P the cataract. The steam distribution is governed by four double-

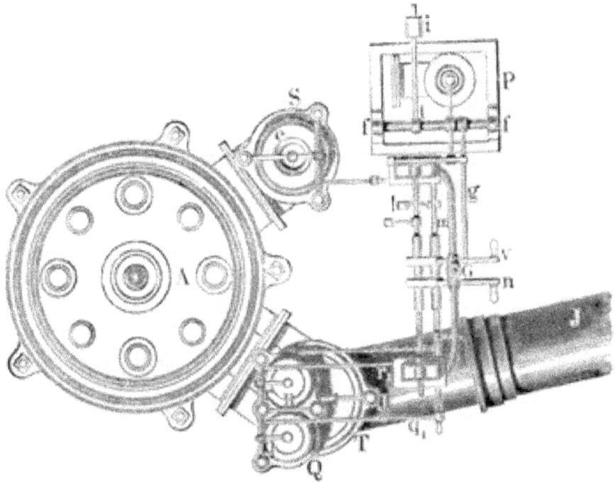

Fig. 136.

seated valves, whose position, with reference to the steam cylinder A, may be better understood from the plan in Fig. 136. These valves are

1. The throttle valve Q.
2. The admission valve R.
3. The equalising valve S, and
4. The release or condenser valve T.

The throttle valve Q, to which steam is admitted from below, is suspended from the lever qq_1, and may be adjusted according to the demand for steam by means of the rod q_2, which at one end is provided with a screw mechanism. By way of this valve steam reaches the admission valve R, which is suspended from another lever rr_1, and, after being opened by the fall of a weight, allows the steam to enter at the top of the cylinder and force the piston downwards. When the latter has passed through a portion of its stroke this valve is closed, and the remainder of the stroke is completed under influence of the expanding steam. The equalising valve S now opens, placing the upper and lower ends of the cylinder in communication with each other through the vertical pipe H, and allowing the piston to ascend under the action of the weight of the pump rod, the steam above the piston being at the same time forced into the space below it. Finally, by the dropping of a weight the release valve T is opened, discharging the steam through the exhaust pipe JJ to the condenser K. The admission valve is now again opened and a new stroke begins.

The general arrangement of the outside valve motion and cataract, and the method of operating the valves by means of levers, rods, tappets, weights, etc., will be explained below for a few special constructions.

In *Beam engines* it is quite necessary that the cylinder should be firmly secured to the foundation, owing to the tendency of the steam pressure to lift it; and besides, on account of the great oscillating mass, this type does not allow of the same velocity of running as the *direct-acting* pumping engines. For these reasons, and because of their simpler construction, the latter are employed to advantage when local conditions admit of placing the cylinder directly above the shaft. A good idea of the arrangement and working of a direct-acting pumping engine may be obtained from the annexed illustrations of a machine erected at the coal mine "Laumonier" near Liege.[1]

Fig. 137 represents a side elevation showing the steam cylinder A resting on cast-iron girders, between which the piston rod B passes. The latter is connected to the upper end of the pump rod CDC by means of a joint block provided with gudgeons CC, and a balance beam EFG supported at F is attached to the rod by the link DE. The balance beam is weighted with a sufficient number of cast-iron discs G, and

[1] See *Bulletin de la Société de l'industrie minérale*, Tome i., St. Etienne, 1855 et 1856.

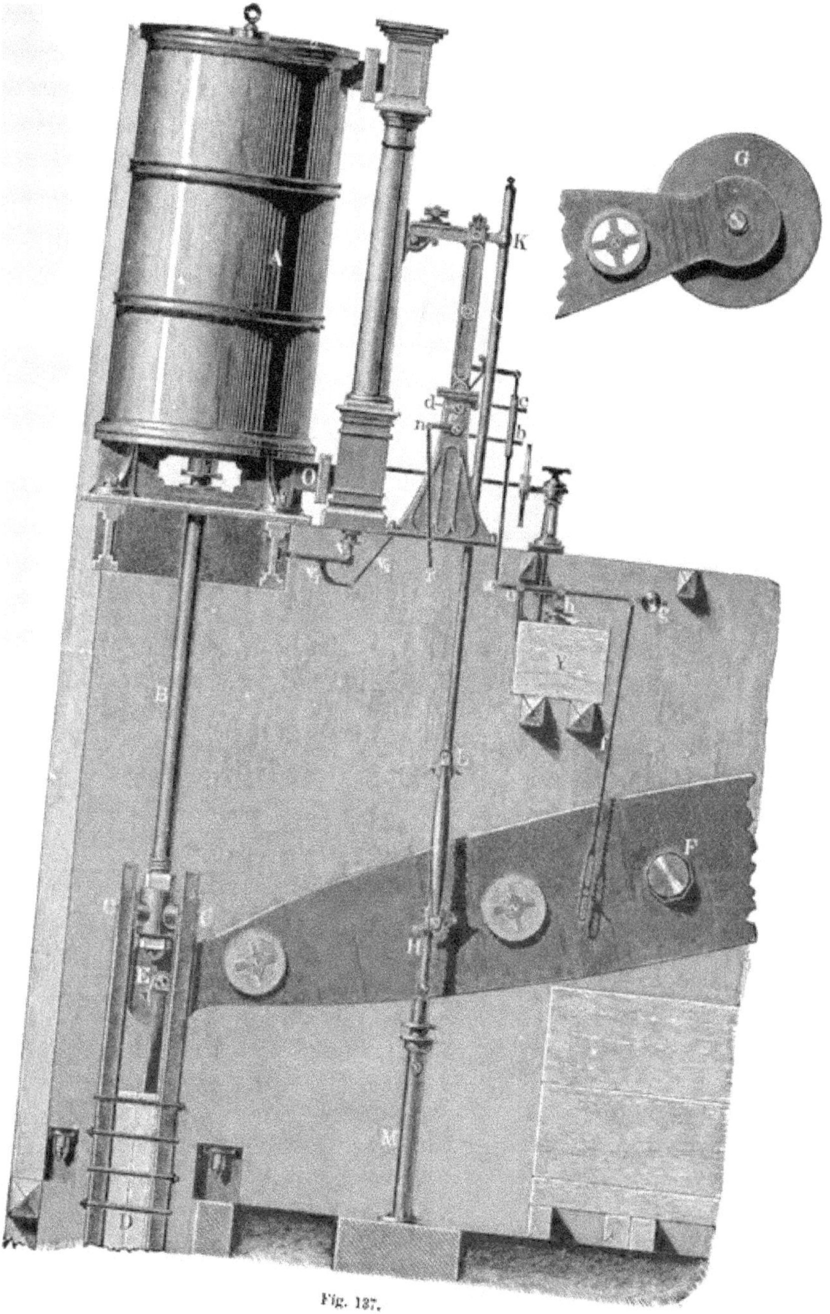

Fig. 137.

through the connecting rod HL gives motion to the valve rod KL, to which the piston of the feed-pump M is also attached. In Fig. 138 a vertical section through the valve chest may be seen, showing the two double-seated valves U and V. Steam is supplied from the boiler through the pipe N, and, when the admission valve U is opened, enters the valve chest UV, and thence the cylinder through the inlet pipe O, forcing up the piston together with the pump rod. During the ascent of the piston the valve U is pressed down, and at the end of the stroke the release and equalising valve V is opened, allowing the previously active steam to return through O, and through V reach the space W, which, by means of the pipe Q, communicates with the upper end of the steam cylinder, and by the regulating valve X and the pipe R with the outside atmosphere. In consequence of this arrangement the exhausting steam is partly utilised to keep the cylinder at a somewhat higher temperature, and thus further the action of the live steam.

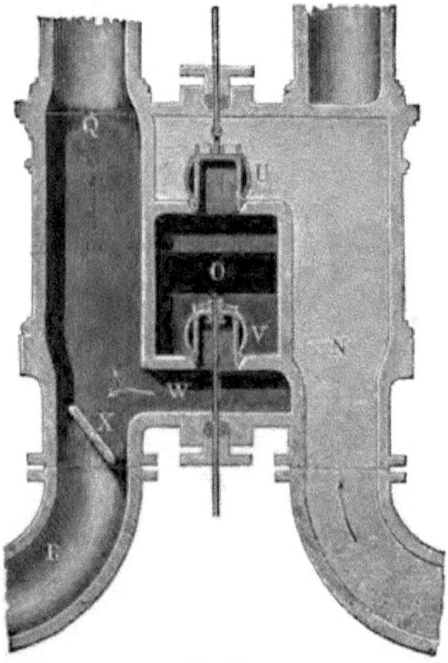

Fig. 138.

The proper opening and closing of the valves U and V is accomplished by weights, and is governed by tappets S and T attached to the valve rod LK (Fig. 139), the mechanism and mode of action being as follows. The admission valve U is connected to the arm *dk* of the valve shaft *d* by means of the rods u and u_2 and the two-armed lever u_1, and similarly the release valve V is connected to the arm *el* of the valve shaft *e* by the rods v and v_2 and the one-armed lever v_1. Further, the shaft *d* carries a rocker arm *s* which is pressed upward by the tappet S, and the shaft *e* an arm *t* which is depressed by the tappet T, the former shaft being thus revolved through a

Fig. 139.

certain angle to the left, and the latter through an angle to the right. Two weights *p* and *r*, the former acting on the arm *dm* of the shaft *d*, and the latter on the arm *en* of the shaft e, serve to give opposite rotations to the shafts. Finally, there are attached to the shafts the two stops d_1 and e_1, which engage the pawls wx and yz (Fig. 140), and thus counteract the tendency of the weights to rotate the shafts. The release of the pawls is effected by the cataract Y (Fig. 137), which essentially consists of a force pump. During the ascent of the pump rod the weighted lever aog, having its fulcrum at o, and being connected to pump h of the cataract, is pushed upward at the right by the balance beam EFG through the slotted connection f, while at the end of the up stroke it is pressed downward by the weight g; the rod ac, which is attached to the lever aog, is thus first pulled downward and subsequently pushed upward, which causes the pawls yz and xw to be alternately disengaged from the stops e_1 and d_1 by the action of the studs b and c (Fig. 140) secured to the rod.

The valve motion operates as follows. Suppose the steam piston, and consequently also the pump rod, to be in their lowest position and both valves to be closed. The weight g of the cataract now descends, and, pushing the rod ac upward, releases the pawl xw from the stop d_1, allowing the weight p suspended from the valve shaft d to turn the latter to the right and open the admission valve U by means of the arm dk, etc. When the ascending steam piston has completed a certain portion of its stroke, the rocker arm s is raised by the tappet S, which causes the admission valve U to close and the pawl wx to drop back against the stop d_1.

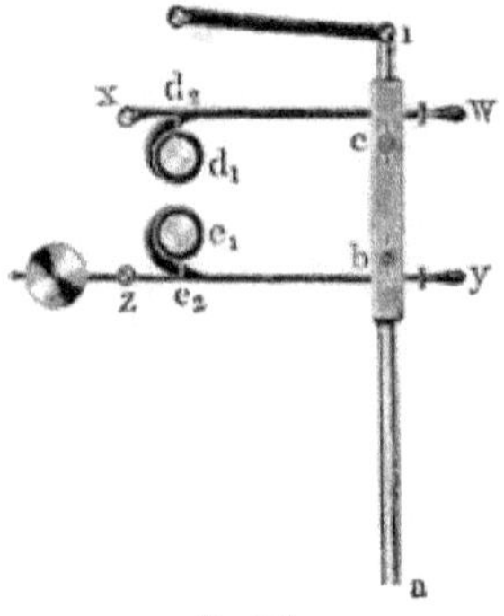

Fig. 140.

During the remainder of the stroke the steam acts expansively, and the balance beam raises the right arm og of the lever ag together with the weight g and the plunger h, thus evidently pulling the rod ac downward. At the end of the up stroke, therefore, the stud b disengages the pawl yz from the stop e_1, releasing the shaft e and causing the latter to be turned to the left by the weight r; the release valve V, which is connected to the arm el of the shaft is thus opened, and consequently the piston descends together with the pump rod and valve rod LK. When, finally, the rocker arm t is depressed by the tappet T, the valve V will close, the pawl yz will again drop against the stop c_1, and a double stroke will be completed. The pause which now occurs is evidently dependent on the time required by the plunger h to descend, and this time may be regulated by varying the discharge orifice of the plunger cylinder by the aid of a cock.

The above pumping engine has a steam cylinder of 1·88 metre [6·17 ft.] and pump cylinders of 0·5 m. [1·64 ft.] diameter, the stroke is 3 m. [9·84 ft.], and the machine makes from 6 to 8 double strokes per minute.

§ 41. **Compound Pumping Engines.**—Since a high degree of expansion is favourable to the action of steam in steam engines, pumping engines ever since the days of *Watt* have been designed to use the steam expansively, and it is essentially to the high ratio of expansion (sometimes twelve-fold) employed in the Cornish engines that we must attribute the economical performance of this type. This was made possible by the use of high-pressure steam (of 4 to 5 atm.) and by the employment of a condenser. It must be borne in mind, however, that serious disadvantages accompany the use of great expansion in pumping engines, capable at times not only of disturbing the proper action of the machine but also of causing its entire destruction, as has been the case in some instances. These disadvantages are due to the great variation of steam pressure on the piston, which evidently increases with increased expansion. The following may serve to illustrate this point. Let F denote the sectional area of the piston, p the pressure of steam per unit of area, and consequently $Fp = P$ the driving force, then, in a non-expansive engine the total resistance W, including friction, opposing the motion of the pump rod may be placed at the same amount. The pump rod in this case ascends with a practically uniform velocity, the rate of which may be easily governed by means of the throttle valve.

On the other hand, when admission of steam is cut off when the piston has completed a certain fraction $\frac{1}{\epsilon}l$ of its total stroke l, that is, when the expansion is ϵ-fold, then the previously constant pressure Fp will gradually decrease until at the end of the stroke, according to *Mariotte's* law, it will become $P_1 = Fp_1 = F\frac{p}{\epsilon}$. The resistance W of the pump rod can therefore no longer be taken at the maximum value P of the original steam pressure, but will be expressed by a mean value between the initial pressure P and the final P_1, being obtained by placing the work Wl performed by this resistance equal to work done by the steam. Basing our calculation on *Mariotte's* law, it is well known that the latter, for the volume $V = F\frac{l}{\epsilon}$ and the pressure p of the steam and the ratio of expansion $\frac{p}{p_1} = \epsilon$, will be given by

$$Vp\left(1 + \text{hyp log}\frac{p}{p_1}\right) = F\frac{lp}{\epsilon}(1 + \text{hyp log}\,\epsilon).$$

After placing the two expressions equal, or

$$Wl = F\frac{lp}{\epsilon}(1 + \text{hyp log}\,\epsilon),$$

we therefore obtain for the uniform resistance W

$$W = F\frac{p}{\epsilon}(1 + \text{hyp log}\,\epsilon) = Fp_1(1 + \text{hyp log}\,\epsilon).$$

A clear picture of the above is obtained from the well-known diagram, Fig. 141, in which above the base line $AB = l$ the pressures on the piston at different points are drawn as ordinates. While the pressure is constant from A to D_1 through the distance $\frac{1}{\epsilon}l$, and here equals $AC = D_1D = Fp$, it has at B dropped to the value $BE = F\frac{1}{\epsilon}p = Fp_1$, the area ACDEBA furnishing a measure of the work done by the steam during one stroke of the piston. If we further set off AK $= Fp_0$ to represent the pressure in the condenser, or, for non-condensing engines, the pressure of the atmosphere on the back of the piston, and if we make KF equal to the constant resistance Q of the rod, we evidently obtain in the rectangle AG the amount of work consumed by the back pressure and resistance. The point of intersection H determines the piston position in which the driving force is exactly equal to the resistances (including back pressure), and we see therefore that while the work A represented by the four-sided figure FCDH is spent in accelerating the masses, an equal amount of energy, given by the area HGE, must be derived from their retardation during the remainder of the stroke in order that the rod may continue to ascend.

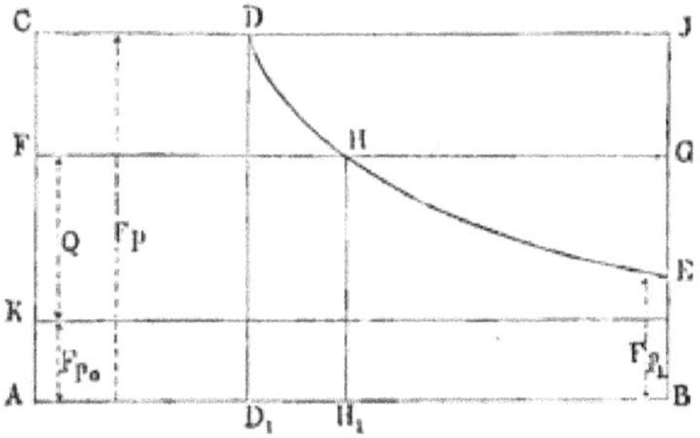

Fig. 141.

From the above it is apparent that the pump rod which begins its motion at A and concludes it at B with no velocity will reach its maximum velocity v at H. Since this velocity is produced by the work A = FCDH which has been expended on the rod, we can calculate it from the relation

$$G \frac{v^2}{2g} = A,$$

where G is the weight of the rod including counterpoises, balance lever, etc., which take part in the motion, the latter parts being reduced to the axis of the rod.

This equation shows that the maximum velocity v becomes greater as the weight G decreases and as the quantity A increases; the latter quantity evidently increases with the ratio of expansion, and becomes zero when the steam is used non-expansively. It is therefore apparent that when the weight G of the rod is very great, the value of A may be taken correspondingly large without danger of excessively increasing the velocity v.

This explains why such a high degree of expansion was possible in the deep mines of Cornwall, where the weight of the pump rods is very considerable. When, however, it was desired to use great expansion also with lighter pump rods, the velocity v of the latter increased considerably, and serious accidents have been caused by such arrangements, as, for instance, that the piston has been thrown with great force against the cylinder head, shattering it into fragments. Experience has proved that it is inadvisable to allow the velocity of the rod to exceed 2 metres [6·56 ft.] By thus limiting the maximum velocity v and deducing a value for the energy A spent in accelerating the masses with a certain ratio of expansion ϵ, the required weight G of the masses may be calculated from the equation $G\frac{v^2}{2g} = A$; in case the pump rod has a smaller weight G_1, the deficiency $G - G_1$ may be made up by adding one-half of it, or $\frac{G - G_1}{2}$, directly to the rod, and allowing the other half to act on a balance lever so as to retain the excess of weight G_1 of the rod. It is evident that this arrangement subjects the rod to a greater pull and adds to the expense of the whole construction, and this is the reason why pumping engines which operate with pump rods are often allowed to work without or with but little expansion, a greater expenditure of fuel being preferred to the above objections.

There is another disadvantage connected with the expansive use of steam in the pumping engines referred to, resulting from the variation of the steam pressure on the piston. This is owing to the fact that at the beginning of the stroke the steam pressure Fp, which in the figure is represented by AC, exceeds the back pressure, together with the resistance of the rod AF, by an amount $FC = F(p - p_0) - Q$, which causes the rod to begin its motion with the corresponding acceleration $\frac{F(p - p_0) - Q}{G}g$. As a result a *separation* of the water will occur in the lowest pump, which is always arranged for suction, and consequently a shock or water-blow will take place unless the water in the suction pipe is correspondingly accelerated by the pressure of the atmosphere, as was further explained for pumps driven by cranks in § 26. Denoting by h_1 the height of suction and length of suction pipe, and by b the height of the water barometer, we have for the maximum acceleration of the water in the suction pipe the expression

$$\frac{b - h_1}{h_1}g,$$

and consequently, under the assumption that the pump barrel

and suction pipe have the same sectional area, the following condition must be satisfied:

$$\frac{h - h_1}{h_1} > \frac{F(p - p_0) - Q}{G}.$$

As from practical considerations the height of suction h_1 is as a rule considerable, amounting to between 5 and 8 metres [16 to 26 ft.], it will be seen that a water-blow may easily occur for high degrees of expansion when the rod is considerably accelerated through the excess of pressure during the period of admission.

To avoid the above objections to a certain extent, and yet gain economy by the use of a high degree of expansion, two cylinders have been employed in the same manner as in *Woolf's* steam engine (Weisbach's *Mechanics*, vol. ii., on Steam Engines), with simultaneous expansion in both cylinders, the results obtained being very favourable. Already in the last century two cylinders were employed by *Hornblower* [1] for the same purpose in Cornish pumping engines, using low-pressure steam, but the plan was abandoned and was not again introduced until in modern times *Kley* applied it to high-pressure steam, and demonstrated the value of the principle. This may be explained by Fig. 142, which represents the essential features of one of the two single-acting pumping engines erected by *Kley* in 1861 and 1862 at the mine "Altenberg" near Aachen. The two steam cylinders A and B are placed side by side above the shaft in such a manner that the pump rod G is attached directly to the piston rod k_2 of the larger cylinder B. As usual, the weight of the pump rod while the latter descends presses up the water from the force pumps, the excess of weight being balanced by means of the counterpoise lever C attached to the pump rod by the connecting rod c_1c_2. The piston rod k_1 of the smaller cylinder A is likewise connected to the balance lever by c_3c_4, and the pump rod is therefore raised by the combined action of both pistons in their ascent, together with the force exerted by the counterweight. The steam distribution is effected by the five valves α_1, α_2, β_1, β_2, and γ, of which α_1 and β_1 are admission valves letting in steam under the pistons A and B, while α_2

[1] See Kley, *Die einfach- und directwirkenden Woolf'schen Wasserhaltungsmaschinen der Grube "Altenberg" bei Aachen*, 1865.

and β_2 are the corresponding release valves. The valve γ finally allows the escaping steam to reach the condenser through the pipe R, the air pump L being driven from c_4.

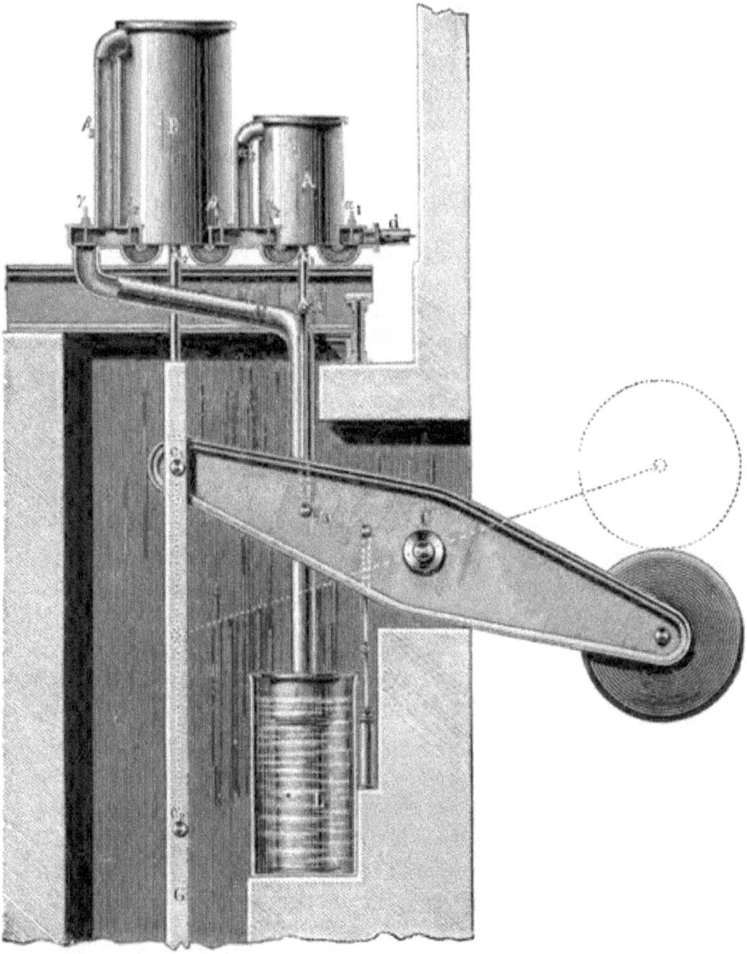

Fig. 142.

From the figure it is, in the first place, evident that if, during the descent of the rod, only the release valves a_2 and β_2 are open, the steam contained below each piston will simply escape through the communicating pipes a_3 and β_3 to the spaces above the pistons, thus establishing the same pressure on both sides of the latter and allowing the pump rod to sink under influence of its net weight. But when the ascent is to commence the valves a_2 and β_2 are closed, whereas the other three valves a_1, β_1, and γ are opened, and consequently fresh steam from the boiler will be supplied through the pipe *d* and valve a_1 to the under side of the smaller piston; moreover, the steam contained above the latter escapes through β_1 to the lower side of the larger piston B, and at the same time the space above this piston will be in communication with the condenser. From the theory of compound engines it is now apparent that the work obtained from the steam which acts on the upper side of the smaller piston and on the lower side of the larger one will equal the expansive action of the steam which originally fills the small cylinder and finally occupies the volume of the large one. Since in the engines at Altenberg the ratio of piston areas f and F are as 1 : 2, and the same ratio obtains for the length of strokes, we perceive that the steam expands four times, or in general the ratio of expansion will be ϕ, if $\phi = \frac{FL}{fl}$ denotes the ratio of cylinder volumes. If the admission valve a_1 is closed before the end of the up stroke so as to give a ratio of expansion ν already in the small cylinder, the total expansion will be given by $\epsilon = \nu\phi$. In the Altenberg engines there is no expansion in the small cylinder, and consequently, since $\nu = 1$, we have $\epsilon = \phi = 4$.

The valves are operated in the same manner as in the single-cylinder Cornish engines (by pawls, tappets, and cataracts), this being easily accomplished owing to the fact that the opening of the admission valve a_1 always coincides with that of β_1, and that a_2 and β_2 likewise always move together. It is only for the case that a ν-fold expansion is desired in the smaller cylinder that the valves a_1 and β_1 do not close simultaneously. The arrangement of the valve motion is therefore practically no more complicated than for single cylinder engines. From the diagram we can readily see that there is much less irregularity in the action of the driving force when compounding is adopted than for the same degree of expansion in one cylinder.

To make this clear, let us assume the action of the small piston A to be transferred to the point of the balance lever where the large piston B acts. This can be done by assuming at this point a piston of the area $f_1 = \frac{fl}{L}$ working in a cylinder having a volume f_1L equal to that of the smaller cylinder, or fl; we then have $\frac{F}{f_1} = \frac{FL}{f_1L} = \phi$. By this assumption the action of the steam will be unaltered, provided that the expansion in this cylinder is retained the same as before, or ν-fold, that is, that the admitted volume of steam $V = f_1 \frac{1}{\nu} L = f \frac{1}{\nu} l$ remains unaltered. We can now construct the diagrams for both pistons on the same base line $OO_1 = L$, Fig. 143.

Let p be the initial steam pressure, and $p_1 = \frac{1}{\nu} p$ the pressure below the smaller piston at the end of the stroke, then the line A will give a representation of the work done by the steam acting on the under-side of the smaller piston, provided that

$$OA = f_1 p,$$

$$O_1A_1 = f_1 p_1 = f_1 \frac{1}{\nu} p,$$

and

$$AA_2 = \frac{1}{\nu} L = \frac{1}{\nu} OO_1,$$

and that the curve A_2A_1 is constructed according to *Mariotte's* law. During this stroke the steam which was exhausted from the small cylinder during the previous stroke exerts a driving action on the large piston and a resisting effort on the upper side of the smaller one. Since the pressure of this steam is p_1 in the beginning, the initial driving force on the large piston will be $OB = Fp_1$, and the force resisting the small piston will be $Oa = f_1 p_1$. At the end of the stroke this steam has expanded from the volume f_1L to FL, or in the ratio $\frac{F}{f_1} = \phi$, the pressure therefore having dropped to $p_2 = \frac{1}{\phi} p_1 = \frac{1}{\nu\phi} p = \frac{1}{\epsilon} p$. The pressure on the piston at the end of the stroke will therefore be $Fp_2 = f_1 p_1 = O_1A_1$, so that the work done by the expanding steam on the large piston will be obtained from the line BA_1, and that done on the upper side of the small piston from the line aa_1, if we make $O_1a_1 = f_1 p_2$. These lines BA_1 and

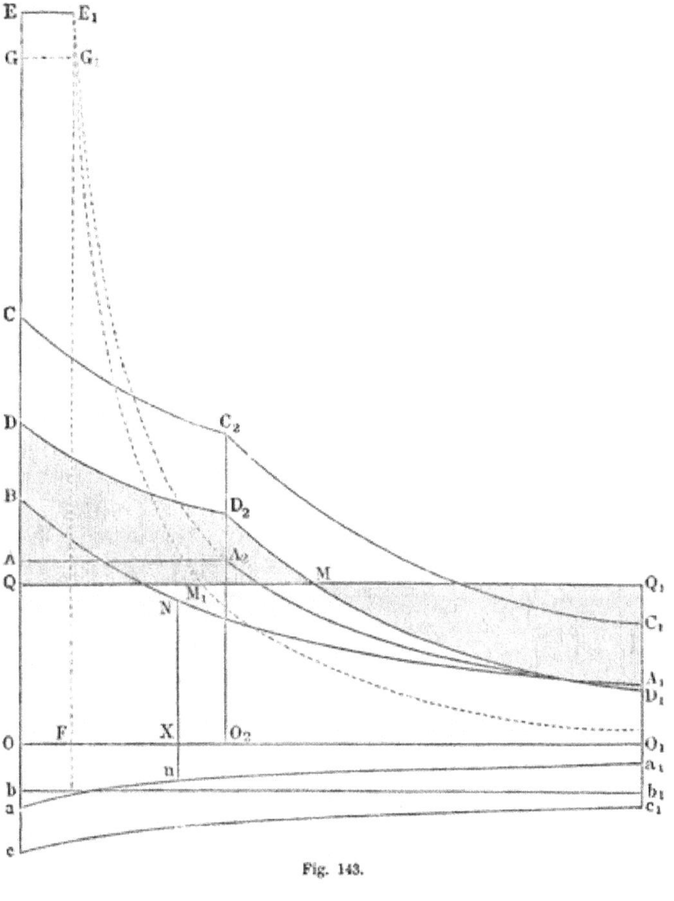

Fig. 143.

aa_1 may be easily constructed by the aid of *Mariotte's* law, as the pressure of the steam p_x for any piston position X, at the distance x from the beginning of the stroke, may be found from the equation

$$f_1 L p_1 = [Fx + f_1(L - x)]\,p_x,$$

and the ordinates

$$XN = Fp_x \text{ and } Xn = f_1 p_x$$

drawn accordingly. If, finally, p_0 denotes the pressure in the condenser, the straight line bb_1 at a distance $Ob = Fp_0$ from the base line O will represent the back pressures on the large piston. By combining all these lines in such a manner that the lines C will represent the sum of A and B, and c the sum of a and b, and by subtracting the ordinates of the resulting back-pressure curve c from those of the resulting driving-pressure curve C, we finally obtain in the area $ODD_2D_1O_1$, determined by the curve D, the total work done during one stroke of the pistons. If we now draw, parallel to the base line, the straight line QQ_1 representing the average pump resistance, we find at the point of intersection M the position where the pistons have their maximum velocity, and in the cross-hatched surface QDD_2M we have a measure of the work expended in accelerating the masses. While the average resistance of the pump rod is given by OQ, there will be a maximum driving force OD applied to it at the beginning of the stroke, so that the accelerating force is given by QD. It is now easily seen that both this excess of *force* and the *work* expended in accelerating the rod will, in the case under consideration, be considerably smaller than in a single-cylinder engine of equal power, that is, one in which the same volume of steam

$$V = f_1 \frac{1}{\nu} L$$

is allowed to expand in the same ratio $\epsilon = \nu\phi$. Such an engine must evidently have a cylinder volume

$$\epsilon V = \nu\phi f_1 \frac{1}{\nu} L = \phi f_1 L = FL,$$

that is, equal to the large cylinder, and the steam must be cut off when the piston has passed through the distance $\frac{1}{\epsilon} L$. The initial pressure, therefore, would be given by $OE = Fp$, and would remain constant throughout the distance $OF = \frac{1}{\epsilon} L$. The diameter corresponding to such an engine is represented by the dotted line $EE_1A_2A_1$ in the figure, and it clearly shows not only that the initial pressure to a higher degree exceeds the average resistance, but also that a greater amount of work, or that represented by the area QGG_1M_1, is expended in accelerating the rod than is the case in an engine constructed on *Woolf's* compound principle.

Owing to this fact, and also on account of their great economy of fuel, the compound types of pumping engines have been largely introduced in recent times, both as single and double-acting.[1] In regard to *Kley's* pumping engines at Altenberg, it may be mentioned that their steam cylinders are 1·70 m. and 1·20 m. [5·58 and 3·94 ft.] in diameter, and the stroke of the larger cylinder is 2·8 m. [9·19 ft.], and that of the smaller one half of the former. The plungers are 0·55 m. [1·80 ft.] in diameter, and their maximum velocity is 0·84 m. [2·76 ft.] when making nine double strokes per minute. Their coal consumption is 2·4 kilograms [5·3 pounds] per horse-power per hour.

§ 42. **Double-Acting Pumping Engines.**—All of the older pumping engines with pump rods were single-acting, that is to say, the steam power was utilised merely to raise the rods, which latter forced the water upward on their down stroke by the action of their weight.

In order, however, to reduce the dimensions of the steam cylinders, double-acting engines have been constructed in recent times using steam power on both strokes. There are chiefly two distinct arrangements adopted with this object in view. One is to balance the weight of the pump rod to such an extent by means of a counterpoise G (Figs. 144 and 145), that the excess of weight will be equal to one-half only of the total resistance, or $\frac{1}{2}Q$. The steam pressure required on the piston for raising the rod will therefore also be equal to $\frac{1}{2}Q = D$, and since the piston on the return stroke likewise acts with the same force to lift the counterpoise, the total force acting on the pump rod will equal the resistance Q. This arrangement was only possible with heavy wooden rods, owing to the compression to which the rod is subjected on the down stroke by the action of its weight, wrought-iron rods made of

[1] Concerning the *Woolf* engines erected by *Kley* at Saarbrücken (single-acting) and at Rüdersdorf (double-acting), see Hörmann's pamphlet, entitled *Die neuen Wasserhaltungsmaschinen*, etc.

round iron being generally of so small diameter that bending would result were they to be exposed to compression. For

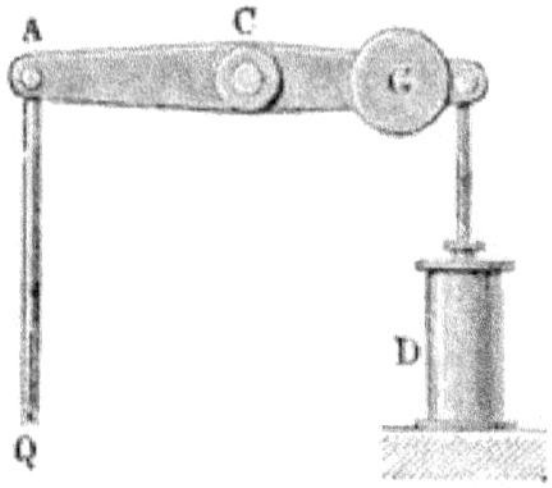

Fig. 144.

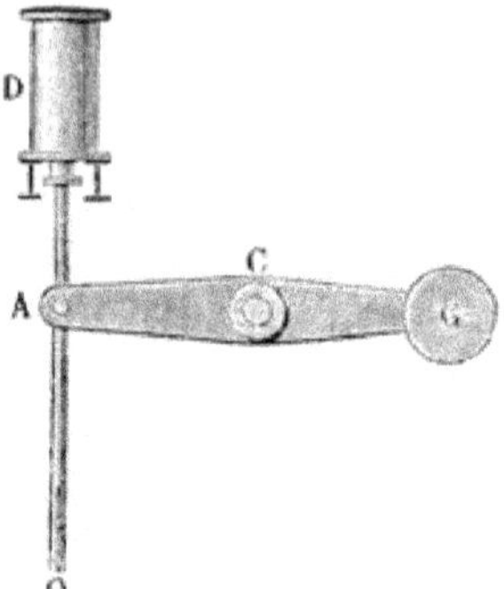

Fig. 145.

this reason *Ehrhardt* attempted, by the use of a cross-section, as shown in Fig. 146, to make wrought-iron rods of sufficient stiffness to permit of their being subjected to compression not only by the action of their weight, but also by the steam pressure from above. By this method the rod obtains a weight equal to one-half of the pump resistance only, or $\frac{1}{2}Q$, and is forced down by the equally great steam pressure. Pump rods have also been constructed in the shape of wrought-iron pipes (*Rittinger's* system), serving at the same time as delivery pipes for the discharging water.

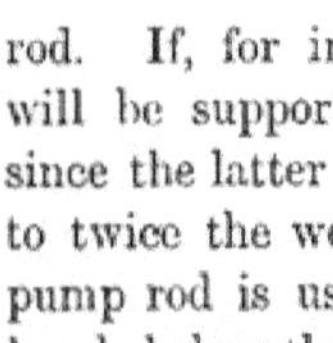

Fig. 146.

Besides dispensing with the counterpoise, the above arrangement offers a special advantage due to the slight weight of the pump rod. If, for instance, the rod should break by accident, it will be supported by the water column in the delivery pipe, since the latter exerts an upward pressure on the plunger equal to twice the weight of the rod. On the other hand, when a pump rod is used which is heavier than the water column, a break below the balance lever would cause the rod to fall with an acceleration, which might be detrimental to the pumps. The continual change from tension to compression in the rod has caused a number of breakages in spite of the greatest care exercised in the construction, and for this reason the plan of allowing the steam to act on the rod from above has been abandoned in modern practice, the use of a counterweight being preferred.

The pumping engines described above, both the direct and indirect acting, belong to the *reciprocating* class, no rotary motion whatever being employed. As a result, these engines are objectionable in some respects. In the first place, the reversal of the piston does not take place smoothly with a gradually decreasing and increasing velocity, as in the machines driven by a crank, the action being instead sudden and jerky, thus causing severe stresses in the parts. Neither is the stroke constant, but varying with the fluctuations in steam pressure, especially when expansion is employed. It was shown above that, as the steam expands a certain fixed mass must be accelerated in order that too great a velocity of the rod may be prevented. It is therefore evident that when the steam pressure is somewhat low the stroke easily falls short of the intended length, and that an increase of pressure above a certain limit may cause the piston to strike against and blow out a cylinder head. Such accidents have often occurred, and could not possibly have been avoided in some instances, when the pump rods have broken, causing the unloaded piston to be sent forward by the mighty force of steam at a velocity against which even the most powerful catch pins would offer no protection. Also in starting, when the delivery pipes are not yet filled with water, or when the pumps "snift," that is, draw in air in place of water, which happens from lack of water in the pump well, similar accidents may occur owing to the reduced load. As a safeguard, the clearance between piston and cylinder heads is made large, which naturally interferes with economy of steam; moreover, from considerations given above, the expansion in most cases cannot be carried very far, and for this reason the coal consumption is considerable.

Fly-wheel shafts have therefore been largely introduced in pumping engines, connection to the walking beam being brought about by the usual crank and connecting rod. With this arrangement the changes of direction at the dead centres evidently take place gradually and without shocks, and besides, the stroke being fixed once for all, danger of the piston striking against the cylinder heads is prevented. In consequence, these *rotative*, or so-called *crank and fly-wheel* engines may be run at a greater velocity, and the energy stored up in the fly-wheel, admitting as it does of the use of a higher degree of expansion, materially increases the efficiency of these machines. Their construction, however, is not so simple as that of the reciprocating type, their first cost is greater, and they are not possessed of one advantage peculiar to the latter type, namely, to admit of reducing the speed at will when the water supply is scant, the cataract furnishing a means of varying the intervals between the strokes.

On the other hand, in crank and fly-wheel engines a certain velocity of rotation depending on the mass of the wheel is required to pass the dead centres, and experience has proved that even if the lower limit of speed is placed at four revolutions per minute, the fly-wheel required will be very large. This is of the greatest importance in pumping plants for mines, where, as a rule, the water supply is variable, and where, in case of failing supply, when a crank and fly-wheel engine is used, it would be necessary to run it periodically and to construct large and expensive sumps for the water to collect in during the periods of rest.

It is owing to the above facts that these engines, in spite of their great advantages, have met with considerable opposition, and they are principally employed only in cases where the water supply is abundant and steady. The results obtained in such cases have been very satisfactory. The arrangement of a compound crank and fly-wheel pumping engine of 700 h.p. constructed at *Hoppe's* machine works in Berlin for the coal mine "Ferdinand" near Kattowitz[1] is indicated by Fig. 147. A and B are the steam cylinders of diameters 1·491 m. [4·89 ft.] and 2·040 m. [6·69 ft.] respectively, the piston rods being connected to the walking beam FED by means of a parallel motion; the fulcrum of the beam is at C, and the pump rod is attached at the end D. The connecting rod JK and crank HK rotate the fly-wheel shaft H, which, besides the fly-wheel S, carries the eccentric for operating the valves of the steam cylinders. The air pump for the condenser is driven by a separate steam engine. By the aid of two sets of suction pumps [of 36·1 metres or 118·44 ft. lift] and three sets

[1] See Hörmann, *Die neuen Wasserhaltungsmaschinen auf den*, etc.

of force pumps [272·9 m. or 895·3 ft.] the water is raised to a total height of 309 metres [1013·8 ft.] The stroke of the large cylinder B is 3·452 m. [11·33 ft.], that of the small cylinder 2·432 m. [7·97 ft.], and that of the pump 1·726 m. [5·66 ft.], or exactly one-half of that of the large cylinder. From these figures we obtain a ratio of cylinder volumes of 1 : 2·66; the steam pressure is three atmospheres by gauge and the total expansion is six-fold. To prevent the upper pumps from drawing in air, the plunger diameters are made to increase slightly downward, those of the upper pumps being

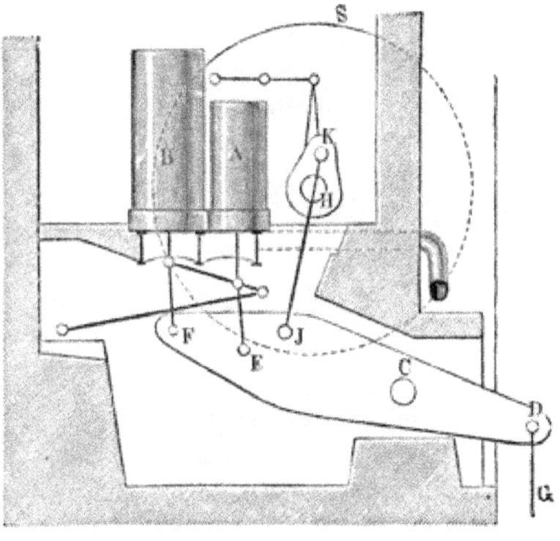

Fig. 147.

0·628 m. [2·06 ft.], and those of the lowest 0·642 m. [2·11 ft.], so that at a speed of 15 revolutions per minute a delivery of 7·420 cub. m. [262 cub. ft.] may be counted on. The pump rod is made cylindrical, of wrought-iron, its weight being balanced by the beam in such a manner that the force required to raise it is the same as that needed to depress it; the rod is therefore subjected alternately to tension and compression.

With a view to combining the above-mentioned advantages of both the reciprocating and crank and fly-wheel engines, and at the same time avoid their objectionable features, Kley has invented an arrangement which has been extensively introduced, and must be considered one of the most important advances in this field. The engine is provided with a walking beam DCE (Fig. 148) and a fly-wheel shaft H; the

valve motion is independent of this shaft, however, being operated from the rod *s* by means of cataracts, as in the Cornish engines. The steam piston A acts on the walking beam at B, the pump rod being attached at D and the necessary counterpoise at G. For operating the valve rod a lever CK is secured to the beam shaft C and acts on the auxiliary lever *kct*. In consequence of this arrangement it is possible to limit the number of strokes at will down to one per minute by lengthening the intervals between strokes by means of the cataract, while by dispensing with the intervals the engine may be run at as quick a rate as is consistent with its proper working. In the latter case the fly-wheel revolves continuously in the same direction, while in slow running the valve motion is so adjusted that the interval occurs when the crank stands either a little in *front* of or a little *beyond* the dead centre. The two cases differ in that in the former instance the fly-wheel on

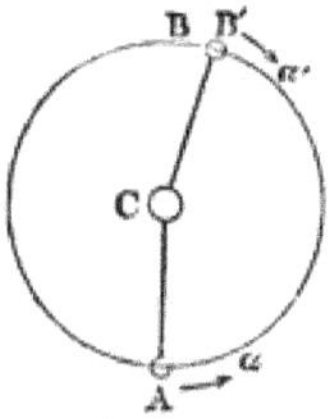

Fig. 149.

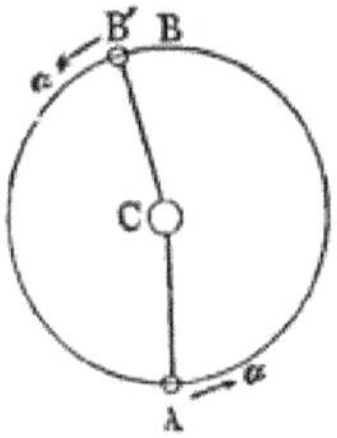

Fig. 150.

the next stroke will move in a direction opposite to its previous motion, while in the latter it will move in the same direction. For instance, if the crank on its up stroke proceeds from the position A near the lower dead centre (Fig. 149) in the direction of the arrow *a*, and reaches the point B′ only, in *front* of the upper dead centre, then the motion during the down stroke succeeding the pause will be in the direction of the arrow *a′*. On the other hand, when the crank is not brought to rest until it reaches the point B″ (Fig. 150), on the *other* side of the centre, the motion *a″* on the down stroke will be in the same direction as *a*. By adjusting the valve tappets, either one or the other of these two conditions may be brought about.

In addition, these engines are especially safe in case of a breakage in the pump rod, or when the greater portion of the load is suddenly thrown off for some other cause. The whole force acting on the piston will then be expended in accelerating the fly-wheel, and consequently the latter will be thrown considerably beyond the dead centre, causing the plunger of the cataract, which in its descent was to open the admission valve in preparation for the following stroke, to ascend again before it has had time to act, thus throwing the valve motion out of action and bringing the engine to a stand-still. These pumping engines are double-acting, have light fly-wheels, are condensing and work with a high rate of expansion, their performance therefore being very economical. The larger engines, above 200 horse-power, are compounded.

§ 43. **Pumps for Waterworks.**—The purpose of a pump is not always merely to raise water to a greater elevation, but to force it into some space which is already filled with air, steam, or water under a higher pressure. This is the case not only with boiler feed-pumps but also with the majority of pumps used for supplying cities with water. In order to prevent, in the latter instance, the irregular motion of the pumped water from being transmitted to the water in the pipe conduits, which would interfere with the working of the machine and unduly strain the pipes,[1] it is often necessary to collect the water in the first place in a reservoir located near the pump, and thence distribute it to various points according to demand. Such reservoirs must be located at a great height, or be made in the shape of a tower or column in order to supply water to the upper stories of high buildings. At the present day they are often made of iron plates forming so-called *stand pipes* of 1 to 2 metres [3¼ to 6½ ft.] diameter, and of a height not infrequently reaching 50 or 60 metres [160 to 200 ft.] Occasionally two stand pipes are erected side by side, the water being pumped into one of these, from which it is discharged through a side pipe at the top into the other stand pipe, whence it descends into the main conduit.

In place of high stand pipes large column-shaped *air chambers* have been introduced (in France); small air pumps are then made use of for replacing the air which leaks through the seams of the vessels or is carried away with the water.

The arrangement of the pump end of a waterworks pump

[1] Direct pumping into the water mains without the use of a reservoir or stand pipes is frequently resorted to, and will be commented on below.—TRANSLATOR'S REMARK.

for city supply is shown in Fig. 151, which represents a vertical section through a portion of the Cornish pumping

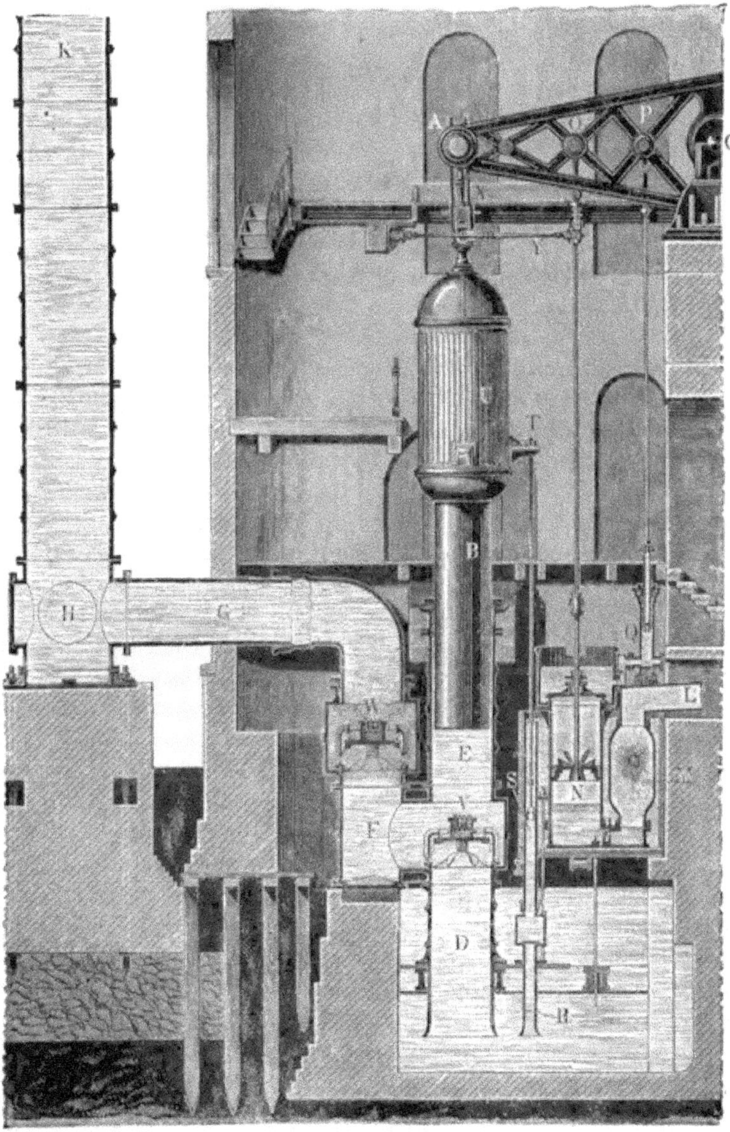

Fig. 151.

engine at East London Waterworks, Old Ford, England.[1] CA is the left half of the walking beam, which is oscillated by a single-acting Cornish steam engine coupled to it at the right by means of a *Watt's* parallel motion; further, B is the plunger with the case U for reception of the balance weights. At E the pump barrel is visible, at HK a portion of the stand pipe, at F and G pipes which connect the pump barrel with the stand pipe, at V the suction valve is shown, and at W the delivery valve. The plunger is on its down stroke, and consequently V is closed and W open. From the walking beam there are further suspended the piston rods of the air pump N and the boiler feed-pump Q, while that of the cold-water pump RS, which, like the main pump, takes its water from a well built of masonry, is attached to the plunger B, and consequently has the same stroke as the latter. Finally, we see in L the exhaust pipe for the steam, which is condensed in the condenser M.

The diameter of the steam piston is 2·04 m. [6·72 ft.], that of the plunger 1·04 m. [3·44 ft.], that of the pump barrel and pipes 1·093 m. [3·59 ft.], and that of the 38-m. [125 ft.] stand pipe is 1·27 m. [4·17 ft.] The steam piston has a stroke of 3·15 m. [10·33 ft.], and the plunger B one of 2·90 m. [9·51 ft.] The high-pressure steam with which the engine works is generated in four cylindrical boilers, as shown in Fig. 152, which represents a plan of the whole plant, including boilers and boiler house, only omitting the walking beam and valve motion. As in the preceding illustration, E here represents the pump barrel, FG the communicating pipe, K the stand pipe, L the exhaust pipe, M the condenser, N the air pump, Q the feed pump, and S the cold-water pump. Of the four boilers, D_1, D_2, D_3, and D_4, with inside fire-boxes R, two are shown in longitudinal section. The steam cylinder A has a wooden casing, the space between the latter and the cylinder being filled with ashes. The steam spaces of all four boilers are connected by the steam pipe BB_1B_2'. The feed water is supplied by the feed pump Q through the pipe TT_1T_2 to the heater U, whence it is conducted to the boiler by the pipe V. There are further seen in O the equalising valve, in W the release valve, in X_1, X_2 the two cataracts, etc.

[1] See the *Cornish* and *Boulton* and *Watt* engines erected at the East London Waterworks, Old Ford, by Th. Wicksteed, London, 1842.

The object of stand pipes, namely, to produce a nearly uniform pressure in the pipe conduit, can be only partly accomplished by the use of air chambers, owing to the great elasticity of air, but it may be attained in a simpler manner by substituting a loaded plunger for the water column contained in a stand pipe. For, evidently, the same result will be produced whether the water is forced into a reservoir by the pressure of a water column having a height h and a

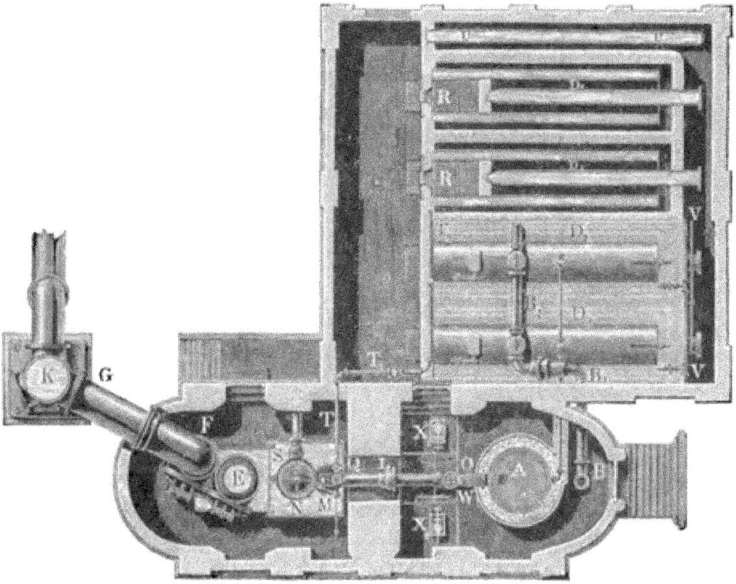

Fig. 152.

sectional area F, or is subjected to the pressure $P = Fh\gamma$ of a plunger having a sectional area F. For instance, if in a given time the water supplied to the main reservoir exceeds or falls short of that drawn from it by a volume V, either the water column or the loaded plunger will rise or fall through a distance $s = \frac{V}{F}$; in both cases the difference in water pressure on the area F will be given by the quantity

$$\Delta P = \pm Fs\gamma = \pm V\gamma,$$

and the pressure per unit of area by

$$\Delta p = \frac{\Delta P}{F} = \pm \frac{V\gamma}{F},$$

the variation being therefore smaller the greater the area F of the water column or loaded plunger is chosen.

The arrangement of these so-called *accumulators*, which were first introduced by *Armstrong* for operating hydraulic cranes, etc., is described in the volume on Hoisting Machinery, § 17, where their advantages in driving machinery which is only intermittently in motion are discussed.

In modern practice the Cornish steam engine is frequently replaced by a crank and fly-wheel engine for driving water-works pumps, the same remarks applying to this arrangement as to the mine pumps treated above. As was previously mentioned, such engines may be run at a greater velocity than the former type, and a more uniform action is attained, especially when they are constructed of the duplex type with cranks at right angles.

A horizontal pumping engine of this type, constructed by *E. A. Cowper* for the Crystal Palace,[1] is shown in Fig. 153. The reciprocating motion of the steam piston K of 0·90 m. [2·95 ft.] diameter is transmitted directly to the plunger L of 0·548 m. [1·80 ft.] diameter by the piston rod, which is common to both; from the crosshead C motion is communicated, through a forked connecting rod, to the crank ED on the fly-wheel shaft D, which is also provided with another crank at right angles to ED at the other end for coupling to the second engine. As in the ordinary double-acting pumps, two suction valves V and two delivery valves W are used, the water lifted from the well U being first forced into the air chamber R of 2·5 m. [8·2 ft.] length and 1·15 m. [3·77 ft.] width, and hence to the pipe X, which is common to both pumps, and leads to a height of 37 m. [121 ft.] The valves are of bronze and double-seated, of the style shown as applied to a piston in Fig. 55, the clear passage area of each valve, corresponding to a lift of 15 mm. [0·59 ins.], being 0·041 sq. m. [0·441 sq. ft.] The manner of operating the cold-water pump O and the air pump P by means of the bell crank MNF, which has

[1] See *The Artisan*, August 1858, and *Civil Eng.*, 1859.

its fulcrum at G and is connected to the crosshead C_1 is evident from the figure. Since the engine makes 15 revolutions per minute, the average piston velocity for 0·915 m. [3 ft.] stroke is

$$\frac{2 \times 0{\cdot}915 \times 15}{60} = 0{\cdot}457 \text{ m. } [1{\cdot}5 \text{ ft.}],$$

with which the water is raised to a height of 37 m. [121 ft.]; the steam has a pressure of 1·2 atm. by gauge, and is allowed to expand to three times its original volume.

Fig. 153.

Also the pumps of the waterworks at Berlin are of the rotative type, the arrangement being illustrated by Fig. 154. Each of the four steam engines originally erected side by side is provided with two cylinders A, the piston rods being attached to equal-armed walking beams BCD by means of parallel motions. Also here the two cranks EF on the fly-wheel shaft are placed at right angles to each other. Each walking beam drives two pumps, of the type shown in Fig. 59, the piston rods k of which are provided with cylindrical plungers O, whose sectional area is equal to half that of the valved piston K. Of the water lifted from the supply pipe S during the ascent of the plunger one-half is therefore discharged on the up stroke and the other half on the down stroke into the air chamber W, whence it flows through the pipe R, which connects with the air chambers of all the pumps. It may yet be noted that in the branch pipes between the air

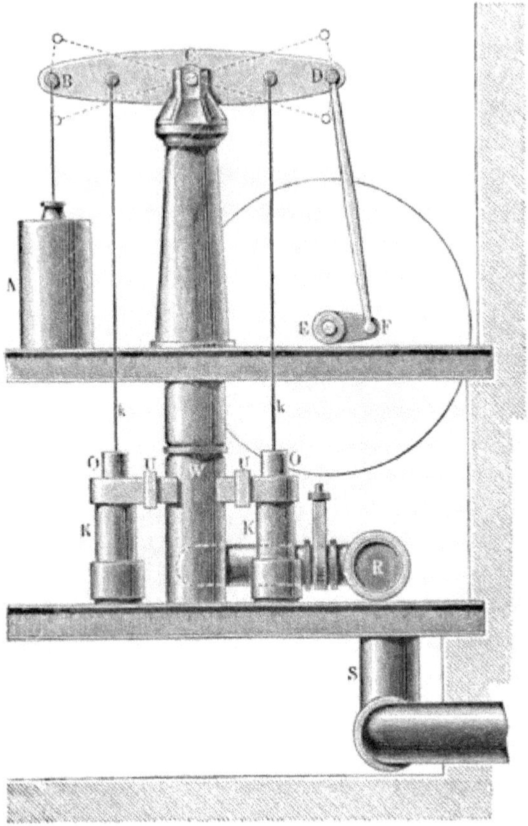

Fig. 154.

chambers and pumps clack-valves U are inserted, which are kept closed by the pressure in the chamber, thus allowing of the pumps being opened for inspection whenever necessary.

The steam cylinders A have 0·942 m. [3·09 ft.] diameter, and 1·255 m. [4·12 ft.] stroke, while the stroke of the plungers is only 0·942 m. [3·09 ft.] As the valved pistons K have a diameter of 0·555 m. [1·82 ft.], and consequently the diameter of the plungers is

$$0{\cdot}707 \times 0{\cdot}555 = 0{\cdot}392 \text{ m. } [1{\cdot}29 \text{ ft.}],$$

the water sucked in by each pump per double stroke, neglecting all losses, will be

$$\frac{3{\cdot}14}{4} \times 0{\cdot}555^2 \times 0{\cdot}942 = 0{\cdot}228 \text{ cub. m. } [8{\cdot}052 \text{ cub. ft.}]$$

In Fig. 155 there is shown one division of a double compound waterworks pump, of a type which has of late years been extensively introduced in the United States, especially for direct pumping into the water mains, the regulation being then effected by means of an automatic regulator which is influenced by the varying pressure in the mains. This style of pump is the design of *Gaskill*, and is manufactured by the *Holly Manufacturing Company* of Lockport, New York.

Each division has one high pressure and one low pressure steam cylinder, and one water cylinder. A (Fig. 155) is the high pressure cylinder arranged on the top of the low pressure cylinder B, which is bolted to the bed plate. The piston rods of the pistons P and Q are attached to crossheads C, running on guides D, and connected by links L to the rocking beam R; the latter is mounted on a shaft S, journalled in heavy supports X, firmly bolted to the bed plate, and rigidly stayed by wrought-iron struts E to the low pressure cylinder B and the water cylinder F. The latter cylinders are in line with each other, the piston rod of the low pressure cylinder B being directly connected to the rod of the pump plunger G. On a level with the centre of the high pressure cylinder A, and directly above the two water cylinders of the pump, a shaft H is arranged, which supports a fly-wheel K, and is connected to the beams R by means of two cranks placed at right angles and connecting rods M. The pillow blocks for this shaft are braced to the high pressure cylinders by heavy cast-iron girders N. The air pump of the condenser is driven by a short arm keyed to the inner end of the beam shaft S. The double-acting plunger G works in a gland O at the centre of the pump barrel, both being accessible on removing the cover U. The suction valves V, communicating with the suction pipe *r*, and the delivery valves (not shown), are arranged in horizontal plates below and above the plunger. The gland O divides the valves of one end of the water cylinder from those of the other end at the centre of the valve plates. I is the air chamber on the delivery pipe T.

The valves of the steam cylinders are operated by means of eccentrics *e* and *c* (Figs. 156 and 157) on a shaft *m* running parallel to the cylinder, and driven by bevel gears from the main shaft H. The admission valves of the high pressure cylinder (Fig. 156) are double-seated (balanced) lift valves which are opened by the eccentrics at the

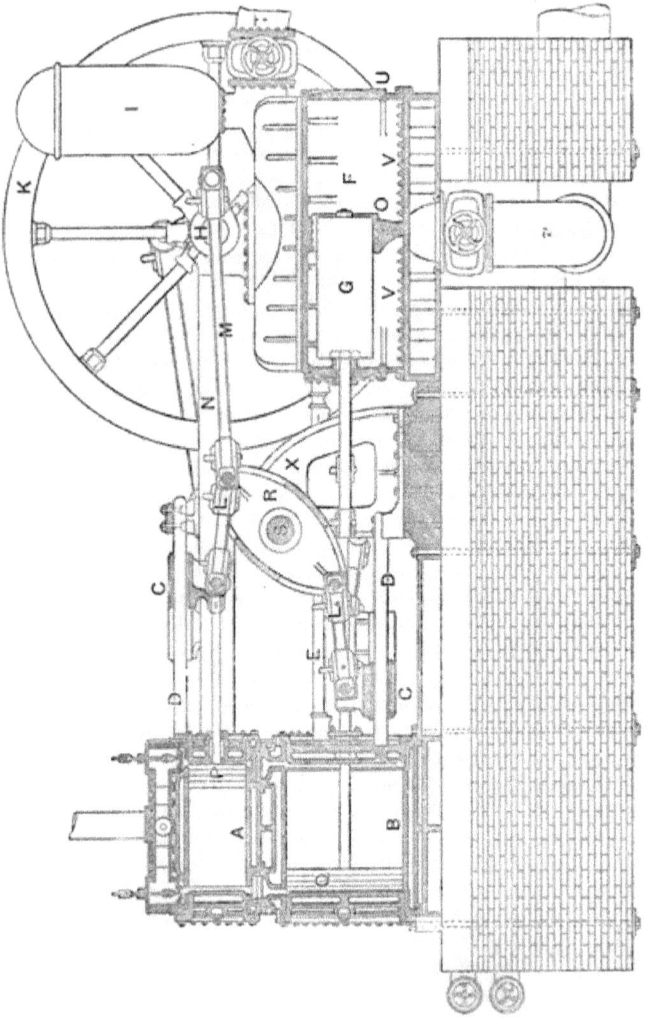

Fig. 155.

proper time, and, by means of the "cut-off" motion shown in Figs. 156 and 157, closed at any desired point of the stroke, the cut-off motion being either regulated by hand or, when pumping is done directly into the mains, by the automatic regulator above referred to, which is connected with the water main. As long as the lug *b* of the eccentric strap *f* rests on the abutment *p*, the valves are moved by the eccentric *c*; when the lug is released from the abutments the straps swing around on the eccentrics, and the valves are closed by the pressure of the springs *d*. The exhaust valves W from the high pressure and Z from the low pressure cylinder are "grid-iron" slide valves, and remain open somewhat less time than is required to complete a stroke; the former are also admission valves to the low pressure cylinder B. Owing to the very close connection between the two cylinders the clearance spaces are com-

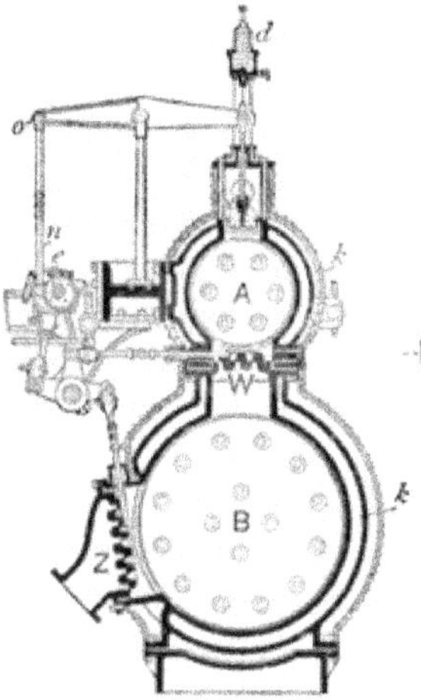

Fig. 156.

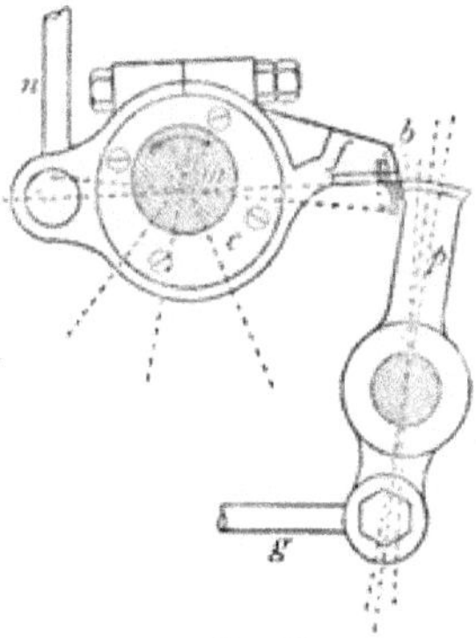

Fig. 157.

paratively small. Steam jackets *k* are provided for the steam cylinders.

The automatic regulator used when pumping directly into the mains consists essentially of a small water cylinder, containing a solid piston acted on by the pressure in the water main, and having a weight attached to it which counterbalances this pressure. This is effected by suspending the weight from a strap passing over a cam that rotates as the pressure changes, thus altering the lever arm of the counterbalance, and keeping it in equilibrium with the water pressure, no matter how much the latter may vary. Whenever the water pressure varies from a given amount, the regulator, by means of a weighted lever, throws into gear a reversible friction clutch which is driven from the engine, and acts to either lengthen or shorten the cut-off of the steam valves by changing the position of the cams or abutments *p* with reference to the eccentric cams *cf* (Fig. 157), which are attached to the shaft *m*, and, by means of the link connections *n* and levers *o*, govern the motion of the spring-actuated admission valves. The rods *q* of each of the abutment pieces *p* are so connected that all four of these pieces will be moved simultaneously.

The shaft on which the counterbalance cam rotates has an index wheel, and the index can be set at any desired water pressure. As long as the water pressure varies one way or the other from the figure at which the index is set, the friction clutch is kept in gear by the weighted lever, and, operating in one direction or the other, acts to change the position of the abutments, and to adjust the cut-off until the required pressure is reached. At this point the index engages with the weighted lever and throws the clutch out of gear. In this manner the pressure is kept practically constant in the system of water pipes. Should the pressure in the latter rise very suddenly, a piston in a safety cylinder raises a lever to which the cut-off gear is connected, and throws the cut-off instantly to zero if necessary.

A horizontal pumping engine of the type just described, erected at the waterworks of South Bethlehem, Pennsylvania, in 1892, for pumping into a reservoir, has a capacity of about 13,150 litres per minute [five million U.S. gallons per twenty-four hours] when operating under a steam pressure at the throttle valve of 5·6 kg. per sq. cm. [80 lbs. per sq. in.], and against a head of 84 metres [276 ft.] The diameter of the two high pressure cylinders is 53·3 cm. [21 ins.], of the two low pressure cylinders 106·6 cm. [42 ins.], and of the two pump plungers 49·5 cm. [19½ ins.]; the stroke of both steam pistons and plungers is 91·4 cm. [36 ins.], and the piston speed, when the pump works at its rated capacity, about 0·58 metres per second [115 ft. per minute]. The duty guaranteed by the builders was 100 million foot pounds per 100 lbs. of coal, which approximately equals 304,670 m. kg. per 1 kg. of coal. A description of a plant for direct pumping into the mains, with a Gaskill engine of the type described above, may be found in *American Machinist*, New York, Nov. 14, 1885.

A most noteworthy departure in recent pumping engineering is the direct-acting "high-duty" pumping engine, designed by *Henry R. Worthington* of New York with a view to securing a high fuel economy with direct-acting pumps, by using the steam expansively in the respective cylinders without the need of employing a fly-wheel for equalising the force exerted by the steam in each stroke. The general arrangement of this type of engine is substantially the same as that of the "duplex" pump shown in Fig. 134, p. 188, with the difference that compounded steam cylinders are used, and a "high-duty" or compensating attachment is introduced which performs the same function as the fly-wheel in the rotative types above described, viz. to store up a portion of the energy given off by the steam during the earlier part of the stroke, and restore it to the pistons in the latter half of the stroke.

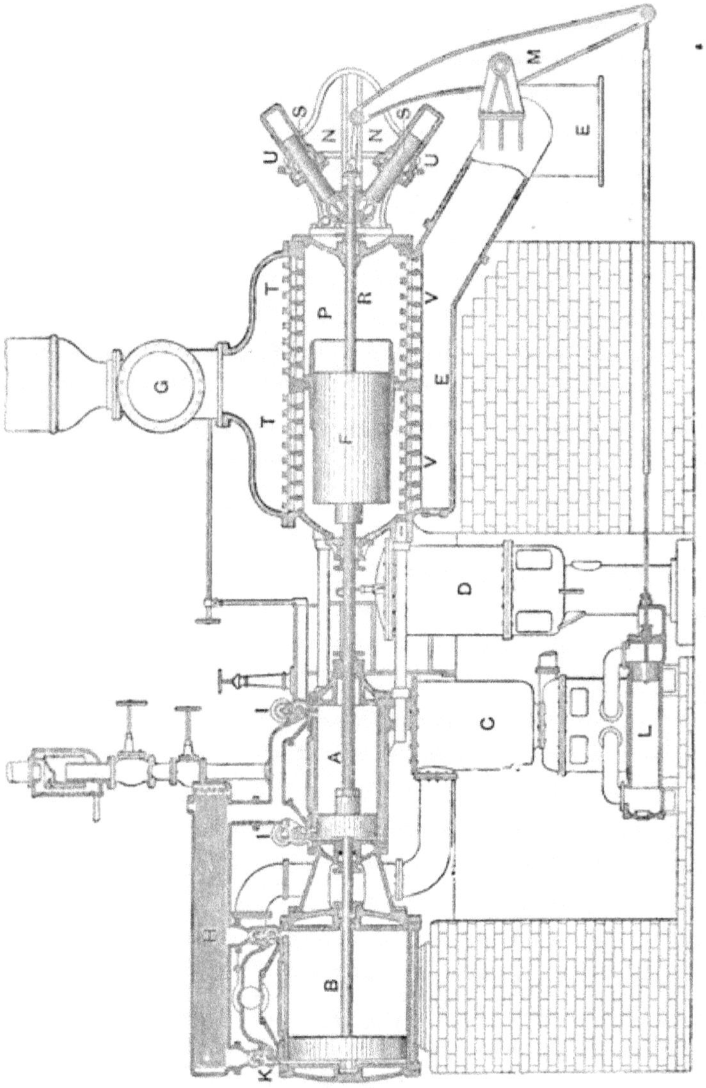

Fig. 158.

A section through one division of a horizontal duplex pumping engine of this description is shown in Fig. 158. Here A is the high pressure and B the low pressure cylinder, both being provided with semi-rotating cylindrical steam and cut-off valves I and K arranged one above the other, the steam valves being operated from the piston rod of the other division of the pump, and the cut-off valves being operated by adjustable link connections from the piston rod of the division to which they are attached. The steam exhausting from the high pressure cylinder through the lower rotary valves passes through a reheater H to the valves of the low pressure cylinder, and after spending its energy in the latter cylinder it is discharged into the jet condenser C, which is connected to the double-acting air pump L, driven by a vibrating lever M attached to the plunger rod R of the pump. P is the pump cylinder with its double-acting plunger F, suction valves V communicating with the suction pipe E, and delivery valves T communicating with the delivery pipe G.

The compensating attachment is arranged as follows:—The plunger rod R is prolonged through the outer end of the pump cylinder, and to the end of this prolongation there is attached a crosshead O moving in guides N secured to the pump. This crosshead is provided with two semi-spherical sockets opposite each other, and into these sockets enter the spherical ends of a pair of plungers S, which can move back and forth through stuffing boxes in a pair of short cylinders U, the "compensating cylinders" being journalled on trunnions in bearings arranged in the guide castings N. The cylinders U, which swing back and forth as the plunger F reciprocates, communicate through their hollow trunnions with an accumulator (compare the volume on *Hoisting Machinery*, § 17). This accumulator may be connected with the delivery pipe, deriving its power from the pressure in the latter.

From the figure it is now evident that at the beginning of each stroke, when the steam pressure on the pistons is at a maximum, and the compensating cylinders U point either as shown in the cut or at the same angle from the opposite end, the plungers S will encounter the greatest resistance from the fluid in the accumulator; as the steam pistons advance, this resistance will gradually decrease as the angle of the compensating plungers with the guides increases, until at the middle of the stroke the resistance will have fallen to zero, since the compensating plungers at this point form right angles with the guides. During the remaining half of the stroke the reverse condition will maintain, inasmuch as the compensating plungers now, instead of opposing the steam pistons, will push in the same direction as the latter move, and consequently the pressure in the accumulator will assist the steam pistons in their motion instead of resisting them as before. Near the middle of the stroke, where the decrease in steam pressure due to expansion is not so great, this assisting action will be comparatively small, and it will gradually increase to a maximum at the end of the stroke, where the expansion has been carried the farthest. By this method the inequalities of driving force due to expansion of the steam in the respective steam cylinders are equalised in a simple, inexpensive, and efficient manner, without the need of cumbersome and heavy fly-wheels with connections. To avoid shocks at the dead centres a throttle valve is introduced in the pipe connecting the compensating cylinders with the accumulator, this valve being throttled at the dead centres by means of a lever swinging with the oscillating cylinders.[1] When an engine provided with an attachment of this kind pumps directly into the water main, the compensating cylinders being connected with the latter, the variation in the water pressure increases or diminishes the load on the compensators, and the engine at once responds automatically by running slower or faster, until the water pressure has been restored to its normal point. In case of a break in the main, the load on the compensators would be removed, and inasmuch as there will be no stored energy to dispose of, and the steam is cut off early in the stroke, there will be no danger of injury to the engine through "racing," as is liable to be the case in pumping engines provided with a fly-wheel when the latter is suddenly deprived of its normal resistance. Descriptions and illustrations of a Worthington pumping engine of the type just commented on, as erected on *Quai d'Orsay* at Paris in 1889 for the Eiffel Tower hydraulic plant, will be found in *Engineering*, London, May 3, 1889; diagrams and accounts of tests of Worthington high-duty engines by *Mair* and *Unwin* are given in the same journal of October 1, 1886, and December 7, 1888.

In Fig. 159 there is shown a perspective view of three vertical Worthington duplex high-duty pumping engines[2] erected in 1891 at the waterworks of the Artesian Water Company in Memphis, Tennessee, each engine having a rated capacity of 26,300 litres per minute [ten millions U.S. gallons of water per twenty-four hours] when working against a pressure of 76·2 metres [250 ft.], the maximum plunger velocity being 0·76 metres per second [150 ft. per minute]. The diameter of the high pressure cylinders is 76·2 cm. [30 ins.], that of the low pressure cylinders 152·4 cm. [60 ins.], and that of the plungers 68·6 cm. [27 ins.]; nominal stroke of plungers and steam pistons is 1·219 metres [4 ft.] At the test prescribed in the contract with the builders, one of these engines is reported to have shown a duty of 117,325,000 ft. lbs. per 1000 lbs. of feed water consumed, which approximately equals 365,680 m. kg. per 10 kg. of feed water (the boiler pressure during the test was 7·7 kg. per sq. cm. = 110 lbs. per sq. in.) From the cut the arrangement of the high and low pressure cylinders A and B and the water

[1] See *Zeitschrift des Vereins Deutscher Ingenieure*, 1888, p. 736, and following.

[2] Compare *Engineering*, London, February 12, 1892.

cylinders C is evident, as well as the mode of operating the semi-rotary cut-off valves from the crossheads O on the plunger rods by means of levers and link connections. These crossheads, which are located between the low pressure and water cylinders, and run in guides in the centre castings D, also serve to receive the ends of the

Fig. 159.

plungers of the compensating cylinders U. The latter are filled with water, and communicate with an air tank which is kept filled with air under pressure.

Further particulars regarding modern pumping plants may be obtained from the following articles in technical journals, describing and illustrating a number of notable pumping engines of English, American, and German design:—In *Engineering*, London, sewage pumping engine at Boston, Mass., designed by *Leavitt*, and built at the *Quintard Iron Works* of New York (Dec. 11, 1885; also described in *American Machinist*, New York, May 31, 1884); triple expansion engine for East London Waterworks, built by *T. Richardson and Sons*, Hartlepool, England (Aug. 8, 1890); Worthington pumps at the Chicago Exhibition 1893, built by *Henry R. Worthington*, New York (April 21, 1893); quadruple expansion pumping engine, built by *E. P. Allis and Company*, Milwaukee (June 15 and 22, 1894); Worthington triple expansion pumping engines at Bombay, India, for pumping sixty million gallons of sewage per twenty-four hours, built by *James Simpson and Company*, London (May 4, 1894); Glasgow hydraulic supply pumps (July 13, 1894); Worthington high-duty pumps at Hornsey Sluice, built by *James Simpson and Company*, London (Nov. 30, 1894); triple expansion pumping engine for Glasgow Docks Hydraulic Installation, built by *Fullerton, Hodgart, and Barclay*, Paisley, near Glasgow (Oct. 18, 1895).

In *Zeitschrift des Vereins Deutscher Ingenieure*; Hülsenberg, *Ueber directwirkende Dampfpumpen* (1884 and 1885); Riedler, *Pumpen mit gesteuerte Ventilen* (1888), *Neuere Wasserwerksmaschinen* (1890), *Amerikanische Pumpwerke* (1893), and *Das Wasserwerk in Boston* (1893); Gutermuth, *Direktwirkende Pumpen* (1888); Ballauf, *Die Worthington Pumpen in den charakteristischen Stellungen* (1893).—TRANSLATOR'S REMARKS.

CHAPTER V

ROTARY PUMPS

§ 44. **Rotary Pumps.**—This name is applied to pumps in which a revolving piston is made use of, whose rotation may be either continuous or reciprocating, and which in its action somewhat resembles the piston of ordinary reciprocating pumps. Many arrangements of this kind have been devised, but with all their variety they have this principle in common with each other and with the ordinary piston pumps, namely, that by the relative motion of two bodies a certain inclosed space is alternately increased and diminished. If care be taken to connect this space while it is *enlarging* with the *suction pipe*, and while it is *contracting* with the *delivery pipe*, then suction and forcing of the water will be effected as in the ordinary piston pumps; for, in fact, the action of the latter is simply due to the alternate increase and decrease of the cylinder space occasioned by the motion of the piston.

All rotary pumps have a closed casing through which pass one or two shafts surrounded by stuffing boxes to prevent leakage. On these shafts, and within the casing, there are bodies of suitable form which, while turning, are in contact with the inner circumference of the casing. If there is only one shaft, valves of some kind are needed, but they are unnecessary when the machine has two rotating shafts, as will be seen hereafter.

We shall first consider the *Bramah* pump (Fig. 160), in which the piston is formed by a rectangular plate AB, provided with two valves WW_1, and swinging about the shaft C in a cylindrical casing. Here the suction pipe H terminates in a chamber provided with the clack valves VV_1, and we see how the oscillations of the piston AB caused by the lever DE alternately open and close the pairs of valves VW and V_1W_1. We also see how the water passes from the suction pipe through the open valve V into the part of the casing that is below the piston, while an equal volume of water above the piston is forced by the closed delivery valve W_1 into the air chamber R, and from there into the delivery pipe S. This pump is principally used in fire and garden engines, and, so far as its mode of action is concerned, it can be regarded as a combination of two single-acting suction and lift pumps. The quantity delivered at each oscillation of the lever of course depends, as in all fire engines and hand pumps, on the extent of the swing of the lever.

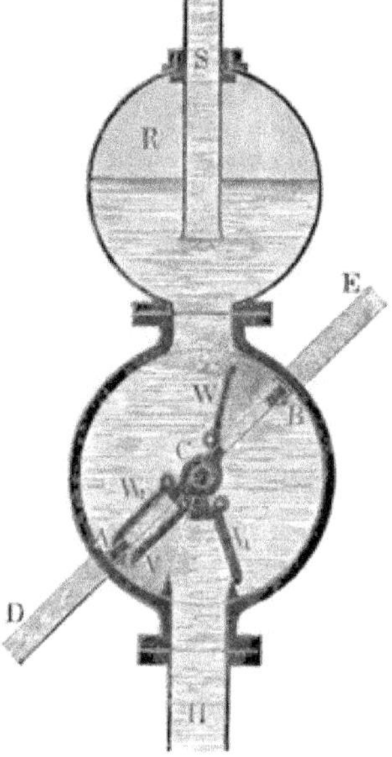

Fig. 160.

Owing to the one-sided pressure on the oscillating piston in the operation of the pump just described, the wear of journal boxes, piston, and casing is quite rapid. This defect is avoided in a Bramah pump, patented by *Abrahamson-Roxendorff*, and manufactured by *G. Allweiler* in Baden, Germany. In this pump[1] two suction and two pressure chambers *g*, *g* and *h*, *h* are provided, as will be seen from Fig. 161, which represents vertical sections of the pump. Two suction valves *n*, *o*, and two delivery valves *p*, *q*, are made use of, these valves being attached to the casing, and no valves being needed in the piston *m*. The latter is provided with two independent passages *s*, *s* and *b*, *b*, the former effecting the communication between the chambers *h*, *h*, and the latter between the chambers *g*, *g*. As the lever is oscillated, the spaces *g*, *g* and *h*, *h* will alternately become pressure and suction chambers, and as the pressure due to suction and forcing at the upper side of one wing of the piston is always counteracted by an equal and opposite pressure at the lower side of the other wing, it is evident that the wear as well as the power lost in friction will be very slight. On account of their double suction and double forcing

[1] *Zeitschr. d. Ingenieure*, 1892, p. 1021.

action, these pumps may be made about one-half the size for the same capacity as the older types of Bramah pumps. At *d* is shown a small pocket just below the delivery pipe *a'*, this pocket serving to receive sand and impurities which are drawn in with the water.

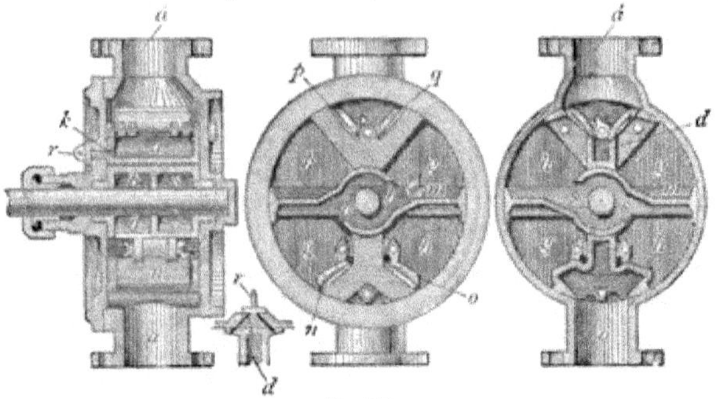

Fig. 161.

The pocket *d* is provided with a clean-out hole *k*, ordinarily closed by a screw *r*, and through which the water may also be drawn from the delivery pipe to prevent freezing. These improved Bramah pumps, which are in extended use as hand pumps, are to be regarded as very efficient and durable machines for pumping a variety of liquids.—TRANSLATOR'S REMARK.

Another pump with a single shaft but continuous rotation is that originated by *Dietz*, Fig. 162; the shaft W is located in the centre of the cylindrical casing BB, and has rigidly attached to it a rim AA which is rotated by it. In this rim A there are slits in which four floats C can move radially, their motion being effected, as shown in the figure, by a suitably curved groove, bounded on the inside by the cam-shaped piece E rigidly secured to the casing, and on the outside by the casing B itself.

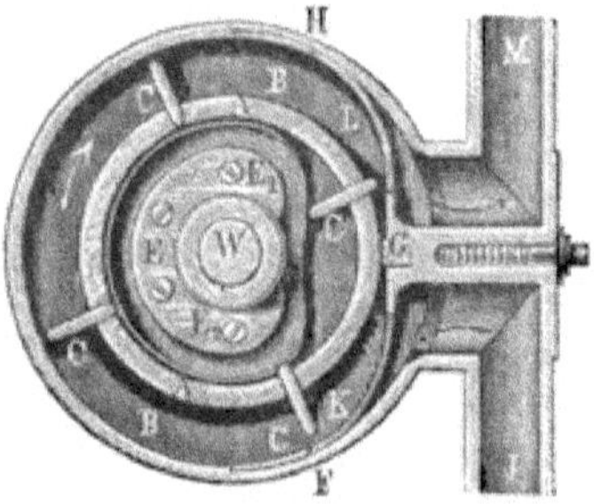

Fig. 162.

Opposite the flat portion of the cam plate E is placed a guiding piece FGH of such a form that the radial width of the groove is everywhere equal to the width of the slides C. It is easy to see that as the shaft W turns with the rim A in the direction of the arrow, the slides will be moved inward by the pressure of the guide L while passing from H to G, and outward by the cam plate E while they are passing from G to F. If the guide is provided with slots L and K, it follows that, owing to the increase of volume at K, water will be sucked in from the pipe J, and in consequence of the contraction of the space between H and G water will be forced through the pipe M. If the shaft were turned in the opposite direction, water would be drawn from the pipe M and forced into J.

In another construction (Fig. 163) the shaft A carries an eccentric disc E, which touches the cylindrical casing G at a

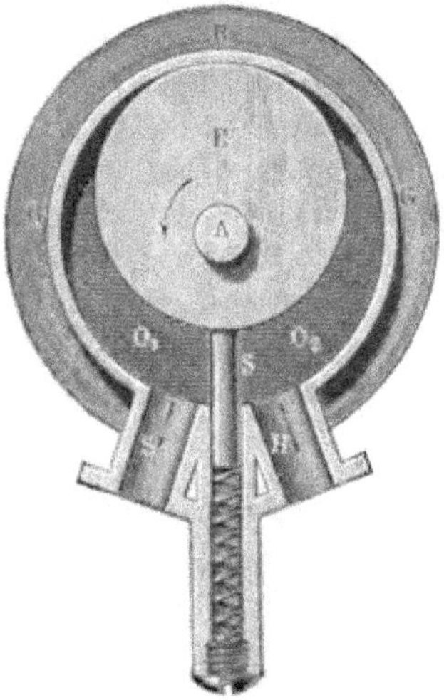

Fig. 163.

Fig. 164.

point B; this disc is always in contact with a slide S, which is pressed against it by a spring, and can move radially in a recess in the casing. The two spaces O_1 and O_2, included between the casing and eccentric disc, are therefore subject to continual variation, O_1 becoming smaller and O_2 larger when rotation takes place in the direction of the arrow. Water must therefore be continually drawn in from the pipe H and forced out through S.

As a rule, every *rotary* steam engine or water-pressure engine can act as a rotary pump, as, for instance, the water-pressure wheel (Fig. 164) described in the volume on Hydraulics.

If in this arrangement the hollow shaft O and its pistons AA be revolved in the direction of the arrow, the water in the spaces O_1 will be forced through the oblong openings C into the central hole O of the shaft, from which it is carried through a stuffing-box into the delivery pipe. On the other hand, the spaces O_2 are filled by the side channels B, which communicate with the other end of the shaft. Of course the slides F, which separate the suction pipe from the delivery pipe, must be so moved that they will offer no obstruction to the passage of the pistons.

The quantity of water which these pumps with one shaft can deliver per revolution is, in every case, determined by the space left free in the casing by the rotating body; it must be noted, however, that the losses due to leakage are very great.

In the pumps just discussed valves are needed to separate the suction water from that in the delivery pipe, but when there are two rotating shafts in the casing such valves can be dispensed with. The separation of suction water and force water is then effected by keeping the two pistons in continual contact, suitable shapes being for this purpose given to the latter. A large number of rotary pumps of this kind have been invented, a few examples of which will be given in the following.

In Fig. 135 *Repsold's* pump is shown, five different positions of the two rotary pistons C and D being indicated. It is

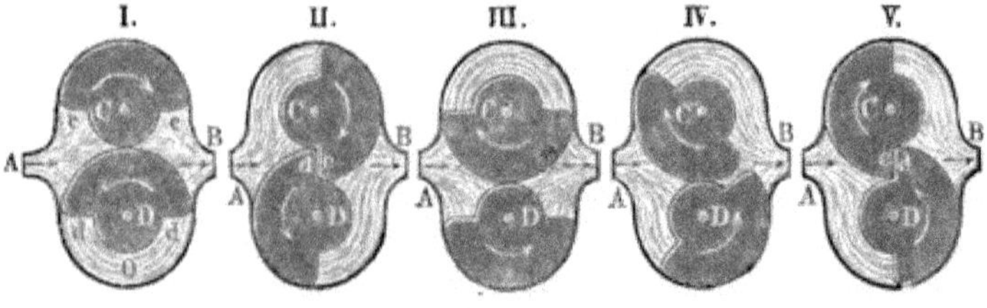

Fig. 165.

evident that if the shafts are revolved at the same rate in opposite directions, as indicated by the arrows, water is sucked in from the pipe A and forced out through the pipe B. Each of the pistons C and D consist of two half cylinders of different diameters, and joined by the steps *c* and *d*. As these steps are brought into contact with each other they must be profiled according to the rules for shaping the teeth of gears (see Weisbach's *Mechanics*, vol. iii. 1, Gearing); as shown in the figure, they are consequently given radial flanks and epicycloidal faces. The pistons C and D may therefore be regarded as *one-toothed* wheels. The shafts receive opposite rotations from two equal spur-wheels on the outside of the casing, which gear with each other and are keyed to the rotating shafts. The figure shows that during one revolution *each piston* delivers a quantity of water equal to the clear space O in the cylindrical portion of the casing which surrounds the piston. This law is generally true for all rotary pumps with two rotating pistons.

The toothed form of the pistons is still more clearly shown

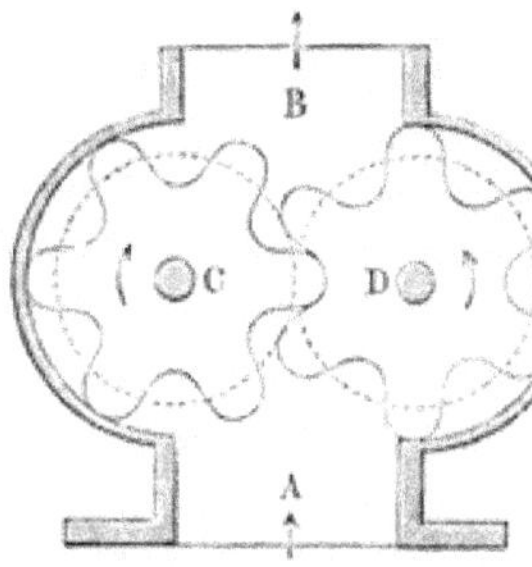

Fig. 166.

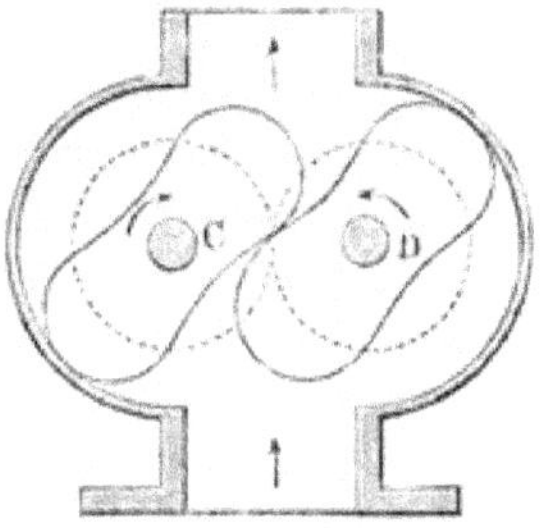

Fig. 167.

in *Pappenheim's* pump (Fig. 166), where the driving-wheels may be dispensed with, for the rotary motion received by one shaft can be communicated to the other by the pistons C and D themselves. It is evident that when rotation takes place in the direction of the arrow, water is sucked in from A and forced out through B, and with an opposite rotation of the shafts the water must move from B to A.

Such pumps have also been employed as blowers, for example *Root's* pump, Fig. 167, where the pistons C and D are to be regarded as *two-toothed* wheels, two outside spur-wheels being here necessary for the transmission of motion. This also applies to the pump shown in Fig. 168, whose mode of action needs no further explanation, after what has been said.

A more detailed discussion of the preceding and other

constructions may be found in Reuleaux's *Kinematics*, chaps. 9 and 10, which treat of *chamber-wheel trains*.

The greatest disadvantage connected with all rotary pumps is the difficulty of maintaining water-tight contact between the casing and the rotating pistons. This difficulty is most marked at the flat surfaces or ends of the casing, for here the relative velocity, and consequently the wear, differs in amount for different points, according to their distance from the axis of rotation. The consequence is that after these pumps have been in use but a short time, the loss of water from leakage is considerable, especially when the lift is great. This is probably the principal reason why these rotary pumps, notwithstanding their comparative simplicity of construction and mode of action, are but little used. They are chiefly adapted for lifting thick, pasty fluids with which valves cannot readily be used and water-tight contacts are less important. In such cases, however, it is more common to make use of centrifugal pumps, which we will now consider.

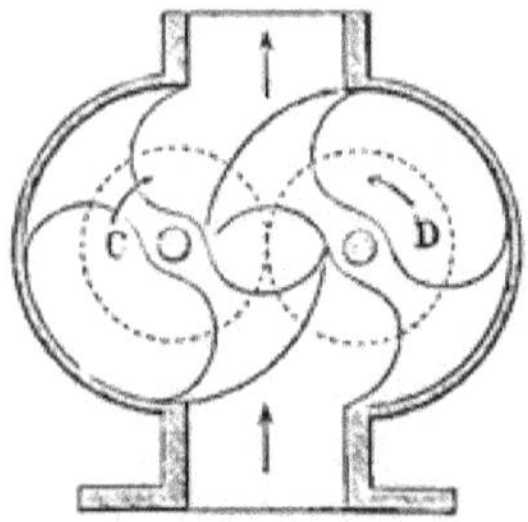

Fig. 168.

§ 45. **Centrifugal Pumps.**—These water-raising machines, in which, as their name indicates, it is principally the centrifugal force which is brought into action, generate the energy needed to lift the water by the *living force* imparted to the latter by a rapidly rotating wheel. For this purpose the wheel is provided with floats like a turbine wheel, and is enclosed in a casing so arranged that the water can be admitted to the wheel at one side and discharged at another. As a matter of fact, these water-raising machines are reversed reaction turbines, and their effect may therefore be estimated in nearly the same manner as that of turbines. Centrifugal pumps may be allowed to exert a suction, though not to the same extent as piston pumps, since experience has shown that their efficiency is reduced thereby. They are likewise little suited for forcing water to great heights, and the lift is therefore seldom taken greater than, say, 15 m. [50 ft.] This is due not only to the fact that the velocity of the wheel would become very great with greater lifts, but also to the increased loss of water through the clearance space necessarily existing between the casing and the rapidly rotating wheel. The efficiency of the best centrifugal pumps is less than that of piston pumps, and seldom exceeds the value $\frac{2}{3}$.

Experiments made by *Morin* with an *Appold* pump gave in the most favourable case an efficiency of 0·68, while other experiments with wheels of unfavourable proportions, for instance with flat floats placed radially, have given much lower efficiencies, down to 20 per cent. Similarly, in the tests made by *Rittinger*,[1] the efficiency did not exceed 35 per cent. But when the wheels are well designed and well made we may in most cases depend on an efficiency of from 60 to 65 per cent. In spite of these low efficiencies, centrifugal pumps are largely used at the present day for moderate lifts or in cases when, as in draining excavations, it is important to procure an apparatus that can be easily set up and started, and may be readily transported. The absence of valves (for at the most only a clack valve at the bottom of the suction pipe is needed) is another advantage which is of particular value when dirty or sandy water is to be lifted. In fact these pumps have been successfully employed in peat beds for removing thick, pasty masses, and they have taken the place of dredging machines[2] when the bottom was of a slimy character.

Centrifugal pumps are made with vertical as well as with horizontal shafts. When the former arrangement is chosen, the wheel is often wholly immersed in the water to be lifted, in order to avoid suction, while pumps with horizontal shafts are always placed above the lower water-level, and consequently must be arranged for suction.

The smaller horizontal pumps are usually driven by belts on account of the great velocity required (up to 2000 revolutions per min.), while the large stationary centrifugal pumps, for example, such as are employed to drain low lands, are usually provided with a vertical shaft and driven by toothed wheels, since, owing to the large diameters (up to 1·5 m. [5 ft.]) and small lifts, a smaller number of revolutions will suffice.

[1] See Rittinger, *Centrifugalventilatoren und Centrifugalpumpen.*
[2] *Zeitschrift des Vereins deutscher Ingenieure*, 1869.

A small centrifugal pump[1] with horizontal shaft is shown in Figs. 169 and 170. The wheel B is fastened to the end of the shaft A and consists of two shroudings *bb*, between which are curved floats *c*; this wheel turns in the spiral-shaped casing H, making 1500 to 2000 revolutions per minute. The water is supplied from the suction pipe C, enters at the centre of the wheel, and, after being caught by the floats, is driven outward to the circumference of the spiral-shaped casing, which guides it to the delivery pipe D, and up the latter to a height corresponding to the imparted velocity or pressure. The pump is driven by a belt pulley

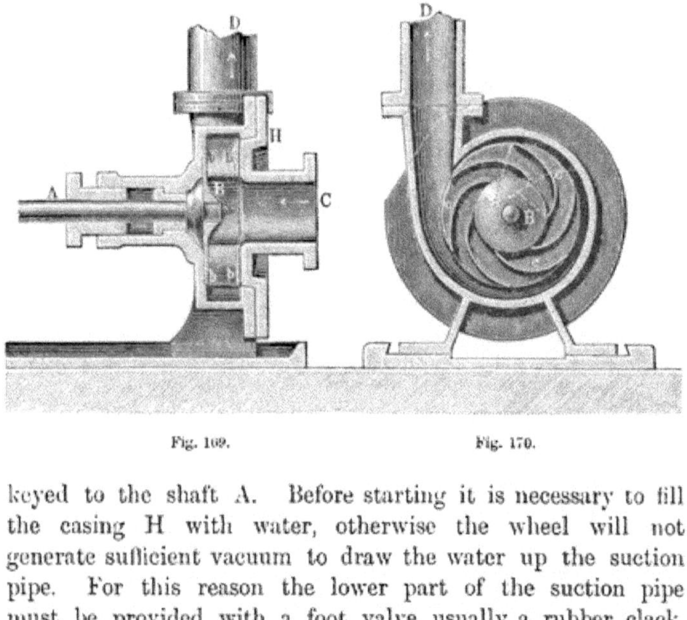

Fig. 169. Fig. 170.

keyed to the shaft A. Before starting it is necessary to fill the casing H with water, otherwise the wheel will not generate sufficient vacuum to draw the water up the suction pipe. For this reason the lower part of the suction pipe must be provided with a foot valve, usually a rubber clack, which remains open when the pump is running, but closes when the latter is brought to rest, thus preventing the water from running out. The pump is filled through a hole that is afterwards stopped up with a screw. In the machine shown the wheel has a diameter of 160 mm. [6·3 ins.], and the quantity discharged is claimed to be 0·45 cub. m. [15·89 cub.

[1] Collection of Drawings by the Society Hütte, 1869, Pl. 9.

ft.] per minute. The lift, as we shall see in the following, depends upon the rotary velocity of the wheel.

A centrifugal pump built by *Henschel and Son* of Cassel is shown in Figs. 171 and 172; it differs from that just

Fig. 171.

described principally in that the wheel B is here composed of a single central plate, with floats on both sides, the casing being made with branch pipes E leading from the suction pipe C for admitting water on each side of the wheel. At the

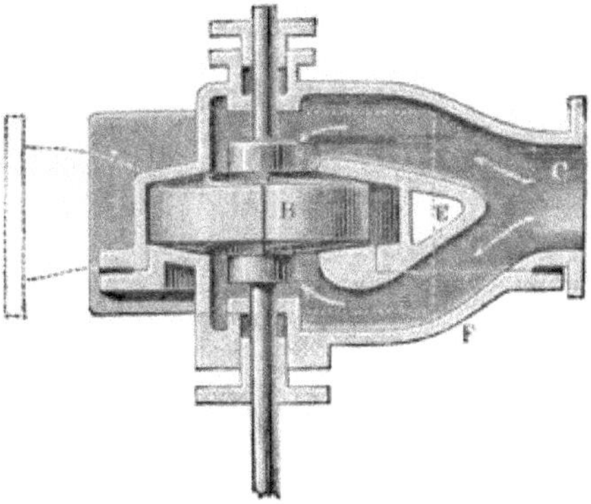

Fig. 172.

same time the cover F is so formed that it can be easily removed and the interior rendered accessible without stirring the casing H from its foundation.

As an example of a centrifugal pump with a vertical shaft,

that designed by *Schwartzkopff* is shown in Figs. 173 and 174.[1] Here the vertical shaft A carries at its lower end

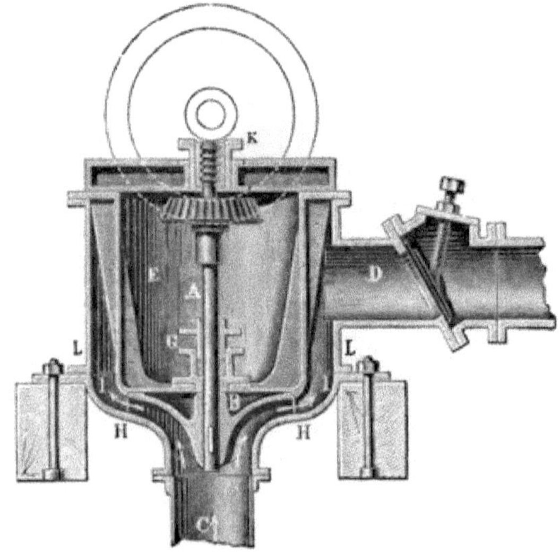

Fig. 173.

the wheel B, which is shaped like a conoid, and on its lower surface is provided with blades of the form shown in

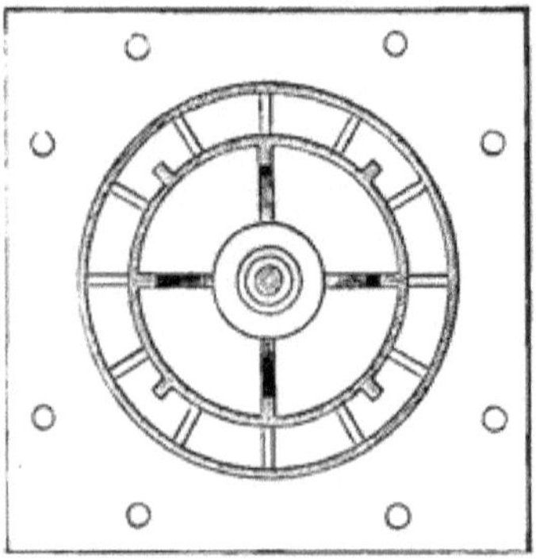

Fig. 174.

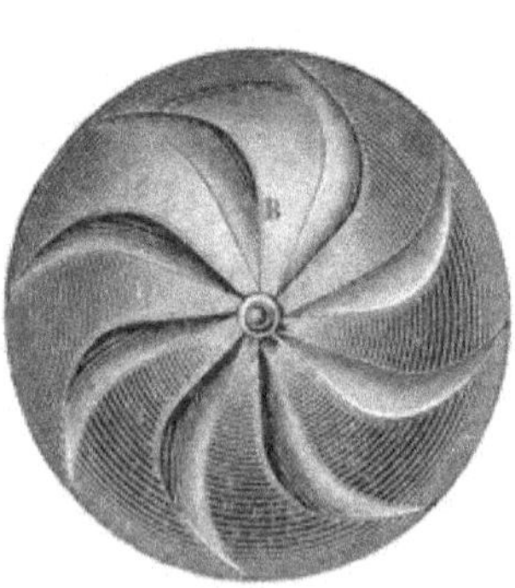

Fig. 175.

Fig. 175, which represents a view of the wheel from below. The latter is surrounded by a bell-shaped casing H, which at

[1] *Verhandlgn. d. Ver. zur Beförderung des Gewerbfl. in Preussen*, 1865.

its upper end L communicates with the delivery pipe D. In L is suspended a bowl-shaped casting E, provided at the bottom with a stuffing box G for the shaft, which is supported by a collar bearing at K. The water which passes from the suction pipe C into the casing is flung outward by the wheel, and flows along the curved surfaces of H into the delivery pipe, where it rises to a height corresponding to its velocity and hydraulic pressure. The upper part of the casing H is provided on the inside with a number of guide blades *l*, which are given a gradual curvature, so as to guide the water, which has received a rotary motion from the wheel, without shock into the radial direction, and thus prevent the formation of eddies and consequent loss of efficiency, which otherwise would occur. In some instances guide blades have been applied, as in turbines, to lead the water into the wheel, but such is not the case in the majority of centrifugal pumps; neither are, as a rule, such guides employed for the discharge in the horizontal types, inasmuch as the latter usually have a casing of such a form that it may be regarded as a single guide blade.

In centrifugal pumps with a vertical shaft the foot valve may be dispensed with, provided the wheel is placed below the lower water-level. It must be noted, however, that for great lifts difficulties may be encountered in the erection of the long shaft and in retaining it in its vertical position, particularly as experience shows that the ground below the suction pipe is considerably loosened by the violent flow of water to the pump when the latter is in vigorous operation.

We may discuss the action of a centrifugal wheel on the water in a manner similar to that employed in tracing the effect of the driving water on the floats of turbines (see vol. on Hydraulics). For this purpose let AB (Fig. 176) be a blade which makes, with the inner circle of radius r_1 and the outer circle of radius r_2, the angles α and β respectively. Let us now assume the ordinary case that water is led to the inner circle without guide blades, that is, in a radial direction, and let us represent the absolute entrance velocity AE of the water by v_1, and the circumferential velocity AF of the inner circle A by u_1. Then, in order that a shock may be prevented, the first element of the blade at A must have a velocity in the direction AL equal to the initial velocity of the water relatively to the rotating wheel. This condition is expressed by the equation

$$c_1^2 = v_1^2 + u_1^2 \quad . \tag{1}$$

In consequence of the initial velocity v_1 of the water and the simultaneous rotation of the wheel, a particle of water entering at A will traverse an absolute path represented, say, by the curve AD, which cuts the outer circumference at the angle KDO = δ.

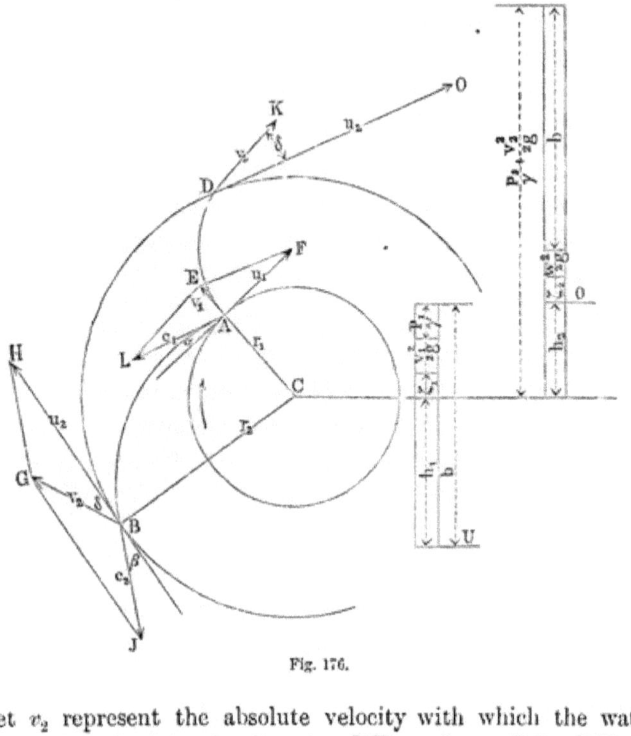

Fig. 176.

Let v_2 represent the absolute velocity with which the water leaves the wheel in the direction DK, and u_2 = DO = BH the tangential velocity of the wheel at the outer circumference; then the relative velocity c_2 with which the water moves along the last element B of the blade AB will be the resultant BJ of the absolute velocity BG = v_2, and the velocity GJ = HB = $-u_2$ equal and opposite to that of the wheel. The triangle GBJ therefore gives the equation

$$c_2^2 = v_2^2 + u_2^2 - 2v_2u_2 \cos \delta \quad . \tag{2}$$

In addition to its velocity v_1 the water entering the wheel at A has a certain hydraulic pressure p_1, which may be represented by a water column of the height $\frac{p_1}{\gamma}$; in like manner $\frac{p_2}{\gamma}$ may represent the hydraulic pressure of the water leaving the wheel at B with the absolute velocity v_2. These pressures may be easily determined. For, let b be the height of the water barometer, h_1 the height of suction estimated from the lower water-level up to the axis C, h_2 the height of delivery measured from C to the upper water-level O, and let ζ_1 and ζ_2 be the heads corresponding to the resistances to motion in the suction and delivery pipes respectively. As may be seen from the figure, we then have

$$\frac{p_1}{\gamma}=b-h_1-\zeta_1-\frac{v_1^2}{2g} \quad . \qquad (3).$$

Since the water that leaves the wheel with a velocity v_2 and a hydraulic pressure p_2 must be capable not only of overcoming the delivery head h_2, the atmospheric pressure b, and the resistance ζ_2, but also of maintaining the velocity of discharge w, we further have

$$\frac{p_2}{\gamma}+\frac{v_2^2}{2g}=h_2+b+\zeta_2+\frac{w^2}{2g} \quad . \qquad (4).$$

Now, in order to determine the velocity u_2 of the wheel, which will give to the water the desired efficacy, it should be noted that, owing to the action of the centrifugal force, the water in passing through the wheel from A to B has its living force increased by an amount corresponding to the head

$$\frac{u_2^2}{2g}-\frac{u_1^2}{2g},$$

less the head ζ_r, which represents the resistance between the blades of the wheel. Accordingly, the action of the rotating wheel on the water passing through it will be represented by the following equation:

$$\frac{p_2}{\gamma}+\frac{c_2^2}{2g}-\left(\frac{p_1}{\gamma}+\frac{c_1^2}{2g}\right)=\frac{u_2^2-u_1^2}{2g}-\zeta_r \quad . \qquad (5).$$

If we here substitute the values c_1^2, c_2^2, p_1 and p_2 given by equations (1) to (4), we obtain, after some simple reductions,

$$h_1 + h_2 + \zeta_1 + \zeta_2 + \frac{w^2}{2g} - \frac{2v_2u_2 \cos \delta}{g} = -\zeta_r,$$

or if we let $h = h_1 + h_2$ represent the total lift, and

$$\zeta = \zeta_1 + \zeta_2 + \zeta_r$$

denote the sum of all the heads corresponding to the hydraulic resistances, we get

$$h + \zeta + \frac{w^2}{2g} = \frac{v_2u_2 \cos \delta}{g} \qquad (6).$$

From this relation we can determine the velocity u_2 of the outer circumference of the wheel by combining it with the equation

$$v_2 = u_2 \frac{\sin \beta}{\sin (\beta + \delta)}$$

obtained from the triangle BGH, thus finding

$$u_2 = \sqrt{g\left(h + \zeta + \frac{w^2}{2g}\right)\frac{\sin (\beta + \delta)}{\sin \beta \cos \delta}},$$

$$= \sqrt{g\left(h + \zeta + \frac{w^2}{2g}\right)(1 + \tan \delta \cot \beta)} \qquad (7).$$

From this follows the absolute velocity with which the water flows from the wheel—

$$v_2 = \sqrt{g\left(h + \zeta + \frac{w^2}{2g}\right)\frac{\sin \beta}{\cos \delta \sin (\beta + \delta)}} \qquad (8).$$

Equation (7) will enable us to compute the velocity of the outer circumference of the wheel, when δ, β, ζ, and w are known. According to *Grove*,[1] we may assume for the angles δ and β

$$\tan \delta = 0{\cdot}5 \text{ ; hence } \delta = 26° \, 34',$$

and

$$\tan \beta = 0{\cdot}3 \text{ ; hence } \beta = 16° \, 42'.$$

It is also sufficiently accurate to place the head $\frac{w^2}{2g}$ due to the velocity of discharge w equal to about 3 per cent of the

[1] *Mittheil. des Gew.-Ver. f. Hannover*, 1869, p. 130.

lift h. Moreover, if we assume, in accordance with existing experimental results, the head ζ representing the resistances equal to $0 \cdot 42h$, we shall get for the velocity of the outer part of the wheel

$$u_2 = 1 \cdot 4 \sqrt{2gh},$$

i.e. about 40 per cent greater than the velocity which would be acquired in falling through the height h of the lift. From this follows the absolute velocity with which the water leaves the wheel

$$v_2 = 0 \cdot 6 \sqrt{2gh}.$$

If we denote the inner radius of the wheel by r_1 and the outer one by r_2, the number of revolutions per minute is expressed by

$$n = \frac{30u_2}{\pi r_2} = 9 \cdot 55 \frac{u_2}{r_2}.$$

The clear width b_2 of the wheel at the outer circumference can be found as follows. Let z represent the number of blades, s their normal thickness, and Q the quantity of water that is to be discharged per second, then we have

$$Q = (2\pi r_2 \sin \beta - zs) b_2 v_2.$$

In like manner for the entrance of the water we have the equation

$$Q = \left(2\pi r_1 - \frac{zs}{\sin \alpha}\right) b_1 v_1,$$

from which we may obtain either of the three quantities b_1, v_1, and r_1 when suitable values are assumed for the others. For instance, if with *Grove* we assume a constant value for the radial component of the velocity of the water in its passage through the wheel, *i.e.* that

$$v_1 = v_2 \sin \delta,$$

then, since

$$v_1 = u_1 \tan \alpha = \frac{r_1}{r_2} u_2 \tan \alpha,$$

we shall have

$$v_2 \sin \delta = \frac{r_1}{r_2} u_2 \tan \alpha,$$

which gives for the angle a which the blade makes with the inner circumference

$$\tan a = \frac{r_2}{r_1}\frac{v_2}{u_2}\sin\delta = \frac{r_2}{r_1}\frac{\sin\beta\sin\delta}{\sin(\beta+\delta)}.$$

Hence, if we accept the customary ratio $\frac{r_2}{r_1} = 2$, and make use of the values $\beta = 16° \ 42'$ and $\delta = 26° \ 34'$ given above, we obtain

$$\tan a = 0{\cdot}375, \text{ or } a = 20° \ 30'.$$

We may further make the assumption that the area of the openings through which the water is admitted to the wheel is equal to the sectional area of the supply pipe, and hence, if we neglect the thickness of the blades, we have $\pi r_1^2 = 2\pi r_1 b_1$, which gives for the clear width at the inner circle of the wheel $b_1 = \frac{r_1}{2}$, etc.

As regards the number and thickness of the blades, we may apply the same rules as in the case of turbines, in which the usual number of blades for wheels of moderate size varies from 4 to 10, and the thickness from 5 to 10 mm. [0·2 to 0·4 in.]

After thus computing the angles a and β which the blades make with the inner and outer circumferences, it is necessary to make some definite assumption from which to determine the *form* of the blades. This is due to the fact that, since the resistances experienced by the water in the wheel cannot be represented by analytical expressions, it is impossible to ascertain by calculation the most advantageous form of blade, or that which reduces the internal resistances to a minimum. If no such resistances were encountered, any form of blade intersecting the circumferences at the required angles a and β would evidently answer the purpose.

As a basis on which to proceed in designing the blades, suppositions have been made with respect to the law according to which the velocity of the water should vary in the wheel.

For instance, it has been suggested that the radial component of the velocity of the water should remain constant, and that the tangential velocity should increase according to a certain law,[1] and hence the absolute path of the water and finally the form of the blade has been determined, as in turbines (see vol. on Hydraulics). The arbitrariness of all such suppositions may justify us in assuming directly as simple a form as possible for the blades, say a circular arc that cuts the two circumferences at the desired angles α and β. To draw such an arc,[2] lay off at any point B of the outer circumference (Fig. 177) a line BJ making with the tangent BH the angle HBJ $= \beta$, and at the point C lay off a line CN making

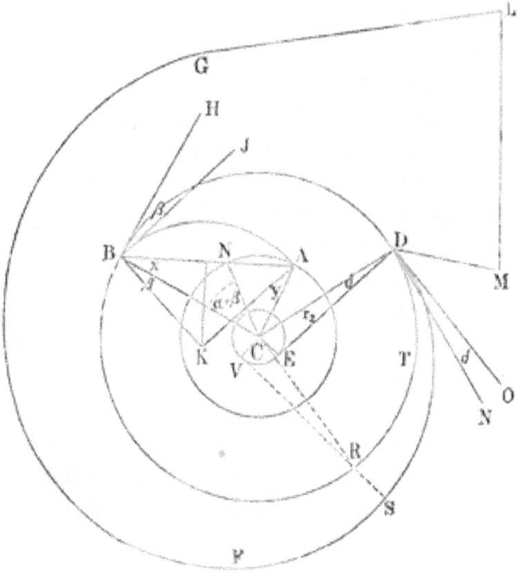

Fig. 177.

with the radius CB an angle BCN $= \alpha + \beta$. This second line cuts the inner circumference at N, and if we connect this intersection with the point B and prolong the line thus obtained until it intersects the inner circle at a second point A, then the circular arc passed through A and B with a centre K lying on the perpendicular to BJ at B will satisfy the required condition, as may be seen from the figure; for, the isosceles triangle CNA gives CNA = CAN or $\alpha + \beta + x = y + \beta + x$, from which we obtain CAK $= y = \alpha$.

The guiding apparatus which receives the water that leaves

[1] See Fink, *Theorie der Centrifugalpumpen Zeitschr. d. Ing.*, 1868, S. 1.

[2] See Article by Grove, *Mittheilungen d. Gew.-Ver. f. Hannover*, 1869.

the circumference of the wheel at the angle δ and gradually leads it into the direction of the delivery pipe, is formed, in horizontal centrifugal pumps, by the casing, whose circumference may be regarded as a single guide blade. This casing is brought almost in contact with the wheel at a point D, and is here given the direction of the issuing water by making the angle NDO equal to δ. A suitable form is obtained for the casing by shaping it as an involute DFG of a circle, which is concentric to the wheel and has the radius $CE = r_2 \sin \delta$. At any point R of the circumference of the wheel the sectional area RS furnished by the casing for the issuing water will then be equal to the area of efflux DTR, which discharges the delivered water through RS. For, if we denote the angle

$$DCR = ECV \text{ by } \phi,$$

then

$$RS = \text{arc } VE = r_2 \sin \delta \,.\, \phi\,;$$

hence the sectional area offered to the water at this place is $b_2 r_2 \phi \sin \delta$, where b_2 is the clear width of the wheel. But this is evidently equal to the area offered by the arc $DTR = r_2\phi$ to the water issuing at the angle δ, the diminution of area due to the thickness of the blades being neglected. The casing is provided with a neck GLMD which connects with the delivery pipe.

The power needed to drive the pump is equal to that required to lift a quantity of water Q per second to the height

$$h + \zeta + \frac{w^2}{2g};$$

expressed in horse-power, this is given by

$$N = \frac{Q\gamma}{75}\left(h + \zeta + \frac{w^2}{2g}\right) = 13{\cdot}33Q\left(h + \zeta + \frac{w^2}{2g}\right)$$

$$\left[N = 0{\cdot}114Q\left(h + \zeta + \frac{w^2}{2g}\right)\right].$$

As the water is actually lifted to the height h only, the efficiency of the pump will be expressed by

$$\eta = \frac{h}{h + \zeta + \frac{w^2}{2g}}.$$

With the above assumed average values of $\zeta = 0{\cdot}42h$ and $\frac{w^2}{2g} = 0{\cdot}03h$, the efficiency of the pump becomes

$$\eta = \frac{h}{1{\cdot}45h} = 0{\cdot}69.$$

This does not include the resistances due to the pull of the belt and the weight of the wheel, which resistances must be specially determined for each particular case.

EXAMPLE.—What dimensions must be given to a centrifugal pump which is to raise 6 cub. m. [211·9 cub. ft.] of water per minute to a height of 5 m. [16·4 ft.]?

If we assume $\beta = 15°$, $\delta = 25°$, and

$$\zeta + \frac{w^2}{2g} = 0{\cdot}5h,$$

we obtain from (7) for the velocity of the outer circumference of the wheel

$$u_2 = \sqrt{g\left(h + \zeta + \frac{w^2}{2g}\right)(1 + \tan\delta\cot\beta)}$$
$$= \sqrt{9{\cdot}81 \times 5 \times 1{\cdot}5(1 + 0{\cdot}466 \times 3{\cdot}73)} = 14{\cdot}20 \text{ m. [46·6 ft.]},$$

and for the absolute velocity of efflux, according to equation (8),

$$v_2 = \sqrt{9{\cdot}81 \times 5 \times 1{\cdot}5 \frac{\sin 15°}{\cos 25° \sin 40°}} = 5{\cdot}72 \text{ m. [18·8 ft.]}$$

The radial component of the velocity of efflux is therefore

$$v_2 \sin\delta = 5{\cdot}72 \times 0{\cdot}423 = 2{\cdot}42 \text{ m. [7·94 ft.]}$$

If we suppose the water to enter the wheel and pass through the suction pipe with this velocity, the diameter d of the latter will be obtained from

$$\frac{\pi d^2}{4} v_2 \sin\delta = Q, \text{ which gives } d = \sqrt{\frac{4}{\pi}\frac{6}{60 \times 2{\cdot}42}} = 0{\cdot}230 \text{ m. [9·06 ins.]}$$

If, therefore, we give to the wheel an inner diameter of 0·24 m. [9·44 ins.], that is, a radius $r_1 = 0{\cdot}12$ m. [4·72 ins.], and assume $r_2 = 2r_1 = 0{\cdot}24$ m. [9·44 ins.], we get for the velocity at the inner circumference

$$u_1 = \frac{r_1}{r_2} u_2 = \frac{1}{2} 14{\cdot}2 = 7{\cdot}1 \text{ m. [23·3 ft.]}.$$

and for the angle α at the inner circumference

$$\tan\alpha = \frac{v_1}{u_1} = \frac{2{\cdot}42}{7{\cdot}1} = 0{\cdot}341,$$

which gives $\alpha = 18° 40'$. Now, assuming 6 blades, each 6 mm. [0·24 in.] thick, we obtain the clear width b_1 at the inner circumference from

$$\left(2\pi r_1 - \frac{zs}{\sin \alpha}\right) b_1 v_1 = Q,$$

or

$$\left(2\pi \times 0{\cdot}12 - \frac{6 \times 0{\cdot}006}{0{\cdot}323}\right) 2{\cdot}42 \times b_1 = \frac{6}{60},$$

which gives $b_1 = 0{\cdot}065$ m. [2·56 ins.]; in like manner the width b_2 at the outer circumference is obtained from

$$Q = \left(2\pi r_2 - \frac{zs}{\sin \beta}\right) v_2 \sin \delta b_2.$$

giving

$$b_2 = \frac{0{\cdot}1}{\left(2\pi \times 0{\cdot}24 - \frac{0{\cdot}036}{0{\cdot}259}\right) 2{\cdot}42} = 0{\cdot}030 \text{ m. [1·18 ins.]}$$

Finally, we have for the number of revolutions of the wheel per minute

$$n = \frac{60 u_2}{2\pi r_2} = \frac{30 \times 14{\cdot}20}{3{\cdot}14 \times 0{\cdot}24} = 564,$$

and for the horse-power required to drive it

$$N = \frac{Q\gamma}{75}\left(h + \zeta + \frac{w^2}{2g}\right) = \frac{6 \times 1000}{60 \times 75} \times 5 \times 1{\cdot}5 = 10.$$

CHAPTER VI

ADDITIONAL WATER-RAISING MACHINES

§ 46. **The Hydraulic Ram.**—Instead of making use of a swiftly rotating wheel for imparting to the water a velocity which will carry it to the required height, we can, when a fall of sufficient head is available, employ for the purpose the living force possessed by the water while in motion in a pipe. It is on this principle that the *hydraulic ram* depends. By this

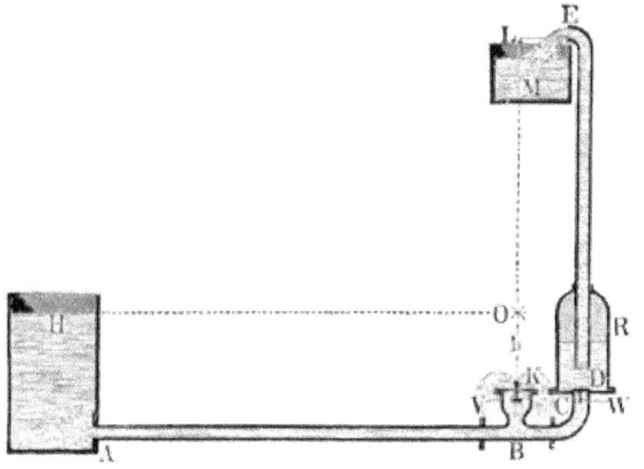

Fig. 178.

apparatus, which was invented by *Montgolfier* in 1796, a portion of the driving water is forced to an elevation which is greater than that from which it has fallen. The arrangement of a hydraulic ram is essentially as follows. The reservoir HA (Fig. 178), from which the water is supplied, connects by means of a pipe ABC with the air chamber R, from which the delivery pipe DE leads to the tank LM, where the lifted water is received. At C, where the supply pipe enters the air chamber, a *delivery valve* W, opening upward, is placed, and in the short branch pipe BK is located a so-called *waste valve* which opens downward.

In order to explain the action of the machine, let us first assume the two valves V and W to be closed, and that the two pipes ABC and DE are full of water, the air chamber R being at the same time filled partly with water and partly with air. Now, if the aperture K is opened by pressing down the valve V, water will discharge through K and a fresh supply will flow from the reservoir AH into the pipe AB. Since the *hydraulic* pressure on the upper surface of the waste valve is less than that on the lower surface, on account of the greater velocity of the water passing through the valve opening, the valve will close again as soon as the velocity of efflux has become sufficient to cause an upward excess of pressure greater than the weight of the valve. This closing of the waste valve causes the water moving in AB to open the delivery valve, and consequently water is forced into the air chamber R till the living force of the water in the supply pipe is wholly destroyed by the operation. The incoming water causes a compression of the air, and compresses the air in the chamber, thus causing a discharge from the orifice E of the delivery pipe DE. After the water has been brought to rest in AB it gradually begins to move in the opposite direction from B to A under the influence of the greater pressure in the air chamber. As this flow soon closes the delivery valve W, water ceases to flow from D, and the atmospheric pressure now becoming greater than the pressure of the water in B, pushes down the waste valve, thus automatically causing a new series of operations to begin. The hydraulic ram represented in Fig. 179 was built by *Montgolfier*, and differs from that just described principally in having two air chambers. AB is the end of the supply pipe, V the waste valve, R the air chamber, and DE the delivery pipe leading from the latter; another air chamber S connects by the pipe C with the valve chamber BK, and by the valves WW with the outer air chamber R. The inner air chamber is employed to diminish the hurtful effects of the shocks that accompany the sudden opening and closing of the waste valve, thus not only diminishing the wear of the machine, but also increasing its efficiency. The water in the chamber R is subjected to a pressure exceeding that of the atmosphere, and consequently it absorbs a portion of the air and carries it along to the delivery pipe. To make good the loss of air thus caused, an air valve H opening inward is inserted at the bottom of the air chamber R. When the return flow takes place in the supply pipe and the pressure in the latter becomes less than the atmosphere, this valve opens and admits, during each series of operations, a small quantity

Fig. 179.

of air into the chambers S and R to replace that which has escaped through the delivery pipe.

The hydraulic ram can also be arranged to lift water by *suction*. A ram of this kind was built by *Boulton*,[1] and was described by *Hachette* in his *Traité Élémentaire des Machines*, under the name of *Bélier aspirateur*. Since that time the Belgian engineer *Leblanc*[2] has constructed a double-acting suction ram for draining excavations. It is shown in Fig. 180. A is the supply reservoir, BCD the driving pipe, SG

[1] *Journal des Mines*, Bd. II.

[2] *Annales des ponts et chaussées*, 3, Ser. 7, année 1858, and *Civil Ingenieur*, Bd. 5.

the upper part of the suction pipe, R an air chamber, and BR a branch pipe joining the chamber with the driving pipe. The waste valve of the ordinary ram is here replaced by the admission valve V, and the delivery valve of the ordinary ram by the suction valve W. When V is open water flows from A into the driving pipe, which leads to the well U. After the water has attained a certain velocity in BCD, the valve V is pressed down by the water above it, and, as the supply is now cut off from A, the pressure at B falls below that of the

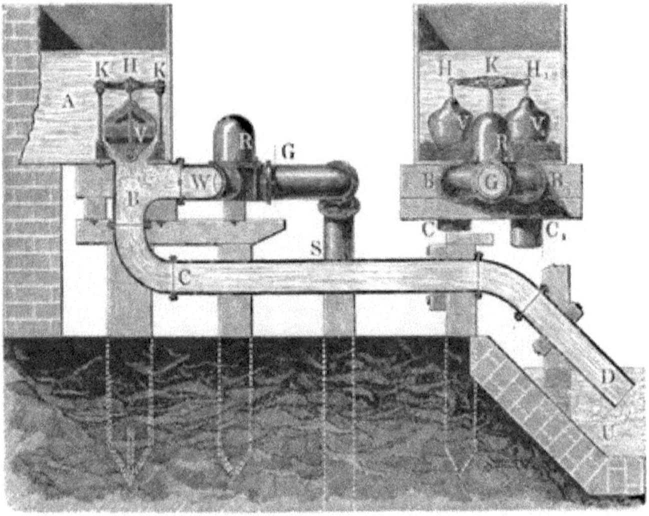

Fig. 180.

atmosphere, with the result that the excess of pressure of the rarefied air filling the chamber opens the suction valve W. Consequently the atmospheric pressure causes water to rise in the pipe SG, from which it passes by the path RWB into the driving pipe, and from there into the discharge well U. This motion lasts but a short time, however, for, as soon as the living force of the water in the driving pipe has been destroyed, the pressure in the pipe BC will overcome that in RW, since the water-level U is higher than the level from which the suction pipe S draws its supply. The result is that the valve W again closes, and after the water in the driving pipe has been brought to rest the valve V opens, and a new series of operations begins.

In this description we have assumed only one driving pipe, one admission valve, etc., while in *Leblanc's* double-acting ram all the parts, with the exception of suction pipe and air chamber, are duplicated, *i.e.* there are two driving pipes attached to the supply reservoir, each having an admission valve, and the air chamber is connected to each driving pipe by a separate branch pipe. To avoid severe shocks, the valves VV_1 and their seats are formed of compressed leather discs, and the suction clacks are also made of leather. Moreover, both of the former are attached by stems to an equal-armed lever HKH_1 attached to the horizontal shaft KK; consequently the closing of one valve will be accompanied by the opening of the other, and the two suction valves W will therefore also be alternately opened and closed.

In the double-acting ram just described the diameter of all the pipes and apertures is 0·2 m. [7·87 ins.], the length of the driving pipe is 3·29 m. [10·79 ft.], the fall is 1·7 m. [5·58 ft.], and the lift 2·25 m. [7·38 ft.] We are in possession of no exact data concerning the performance of this ram, but it is claimed that its capacity equals that of six wooden pumps, each requiring in its operation the combined efforts of twelve labourers.

Eytelwein has made very complete tests of two hydraulic rams of the ordinary type. The results are contained in a paper entitled "Bemerkungen über die Wirkung und vortheilhafte Anwendung des Stoszhebers," von J. A. Eytelwein, Berlin, 1805. The larger of the rams had a horizontal driving pipe 65 mm. [2·56 ins.] in diameter, a delivery pipe 26 mm. [1·02 in.], and a copper air chamber 0·235 m. [9·25 ins.] in diameter and 0·314 m. [12·36 ins.] high. The lengths of the pipes, the fall and the lift, were all frequently changed within wide limits. The smallest length of the driving pipe was 3·6 m. [11·8 ft.], the greatest length 13·65 m. [44·77 ft.], the length of the delivery pipe varied from 10 to 15·5 m. [32·8 to 50·84 ft.], and the lift from 4·7 to 14·8 m. [15·42 to 48·54 ft.] Moreover, in the experiments with the larger ram five different dish-shaped waste valves were employed, only one of which had a passage through it, the area of which was 23·9 sq. cm. [3·7 sq. in.], or nearly equal to the sectional area 25·24 sq. cm. [3·91 sq. in.] of the driving pipe. The delivery valve was of brass, and was either a hanging clack or a disc valve that moved horizontally. The number of operations or strokes per minute ranged from 10 to 180, the corresponding waste water varied from 0·0044 to 0·170 cub. m. [0·155 to 6·004 cub. ft.], and the lifted water from 0·75 to 31 litres [0·0265 to 1·0948 cub. ft.]

Let Q represent the quantity of water flowing through the waste valve, Q_1 the quantity discharged from the delivery pipe, h the fall OK (Fig. 181) or distance of the water-level **H** in the supply reservoir above the outlet K of the waste valve,

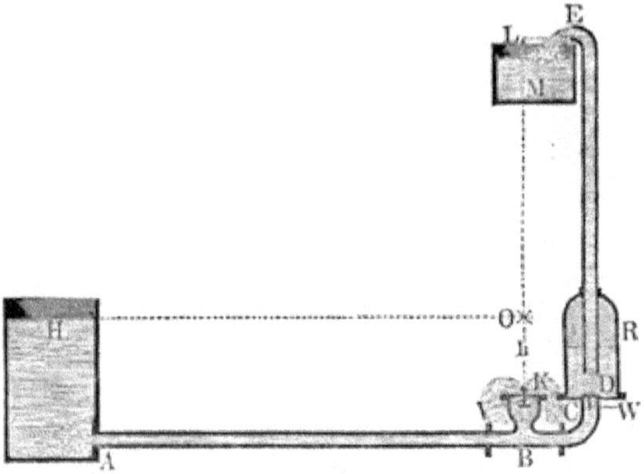

Fig. 181.

and h_1 the lift or distance OL between the water-levels in the upper and lower reservoirs; the efficiency of the ram is then expressed by

$$\eta = \frac{Q_1 h_1}{Q h}.$$

In all, *Eytelwein* made 1123 experiments, partly with the larger and partly with the smaller ram. From these experiments he deduced the formula

$$\eta = 1·12 - 0·2 \sqrt{\frac{h_1}{h}},$$

whence we obtain the following table for the efficiency,

$\frac{h_1}{h}=$	1	2	3	4	5	6	8	10	12	15	20
$\eta=$	0·920	0·837	0·774	0·720	0·673	0·630	0·555	0·488	0·427	0·345	0·226

This shows that for a given fall h the efficiency decreases with an increased lift h_1, and *Eytelwein* therefore suggests that, for great lifts, instead of a single ram several be employed, the first supplying water to the one above it, and so on. He also deduces from his experiments the following important rules.

1. The total quantity $(Q+Q_1)$ of water consumed varies as the square of the diameter (d) of the supply pipe, and if $60(Q+Q_1)$ represents the quantity of water used per minute, in cubic inches, the diameter of this pipe will be given by

$$d=\frac{\sqrt{60(Q+Q_1)}}{21}\text{ inches,}$$

or, when $Q+Q_1$ is expressed in cubic metres, it would be approximately

$$d=300\sqrt{60(Q+Q_1)}\text{ mm.}$$

2. The length l of the supply pipe depends on the delivery head h_1, and may be found from the formula

$$l=h_1+0{\cdot}3\frac{h_1}{h}.$$

3. The diameter d_1 of the delivery pipe has a subordinate influence on the action of the machine; it suffices to take $d_1=\frac{1}{2}d$.

4. The aperture of the waste valve should have the same sectional area as the driving pipe.

5. The weight of the waste valve should be as small as is consistent with sufficient strength.

6. The waste valve may be placed under water without interfering with the effect of the machine.

7. The two valves should be located as closely together as possible.

8. The air chamber diminishes the shocks and increases the effect of the machine. It is sufficient to make the cubic contents of the air chamber equal to that of the delivery pipe.

EXAMPLE.—An hydraulic ram is to be constructed to work with a fall of 2 m. [6·56 ft.], and to raise 30 litres [1·059 cub. ft.] of water per minute to a height of 8 m. [26·25 ft.]

We have $h=2$, $h_1=8$, and consequently the efficiency

$$\eta=1{\cdot}12-0{\cdot}2\sqrt{\frac{h_1}{h}}=1{\cdot}12-0{\cdot}2\sqrt{4}=0{\cdot}72.$$

The required driving water will therefore be

$$60Q = 60\frac{Q_1 h_1}{\eta h} = \frac{0\cdot03 \times 8}{0\cdot72 \times 2} = 0\cdot167 \text{ cub. m. [5·898 cub. ft.]},$$

and the total consumption of water per minute

$$60(Q + Q_1) = 0\cdot03 + 0\cdot167 = 0\cdot197 \text{ cub. m. [6·957 cub. ft.]}$$

The diameters of supply pipe and valve openings should therefore be

$$d = 300\sqrt{60(Q + Q_1)} = 300\sqrt{0\cdot197} = 133 \text{ mm. [5·24 ins.]},$$

while for the delivery pipe $d_1 = \frac{1}{2}d = 66$ mm. [2·62 ins.] will suffice.
For the length of supply pipe we have

$$l = h_1 + 0\cdot3\frac{h_1}{h} = 8 + 0\cdot3\frac{8}{2} = 9\cdot2 \text{ m. [39·18. ft.]},$$

The volume of the air chamber we make equal to that of the delivery pipe, or

$$W = \frac{\pi d_1^2}{4}h_1 = 0\cdot7854 \times 0\cdot66^2 \times 80 = 27\cdot4 \text{ litres [0·968 cub. ft.]},$$

and if we choose the cylindrical form and a diameter of 0·3 m. [0·98 ft.] we obtain for the height of the chamber $\frac{27\cdot4}{7\cdot068} = 0\cdot388$ m. [1·27 ft.]

The general *theory of the hydraulic ram* is quite complicated, and involves the use of unusual analytical methods (see Navier's *Résumé des Leçons sur l'application de la Mécanique*, part ii., and also Venturoli's *Elementi di Meccanica e d'Idraulica*, vol. ii.) Moreover, as it is insufficient for the determination of the effect of the machine, which always requires empirical results, we will in the following present a theory which, though only approximately correct, yet the more closely approaches to the truth the greater the quantity of water in the supply pipe, and the greater the number of revolutions of the ram per minute.

Let F denote the sectional area and l the length of the supply pipe, and h the fall, then the inert quantity of water which must be set in motion when the waste valve opens by the force $Fh\gamma$ will be $\frac{Fl\gamma}{g}$, and consequently its acceleration

$$p = \frac{Fh\gamma}{Fl\gamma}g = \frac{h}{l}g.$$

Owing to this constant acceleration the velocity acquired after t seconds will be

$$v = pt = \frac{h}{l} gt,$$

and if the waste opening has the same area F as the supply pipe, the volume of water which flows out during the time t will be

$$V = F\frac{v}{2}t = \frac{Fl}{h}\frac{v^2}{2g} = \frac{Fh}{l}\frac{gt^2}{2}.$$

Now, if the volume of water in the delivery pipe is small in comparison with that in the supply pipe, the retardation of the latter mass caused by the back pressure $Fh_1\gamma$ in the delivery pipe, when the waste valve closes and the delivery opens, will be given by

$$p_1 = \frac{Fh_1\gamma}{Fl\gamma} g = \frac{h_1}{l} g;$$

and consequently the velocity of the water in the supply pipe will, when t_1 seconds have elapsed after the opening of the delivery valve, amount to

$$v_1 = v - p_1 t_1 = v - \frac{h_1}{l} gt_1.$$

The time t_1 required to bring the whole volume of water to rest will now be obtained by placing $v_1 = 0$, which gives

$$t_1 = \frac{l}{h_1}\frac{v}{g} = \frac{h}{h_1} t,$$

and determines the volume which, in the meantime, has flown into the air chamber to

$$V_1 = F\frac{v}{2}t_1 = \frac{Fl}{h_1}\frac{v^2}{2g} = \frac{Fh^2}{h_1 l}\frac{gt^2}{2}.$$

Assuming the delivery valve to remain open for a short interval t_2 after the subsequent return flow into the supply pipe has commenced, the driving force $Fh_1\gamma$ will give to the volume $Fl\gamma$ a velocity

$$v_2 = \frac{h_1}{l} gt_2,$$

and therefore the volume returned from the air chamber will be

$$V_2 = \frac{Fh_1}{l} \frac{gt_2^2}{2}.$$

Finally, when the delivery valve closes and the waste valve opens, the mass of water $Fl\gamma$ will move with a retardation

$$p_3 = \frac{h}{l} g,$$

and accordingly, after the time t_3, it has attained a velocity

$$v_3 = v_2 - p_3 t_3 = v_2 - \frac{h}{l} g t_3.$$

The volume $Fl\gamma$ is now again brought to rest, that is, v_3 becomes zero, and a new stroke begins after the time

$$t_3 = \frac{l}{h} \frac{v_2}{g} = \frac{h_1}{h} t_2,$$

during which a volume

$$V_3 = \frac{Fv_2 t_3}{2} = \frac{Fv_2}{2} \frac{h_1}{h} t_2 = \frac{Fh_1^2}{hl} \frac{gt_2^2}{2}$$

returns, and an equal quantity of air or water flows in through the waste valve.

The driving or waste water required by the ram per minute is now given by

$$Q = \frac{V - V_3}{t + t_1 + t_2 + t_3},$$

or approximately, when V_3, t_2, and t_3, owing to their insignificance, are neglected,

$$Q = \frac{V}{t + t_1} = \frac{V}{t\left(1 + \frac{h}{h_1}\right)} = \frac{h_1}{h + h_1} \frac{Fv}{2} = \frac{h_1}{h + h_1} \frac{h}{l} F \frac{gt}{2}.$$

Further, the volume of water lifted per second will be

$$Q = \frac{V_1 - V_2}{t + t_1 + t_2 + t_3},$$

or approximately

$$Q_1 = \frac{V_1}{t + t_1} = \frac{h_1}{h + h_1} \frac{V_1}{t} = \frac{h}{h + h_1} \frac{Fv}{2} = \frac{h}{h + h_1} \frac{h}{l} F \frac{gt}{2},$$

and consequently the ratio of the lifted water to the driving or waste water

$$\frac{Q_1}{Q} = \frac{h}{h_1}.$$

The total volume of water consumed will be

$$Q + Q_1 = \left(\frac{h_1}{h + h_1} + \frac{h}{h + h_1}\right)\frac{Fv}{2} = \frac{Fv}{2},$$

which, like the experiments of *Eytelwein,* indicates that the sectional area F of the supply pipe should be proportional to the quantity $Q + Q_1$ of water consumed.

Under the supposition that the volume of water sucked in through the waste valve during the return flow is $V_3 = \frac{Fv_2t_3}{2}$, we obtain for the efficiency of the hydraulic ram

$$\eta = \frac{(V_1 - V_2)h_1}{(V - V_3)h} = \frac{\left(\frac{h^2}{h_1}t^2 - h_1t_2^{\,2}\right)h_1}{\left(ht^2 - \frac{h_1^{\,2}t_2^{\,2}}{h}\right)h} = \frac{h^2t^2 - h_1^{\,2}t_2^{\,2}}{h^2t^2 - h_1^{\,2}t_2^{\,2}} = 1.$$

On the other hand, if there is no such suction through the waste valve, the efficiency will be given by

$$\eta = \frac{(V_1 - V_2)h_1}{Vh} = \frac{h^2t^2 - h_1^{\,2}t_2^{\,2}}{h^2t^2} = 1 - \left(\frac{h_1}{h}\right)^2\left(\frac{t_2}{t}\right)^2,$$

that is to say, that, as has also been proved by experience, it will approach more closely to unity the smaller the ratio $\frac{h_1}{h}$ of delivery height to fall, and the shorter the time during which the delivery valve remains open after the return flow commences.

NOTE.—In regard to the efficiencies $\eta = 0{\cdot}57$ to $0{\cdot}67$ of five hydraulic rams which are in operation in France, see *Formules, Tables,* etc., *par* Claudel, Paris, 1854.

§ 47. **Ejectors and Injectors.**—On the *suction* produced by a jet of water various water-raising contrivances have been based, which may be employed to advantage for some purposes.

To this class belongs the *ejector* designed by *James Thomson*[1] and shown in Fig. 182. The driving water is supplied by the pipe EA, and flows through a conical nozzle into a horizontal discharge pipe which widens toward the end F. Owing to the reduction in pressure which is hereby produced in the chamber D (as in the case of the blast pipe of a locomotive), water will be sucked through the pipe CD, ascending from the reservoir C, and will be discharged, together with the driving water, at F.

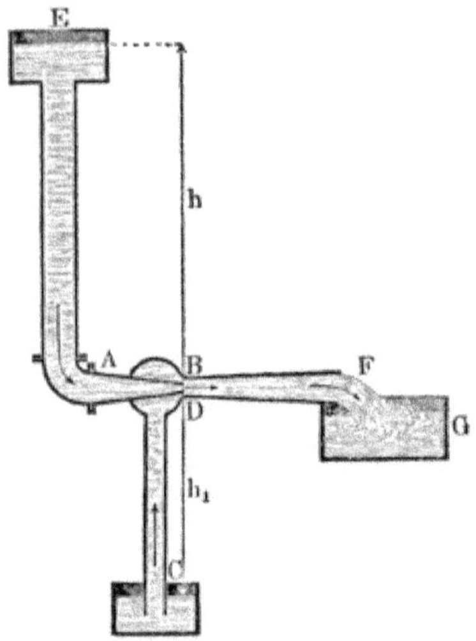

Fig. 182.

According to tests of the apparatus made by *Thomson*, the maximum effect was obtained for a suction $h_1 = 0{\cdot}9h$, where h is the fall of the driving water; for this case the water raised Q_1 equalled $\frac{1}{5}$ of the driving water Q, which performance corresponds to the trifling efficiency of

$$\eta = \frac{1}{5} \times 0{\cdot}9 = 0{\cdot}18.$$

This small efficiency explains why ejectors of this kind have gained but little application, more economical methods being available for the employment of water power. On the other hand, the apparatus constructed by *Nagel*,[2] and based on the same principle, has been used to advantage in the temporary work of draining excavations, etc., the slight efficiency often being of no consequence in such cases. An arrangement used by *Nagel* for draining the excavations for a turbine plant is shown in Figs. 183 and 184. By opening the gate B water is admitted at A to the wooden trough CDE, which is secured to the bottom of the outlet channel FF. The trough is made of rectangular section, and obtains its smallest height at D,

[1] *Report of the British Association*, 1852, and Rankine, *Manual of the Steam Engine*, etc.

[2] *Ztschr. deutsch. Ing.*, 1866, p. 121.

whence it widens towards E, so that it has its minimum cross-section at D. At this point the suction pipe H enters it from above, being carried from the excavation G, which is protected by the dam K. A flap L at the end E of the trough permits of filling the latter with water before starting; when the flap

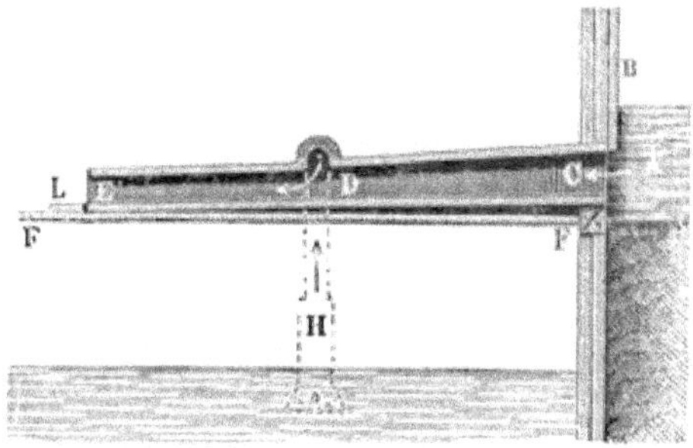

Fig. 183.

is dropped the flow of water through the trough will create a suction in the pipe H. The latter is at its lower end provided with a foot valve, which prevents the water from flowing out of the suction pipe when the action of the apparatus is interrupted by the closing of the gate B. The desired object

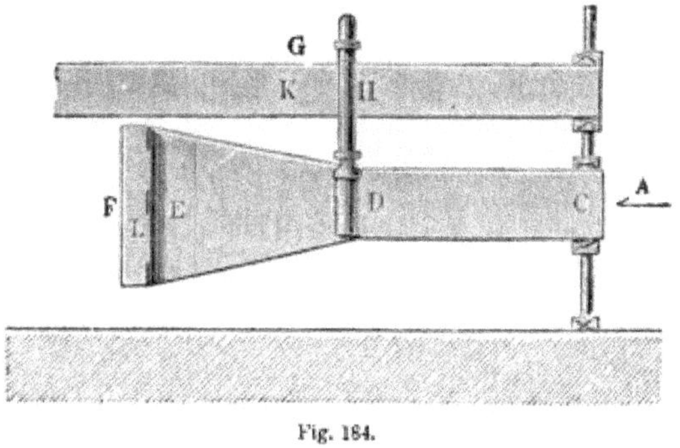

Fig. 184.

was completely achieved in the instance referred to, when the excavation, which measured 24·3 m. [79·7 ft.] in length by 5·6 m. [18·4 ft.] in width, and contained water to a depth of 2·4 m. [7·9 ft,], was emptied in half an hour and kept entirely free from water during the building operations. It is claimed that the apparatus worked equally well when the height of suction considerably exceeded that of the fall.

To discuss the action of this ejector, let h denote the fall of the driving water measured from the upper level E (Fig. 182) to the centre of the nozzle, whose clear aperture will be designated by F; further, let h_1 be the height of suction, and F_1 the sectional area of the suction pipe. Now, if p is the hydraulic pressure in the chamber at the orifice B of the nozzle, and consequently $\frac{p}{\gamma}$ the corresponding water column, then, neglecting frictional resistances, we have for the velocity of efflux v of the driving water at B the equation

$$\frac{v^2}{2g} = b + h - \frac{p}{\gamma} \quad . \quad . \quad . \quad (1),$$

and likewise for the velocity v_1 with which the suction water passes the nozzle

$$\frac{v_1^2}{2g} = b - h - \frac{p}{\gamma} \quad . \quad . \quad . \quad (2).$$

Letting w represent the velocity with which the mixture of driving and suction water Q and Q_1 is discharged, the living force of the latter will be

$$(Q + Q_1)\frac{w^2}{2g}.$$

As the two bodies of water suddenly change their velocities v and v_1 respectively into w, the resulting losses from impact will (according to Weisbach's *Mechanics*, vol. i. § 7, chap. 4) be given by

$$\frac{Q(v-w)^2}{2g} \text{ and } \frac{Q_1(v_1-w)^2}{2g},$$

and consequently we obtain the following equation,

$$Qh = Q_1h_1 + (Q + Q_1)\frac{w^2}{2g} + Q\frac{(v-w)^2}{2g} + Q_1\frac{(v_1-w)^2}{2g},$$

in which the quantity on the left represents the work performed by the driving water in falling from the height h, while Q_1h_1 is the work expended in raising the suction water Q_1 to the height h_1. This equation may also be written thus—

$$Q\left(h - \frac{v^2}{2g} + w\frac{v-w}{g}\right) = Q_1\left(h_1 + \frac{v_1^2}{2g} - w\frac{v_1-w}{g}\right),$$

or, after combining it with (1) and (2),

$$Q\left(\frac{p}{\gamma} - b + w\frac{v-w}{g}\right) = Q_1\left(b - \frac{p}{\gamma} - w\frac{v_1 - w}{g}\right) \qquad (3).$$

The efficiency will be given by

$$\eta = \frac{Q_1 h_1}{Q h} \qquad (4),$$

and for determining the cross-sections we have the relations

$$Q = Fv\gamma\,; \quad Q_1 = F_1 v_1 \gamma\,; \quad Q + Q_1 = (Fv + F_1 v_1)\gamma = Gw\gamma,$$

if G represents the sectional area of the discharge pipe.

Assuming, for instance, $h = 8$ m. [26·25 ft.], $h_1 = 3$ m. [9·84 ft.], and such dimensions that the velocities v_1 and w in suction and discharge pipes will be equal to 5 m. [16·4 ft.], we have

$$\frac{p}{\gamma} = b - h_1 - \frac{v_1^2}{2g} = 10{\cdot}34 - 3 - \frac{5^2}{2 \times 9{\cdot}81} = 6{\cdot}07 \text{ m. [19·92 ft.]},$$

and consequently

$$v = \sqrt{2g\left(b + h - \frac{p}{\gamma}\right)} = \sqrt{2 \times 9{\cdot}81(10{\cdot}34 + 8 - 6{\cdot}07)} = 15{\cdot}5 \text{ m. [50·85 ft.]}$$

Hence we obtain

$$\frac{Q_1}{Q} = \frac{\frac{p}{\gamma} - b + w\frac{v-w}{g}}{b - \frac{p}{\gamma} - w\frac{v_1 - w}{g}} = \frac{6{\cdot}07 - 10{\cdot}34 + 5\,\frac{15{\cdot}5 - 5}{9{\cdot}81}}{10{\cdot}34 - 6{\cdot}07} = \frac{1{\cdot}08}{4{\cdot}27} = 0{\cdot}252,$$

and accordingly the efficiency will be

$$\eta = \frac{Q_1 h_1}{Q h} = 0{\cdot}252\,\frac{3}{8} = 0{\cdot}094.$$

The *Gifford Injector*, which in modern times has gained such an extended application as a means of feeding boilers, also belongs to the class of apparatus which raise water by the action of a jet. Its distinguishing feature is that the driving fluid is steam, which, by virtue of its living force, not only sucks water from a certain depth, but also forces it into the boiler against the pressure in the latter, the action being equivalent to raising the water to a height equal to that of a water column corresponding to the excess of pressure in the boiler. The arrangement and action of this apparatus are essentially as follows. The pipe A (Fig. 185) communicates with the steam space of the boiler, and, when the cock H is open, admits steam through a series of holes into a pipe BC

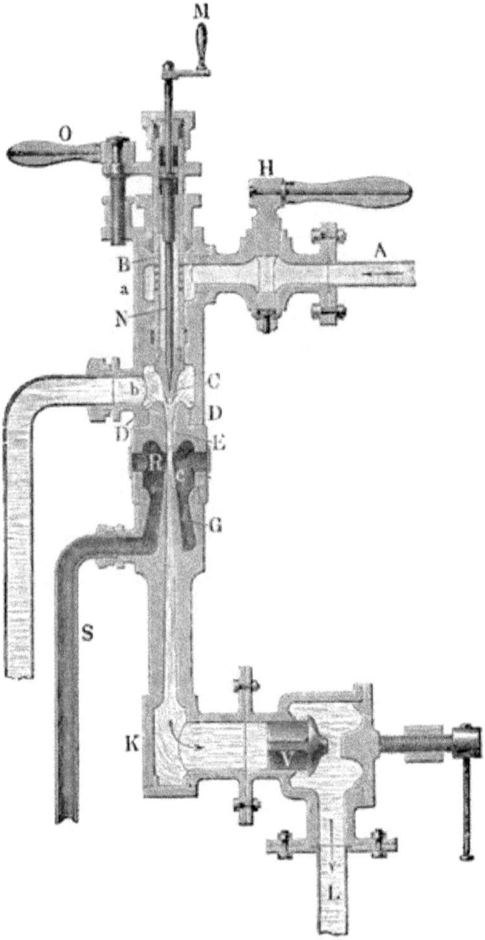

Fig. 185.

provided with a nozzle C. This nozzle terminates in a chamber D, which acts as a condenser, and is connected by the suction pipe *b* with the feed-water supply. This chamber is called the *combining tube*, and has its outlet in a nozzle E, through which flows the mixture of water drawn through *b* and that formed by the condensation of the steam that passes through C. Opposite the tube E is a delivery tube G, which receives the water issuing from E and guides it through the gradually widening pipe K and the check valve V into the pipe L communicating with the water space in the boiler. In this way the steam issuing from C drives a continual stream of water into the boiler. The flow of steam can be regulated by a conical spindle N, which can be moved back and forth by turning the crank M. On the other hand, the amount of feed water can be controlled by moving the receiving tube BC by means of a screw O, thus varying the annular space between the nozzle C and the bottom of the combining tube D. All the water which is not received by the delivery tube G passes into the overflow chamber R, and is discharged by the pipe S. While the apparatus is working normally no water escapes through the pipe S, for overflow occurs only when the apparatus is started, or when the pressure of the steam has fallen below the requisite amount. By the condensation of the steam this feed water is heated, which circumstance, however, involves no loss in the case of boiler feeds, for practically all the heat thus expended is returned to the boiler. But when the injector is employed as a pump or *ejector* for raising water to a great height, this heat produces no useful effect, and is the cause of the low efficiency of steam jet pumps or ejectors when employed for the latter purpose. As the action of the injector depends largely on the condensation of the steam jet, we see that it will be more uncertain the hotter the feed water. Of late certain improvements have been made in the apparatus, however, permitting the employment of feed water of 60° C. [140° Fahr.] The height of suction attainable is usually less than in piston pumps, for there is always in the combining chamber D a pressure corresponding to that of steam of the temperature of the mixture flowing through the chamber. But aside from these defects, which, especially in boiler feeds, are of slight importance, the injector is a most excellent apparatus, and is valued for its uniform action and entire absence of mechanical complications.

In order to determine the action of the injector, approximately, by calculation, let *h* represent the height of a water column corresponding to the excess of pressure (that is, the gauge pressure) of the steam in the boiler, then, for an excess of pressure equal to n atmospheres we have $h = 10{\cdot}34n$ metres [$h = 33{\cdot}9n$ ft.]; in like manner let h_1 represent the height of suction, and h_2 the distance of the combining chamber below the water in the boiler. Moreover, let Q be the weight of the steam flowing per second through the steam nozzle with a velocity v, and Q_1 the weight of the feed water which flows per second into the condensing chamber with a velocity v_1. Now, let p represent the hydraulic pressure in the latter chamber, then we have, for the suction pipe, the equation

$$(1 + \zeta_1)\frac{v_1^2}{2g} = b - h_1 - \frac{p}{\gamma} \quad . \quad (1),$$

where ζ_1 represents the coefficient of resistance for the suction pipe. In like manner we have, for the flow of steam, the approximate equation

$$\frac{v^2}{2g\mu} = \phi h + b - \frac{p}{\gamma} \quad (2),$$

where ϕ is a coefficient less than unity, which takes into account the reduction of steam pressure due to radiation and friction while the steam passes from the boiler to the apparatus, the excess of pressure of the steam before it issues from the nozzle being therefore expressed by ϕh; μ represents the relative volume of this steam, or the ratio of the volume of the steam and that of the water from which it is formed; the living force contained in the steam Q and water Q_1, when flowing into the apparatus, is expressed by

$$Q\frac{v^2}{2g} + Q_1\frac{v_1^2}{2g};$$

a portion of this energy is absorbed by the impact accompanying the change of the velocities v and v_1 into v_2, and the remainder is expended in forcing the mixture $Q + Q_1$ with a velocity v_2 through the nozzle E of the combining chamber. We therefore obtain the relation

$$Q\left(\frac{v^2}{2g} - \frac{(v - v_2)^2}{2g}\right) + Q_1\left(\frac{v_1^2}{2g} - \frac{(v_1 - v_2)^2}{2g}\right) = (Q + Q_1)\frac{v_2^2}{2g},$$

and hence the simple equation

$$Q(v - v_2) = Q_1(v_2 - v_1) \qquad (3).$$

The jet passes from the combining chamber, where the pressure is p, into the delivery tube, whose orifice is subjected to the atmospheric pressure b in the overflow chamber R; accordingly, if we neglect the resistances at the entrance, we find the velocity w with which the water enters the delivery tube from

$$\frac{v_2^2}{2g} + \frac{p}{\gamma} = \frac{w^2}{2g} + b,$$

or

$$\frac{w^2}{2g} = \frac{v_2^2}{2g} + \frac{p}{\gamma} - b \ . \qquad (4).$$

This velocity w, with which the jet enters the delivery tube, must be capable not only of forcing the water into the boiler—*i.e.* of lifting it from a height b corresponding to the atmospheric pressure, to a height $h_2 + h + b$ corresponding to the pressure in the boiler—but also of overcoming the resistances in the pipe between the delivery tube and the boiler, and discharging the water into the boiler itself with a velocity w_1.

If ζ represents the coefficient of resistance in the pipe between the delivery tube and the boiler, we must have the equation

$$\frac{w^2}{2g} + b = h_2 + h + b + (1 + \zeta)\frac{w_1^2}{2g},$$

or, if we make the sectional area of the delivery pipe m times that of the delivery tube, *i.e.* make $w_1 = \frac{w}{m}$, we obtain

$$\left(1 - \frac{1 + \zeta}{m^2}\right)\frac{w^2}{2g} = h_2 + h \ . \qquad (5).$$

From this equation we first compute w, and then assuming a certain entrance velocity v_1 of the water, determine the hydraulic pressure p from (1); subsequently we find from (4) the value v_2, and from (3) the ratio of the weights Q and Q_1 of the steam and water.

This ratio $\frac{Q_1}{Q} = \nu$ also gives the temperature t_2 with which the water reaches the boiler, for, calling t_1 the temperature of the feed water, we have

$$Q640 + Q_1 t_1 = (Q + Q_1)t_2,$$

or

$$t_2 = \frac{640 + \nu t_1}{1 + \nu} \text{ C.} \left[t_2 = \frac{1152 + \nu(t_1 - 32)}{1 + \nu} + 32 \text{ F.} \right] \quad (6).$$

EXAMPLE.—Determine the dimensions of an injector which is to furnish a boiler carrying four atmospheres pressure [58·8 lbs. gauge pressure] with 10 kg. [22 lbs.] of water per minute, the height of suction being $h_1 = 2$ m. [6·56 ft.], and the injector being $h_2 = 1{\cdot}5$ m. [4·92 ft.] below the water-level of the boilers.

Let us assume that the feed water flows through the annular space around the steam nozzle with a velocity $v_1 = 5$ m. [16·4 ft.], and that the coefficient of resistance for the suction pipe is equal to 0·2, then the hydraulic pressure in the combining chamber is found from equation (1) to be

$$\frac{p}{\gamma} = b - h_1 - (1 + \zeta_1)\frac{v_1^2}{2g} = 10{\cdot}34 - 2 - 1{\cdot}2\frac{5^2}{2 \times 9{\cdot}81} = 6{\cdot}8 \text{ m. [22·3 ft.]}$$

If we also assume $\phi = 0{\cdot}75$, *i.e.* that the steam flows into the apparatus with an excess of pressure of three atmospheres [44·1 lbs. gauge pressure], we must take $\mu = 448$ (see Weisbach's *Mechanics*, vol. ii., Table of Relative Steam Volumes); substituting these values in equation (2) we obtain the velocity of efflux of the steam

$$v = \sqrt{2 \times 9{\cdot}81 \times 448 \times (3 \times 10{\cdot}34 + 10{\cdot}34 - 6{\cdot}8)} = 551 \text{ m. [1807 ft. per sec.]}$$

Now, taking the sectional area of the delivery pipe at the entrance to the boiler to be ten times that of the delivery tube, *i.e.* placing $m = 10$; further, assuming the coefficient of resistance for the delivery pipe equal to 2, for the bends equal to 3, and for the clack valve 10, *i.e.* assuming $\zeta = 15$, then equation (5) will give us the velocity of the water when it enters the delivery pipe

$$w = \sqrt{\frac{1{\cdot}5 + 4 \times 10{\cdot}34}{1 - \frac{1 + 15}{100}} 19{\cdot}62} = 31{\cdot}7 \text{ m. [104 ft.]},$$

and therefore $w_1 = 3{\cdot}17$ m. [10·4 ft.] We can now obtain from (4) the velocity v_2 with which the mixture leaves the combining chamber

$$v_2 = \sqrt{2 \times 9{\cdot}81\left(\frac{31{\cdot}72^2}{2 \times 9{\cdot}81} + 10{\cdot}34 - 6{\cdot}8\right)} = 32{\cdot}7 \text{ m. [107·3 ft.]},$$

and therefore equation (3) gives us the ratio of the weights

$$\nu = \frac{Q_1}{Q} = \frac{v - v_2}{v_2 - v_1} = \frac{551 - 32{\cdot}7}{32{\cdot}7 - 5} = 18{\cdot}7.$$

The weight of steam used per minute is therefore

$$\frac{10}{18\cdot7}=0\cdot535 \text{ kg. } [1\cdot18 \text{ lb.}]$$

The temperature with which the water enters the boiler when the temperature of the feed water is 15° C. [59° F.] is

$$\frac{640+18\cdot7\times15}{1+18\cdot7}=46\cdot7° \text{ C. } [116° \text{ F.}],$$

and consequently the water has been heated about 30° C. [54° F.]

The feed water needed per second is given by

$$Q_1=\frac{10}{60}=0\cdot167 \text{ kg. } [0\cdot368 \text{ lb.}];$$

from this quantity, and the velocities determined above, we can determine the dimensions of the apparatus at various points. The annular area F_1 for the suction water will be

$$F_1=\frac{Q_1}{v_1}=\frac{0\cdot167 \text{ cb.dm.}}{50 \text{ dm.}}=0\cdot334 \text{ sq. cm. } [0\cdot052 \text{ sq. in.}],$$

the area of the orifice of the combining tube

$$F_2=\frac{Q+Q_1}{v_2}=\frac{1+18\cdot7}{18\cdot7}\frac{0\cdot167}{327}=\frac{0\cdot175}{327}=0\cdot0536 \text{ sq. cm. } [0\cdot0083 \text{ sq. in.}],$$

and that of the orifice of the delivery tube

$$G=\frac{Q+Q_1}{w}=\frac{0\cdot175}{317}=0\cdot0552 \text{ sq. cm. } [0\cdot00856 \text{ sq. in.}],$$

while the feed pipe must have ten times this size, or

$$G_1=10G=0\cdot552 \text{ sq. cm. } [0\cdot0856 \text{ sq. in.}]$$

As mentioned above, both the steam nozzle and the entrance for the water may be regulated by adjusting screws.

NOTE.—In the above discussion we have neglected the quantity of heat which was transformed into work during the operation of forcing the water into the boiler. With reference to the application of the mechanical theory of heat to the injector, we refer to Weisbach's *Mechanics*, vol. ii., Boilers.

§ 48. **Spiral Pumps.**—Compressed air may also be employed for raising water, the apparatus made use of being then either *spiral pumps* or the air machine constructed by *Höll*. The method is, however, seldom employed.[1] The two apparatus

[1] Regarding the use in America of an "air lift pump," constructed by *Pohlé* of San Francisco, see *Engineering*, London, 1894, vol. i. pp. 723 and 754, and Uhland's *Technische Rundschau*, 1896, pp. 10 and 46.—TRANSLATOR'S REMARK.

just mentioned differ as regards the means introduced for compressing the air, falling water being in *Höll's* pump the motive agent, while in the spiral pump the air is compressed by revolving a shaft. Like the water screw, the latter consists of a pipe wound spirally around a shaft, only the shaft is not inclined to the horizon, but placed horizontally almost on a level with the surface of the water, and besides the outlet of this pipe does not communicate directly with the outside air, but with the lower end of a vertical pipe into which the lifted water is delivered. In Fig. 186 AB is the shaft, which is connected to the delivery pipe MN by means of a stuffing box B, and is revolved by the crank P; CGL is the thread or *worm* wound around and firmly secured to the shaft; it is provided with an enlarged head or inlet at C, and terminates at L in the hollow end of the shaft.

In order to form a clear idea of the action of this machine, we will assume that the worm is filled with

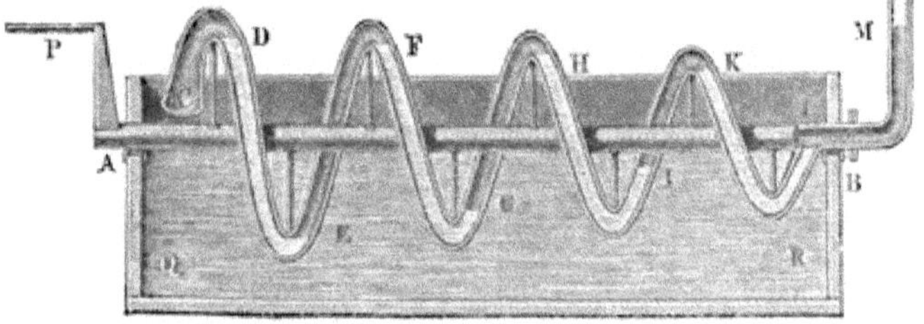

Fig. 186.

water and air occupying alternate arcs, as shown in the figure. At the head CD the air will then have the usual atmospheric pressure, measured by a water column of the height b, while the air enclosed in the arc EF will have a pressure exceeding b by an amount h_1 equal to the height of the water arc DE; further, in the arc GH the pressure will exceed that in EF by the height h_2 of the water column FG, etc. We have, therefore,

the pressure in EF $= b + h_1$

" " GH $= b + h_1 + h_2$

" " JK $= b + h_1 + h_2 + h_3$

" " LM $= b + h_1 + h_2 + h_3 + h_4$, etc.,

if the pressures in the remaining water arcs HJ, KL . . . be denoted by h_3, h_4. . . . In order to keep the air enclosed in the space LBM in equilibrium, it is now necessary that a water column of the height

$$h = h_1 + h_2 + h_3 + h_4 + . . .$$

should be contained in the delivery pipe BMN, since the air will then on both sides be acted on by the same force—

$$b + h = b + h_1 + h_2 + h_3 + h_4 + . . .$$

When the worm is slowly revolved, the air and water arcs gradually advance to the inlet B of the delivery pipe BMN . . ., and after rising in the latter are finally discharged at the top. If the radius of the worm is gradually reduced from the head C to the end L, so as to correspond to the gradually decreasing arcs of air, and if the head C for every revolution is allowed to take in a water arc and a quantity of air, then the equilibrium of the arcs will not be disturbed while the worm revolves slowly, and consequently there will be a uniform discharge of water

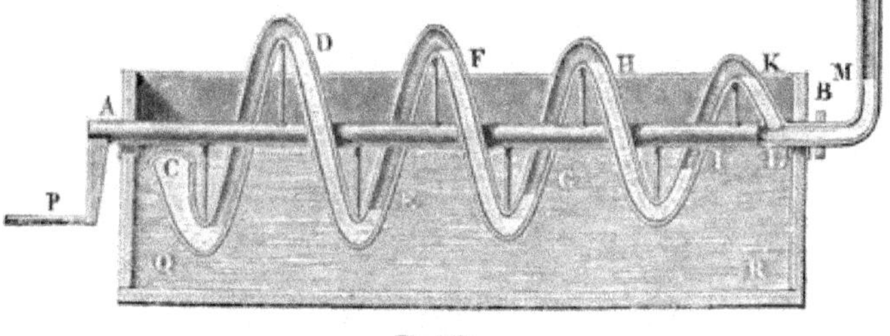

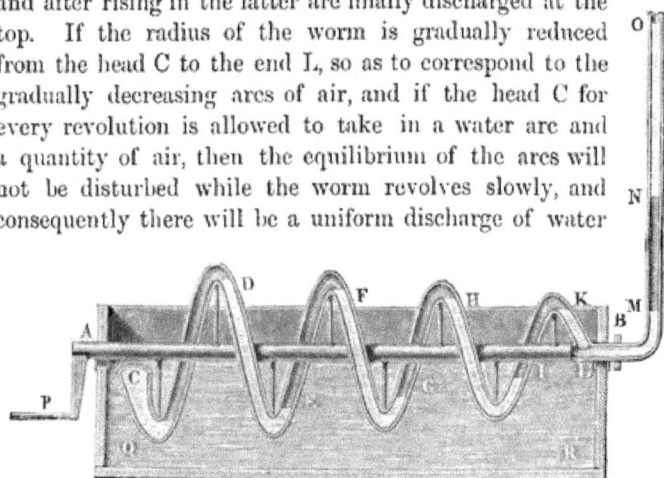

Fig. 187.

at the top of the delivery pipe. For instance, in case the shaft AB has revolved half a turn, then the arcs of water and air will have advanced in the direction from A to B through a distance equal to half the pitch of the screw, as shown in Fig. 187. The forward arc KL will then be partially discharged into the hollow shaft B and its continuation while the head C has taken in a new supply of water.

It is quite essential for the proper working of the machine that the head CD (Fig. 188), although scarcely longer than one quarter of the circumference, should be of sufficient capacity to contain a volume of air which will entirely fill the outer half coil DE. Then, in the opposite position, shown in Fig. 189, it will, although the inlet apparently has described a semicircle in the water, take in an approximately equal amount of water

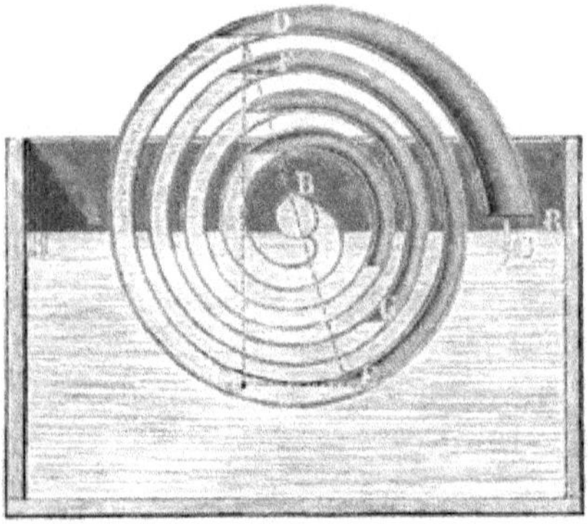

Fig. 188.

CC_1, which, after another half revolution, will fill one half of the outer coil DE, as may be seen in Fig. 188. For the fulfilment of this requirement it is necessary that the centre of the shaft should be located somewhat above the water-level

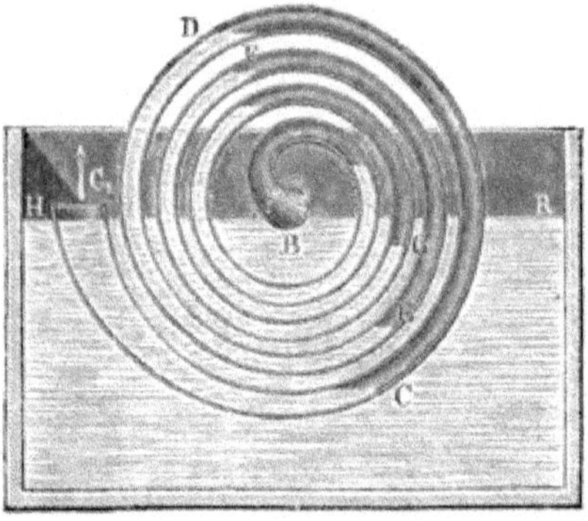

Fig. 189.

HR, and that the mean sectional area $\pi\rho_0^2$ of the head should be equal to twice the sectional area $\pi\rho^2$ of the worm, *i.e.* we should have

$$\pi\rho_0^2 = 2\pi\rho^2,$$

or, in other words, the ratio of the mean radius ρ_0 of the pipe at the head to the radius ρ in the remainder of the worm should be

$$\frac{\rho_0}{\rho} = \sqrt{2} = 1\cdot414,$$

or about as 7 to 5.

The rule according to which the radii of the coils are to be decreased from the head end to the delivery pipe may be deduced as follows by the aid of Mariotte's law. Let r_1 be the mean radius BD = BE of the first coil DEF (Fig. 188), and let ρ be the radius of its cross-section, then the volume of water contained in the arc DE will be

$$V = \pi\rho^2 . \pi r_1 = \pi^2\rho^2 r_1,$$

and the height DQ will be

$$h_1 = 2(r_1 - \rho).$$

Denoting by h the height of the water column or the sum of water columns in the delivery pipe, and by b that of a column which corresponds to the atmospheric pressure, then the volume of the inner arc, if there are in all n arcs, will be

$$V_n = \frac{b}{b+h} V,$$

and consequently its length

$$l_n = \frac{V_n}{\pi\rho^2} = \frac{b}{b+h} \frac{\pi^2\rho^2 r_1}{\pi\rho^2} = \frac{b}{b+h} \pi r_1.$$

If we now add to this length that of a water arc $l = \pi r_1$, we obtain for the total length of the coil n

$$l + l_n = \left(\frac{b}{b+h} + 1\right)\pi r_1,$$

and hence the required radius

$$r_n = \frac{l + l_n}{2\pi} = \left(\frac{b}{b+h} + 1\right)\frac{r_1}{2} = \frac{(2b+h)}{(b+h)}\frac{r_1}{2}.$$

Letting the radii of the coils decrease in arithmetical

progression from the outside, we shall have as the difference between two successive terms

$$d = \frac{r_1 - r_n}{n-1} = \frac{(b+h) - (b + \frac{1}{2}h)}{b+h} \frac{r_1}{n-1}$$
$$= \frac{h}{b+h} \frac{r_1}{2(n-1)},$$

and the corresponding progression for the radii of coils will be

$$r_1, (r_1 - d), (r_1 - 2d) \ldots [r_1 - (n-1)d].$$

For the angle β at the centre, corresponding to the water arc l in the inner coil, we have

$$\frac{\beta^\circ}{360^\circ} = \frac{l}{l + l_n} = \frac{b+h}{2b+h},$$

consequently

$$\beta^\circ = 360^\circ \frac{b+h}{2b+h} = \frac{r_1}{r_n} 180^\circ,$$

and hence we determine the height of water in this coil

$$h_n = r_n - \rho + r_n \cos(180^\circ - \beta) = r_n(1 - \cos\beta) - \rho.$$

The mean height of water in all the turns is

$$\frac{h_1 + h_n}{2} = \left(1 + \frac{2b+h}{b+h} \frac{1 - \cos\beta}{4}\right) r_1 - \frac{3}{2}\rho,$$

and therefore the total number of coils required is

$$n = \frac{2h}{h_1 + h_n}.$$

If the shaft makes u revolutions per minute, the quantity of water raised per second will be

$$Q = \frac{u}{60} V = \frac{u}{60} \pi^2 \rho^2 r_1.$$

Since the machine not only raises water but also compresses air, the work required will be given by

$$L = Qh\gamma + Qb\gamma \text{ hyp log } \frac{b+h}{b} = \left(h + b \text{ hyp log } \frac{b+h}{b}\right) Q\gamma$$

(See Weisbach's *Mechanics*, vol. i. § 6, chap. 4).

If the air in the delivery pipe mixes uniformly with the water and gradually expands during the ascent, then the delivery height will be increased to

$$h + b \text{ hyp log } \frac{b+h}{b};$$

on the other hand, if the air and water remain separated in this pipe, the total delivery height will be equal to the sum of the heights of the air and water columns contained in the latter. Let ρ_1 be the radius of the delivery pipe, then the length of water column produced in the latter by one water arc will be

$$= \frac{l\rho^2}{\rho_1^2},$$

and consequently the number of such columns, as well as columns of air in the delivery pipe,

$$m = h : \frac{l\rho^2}{\rho_1^2} = \frac{h\rho_1^2}{l\rho^2}.$$

The air column at the top is compressed by a water column of a length

$$\frac{l\rho^2}{\rho_1^2} = \frac{h}{m},$$

and therefore its height is

$$\frac{b}{b + \frac{h}{m}} \cdot \frac{h}{m} = \frac{bh}{mb + h};$$

the air column just below it, which is acted on by a water column of a length $\frac{2h}{m}$, has a height

$$\frac{b}{b + \frac{2h}{m}} \cdot \frac{h}{m} = \frac{bh}{mb + 2h},$$

and the next lower air column one of

$$\frac{bh}{mb + 3h}, \text{ etc.}$$

Consequently the total delivery height is given by

$$h+\frac{bh}{mb+h}+\frac{bh}{mb+2h}+\frac{bh}{mb+3h}+\ldots\frac{bh}{m(b+h)}$$
$$=\left[1+b\left(\frac{1}{mb+h}+\frac{1}{mb+2h}+\frac{1}{mb+3h}+\ldots\frac{1}{m(b+h)}\right)\right]h.$$

If a is the lever arm at which the turning force P acts on the shaft, we have

$$P=L:\frac{\pi na}{30}=\left(h+b \text{ hyp log } \frac{b+h}{b}\right)\frac{V\gamma}{2\pi a}.$$

In the calculation of the quantity of water raised per second

$$Q=\frac{u}{60}V=\frac{u}{60}\pi^2\rho^2 r_1,$$

no attention has been paid to the helical shape of the worm, since the pitch of the latter, which is of no consequence, can be assumed quite small. If the diameter of pipe in the worm also be taken quite small in comparison with the radius of coils, the axes of the latter might all be placed in the same plane, and thus be made to form a plane spiral.

NOTE.—Spiral pumps have thus far gained but little application, and have been thoroughly discussed only by *Eytelwein* (see his *Handbuch der Mechanik fester Körper und der Hydraulik*). On the other hand, machines of this class have been more frequently brought into use as blowing engines.

EXAMPLE.—A spiral pump has a worm and delivery pipe of radii $\rho=\rho_1=0{\cdot}08$ m. [3·15 ins.], while the radius of the largest coil is $r_1=1$ m. [3·28 ft.] Determine the radius of the other coils, the height of delivery, and the turning force required, under the supposition that the latter acts with a lever arm of 0·5 m. [1·64 ft.], and that the height of water column in the delivery pipe is $h=12$ m. [39·37 ft.] The maximum volume of water raised per revolution is

$$V=\pi^2\rho^2 r_1=9{\cdot}87\times0{\cdot}08^2\times1=0{\cdot}0632 \text{ cub. m. [2·232 cub. ft.]};$$

further, if we place the pressure of the atmosphere equal to $b=$ 10·34 m. [33·92 ft.] water column, the radius of the smallest coil will be

$$r_n=\frac{(2b+h)r_1}{(b+h)2}=\frac{20{\cdot}68+12}{10{\cdot}34+12}0{\cdot}5=0{\cdot}732 \text{ m. [2·4 ft.]},$$

and the angle at the centre of the corresponding arc

$$\beta^\circ=\frac{r'}{r}180^\circ=\frac{180^\circ}{0{\cdot}732}=246^\circ.$$

The height of the water arc in the largest coil is

$$h_1 = 2(r_1 - \rho) = 2(1 - 0{\cdot}08) = 1{\cdot}84 \text{ m. } [6{\cdot}04 \text{ ft.}],$$

while that of the smallest is

$$h_n = r_n(1 - \cos\beta) - \rho = 0{\cdot}732(1 + 0{\cdot}407) - 0{\cdot}08 = 0{\cdot}95 \text{ m. } [3{\cdot}12 \text{ ft.}],$$

and consequently the mean height of the water arcs

$$\frac{h_1 + h_n}{2} = \frac{2{\cdot}79}{2} = 1{\cdot}395 \text{ m. } [4{\cdot}58 \text{ ft.}]$$

and the number of coils or threads required

$$n = \frac{2h}{h_1 + h_2} = \frac{12}{1{\cdot}395} = 9.$$

Now we have

$$b \text{ hyp log } \frac{b+h}{b} = 10{\cdot}34 \text{ hyp log } \frac{22{\cdot}34}{10{\cdot}34} = 7{\cdot}96 \text{ m. } [26{\cdot}12 \text{ ft.}],$$

and therefore, when the air and water mix uniformly, the maximum delivery height will be

$$h + b \text{ hyp log } \frac{b+h}{b} = 12 + 7{\cdot}96 = 19{\cdot}96 \text{ m. } [65{\cdot}49 \text{ ft.}],$$

and the required driving force

$$P = 19{\cdot}96 \frac{V\gamma}{2\pi a} = 19{\cdot}96 \frac{63{\cdot}2}{6{\cdot}28 \times 0{\cdot}5} = 402 \text{ kg. } [886{\cdot}41 \text{ lbs.}]$$

If the machine is to make twelve revolutions per minute, the quantity of water raised per second will be

$$Q = \frac{n}{60} V = \frac{12}{60} 0{\cdot}0632 \text{ cub. m.} = 12{\cdot}6 \text{ litres } [0{\cdot}445 \text{ cub. ft.}],$$

and, neglecting wasteful resistances, such as friction at the journals and in the pipes, we obtain for the theoretical amount of work expended

$$L = \left(h + b \text{ hyp log } \frac{b+h}{b}\right) Q\gamma = 19{\cdot}96 \times 12{\cdot}6 = 251{\cdot}5 \text{ m. kg. } [1819{\cdot}3 \text{ ft. lbs.}].$$

The length of each water arc or water column in the delivery pipe is

$$l = \pi r_1 = 3{\cdot}142 \text{ m. } [10{\cdot}31 \text{ ft.}],$$

and therefore the number of such columns in the delivery pipe, on an average,

$$m = \frac{h}{l} = \frac{12}{3{\cdot}142} = 3{\cdot}8.$$

The lengths of the air columns in the delivery pipe are as follows :—

$$l_1 = \frac{b}{b+l}\,l = 0{\cdot}767l,$$

$$l_2 = \frac{b}{b+2l}\,l = 0{\cdot}622l,$$

$$l_3 = \frac{b}{b+3l}\,l = 0{\cdot}523l,$$

$$l_4 = \frac{b}{b+4l}\,l = 0{\cdot}451l,$$

which, added together, give

$$2{\cdot}363l = 2{\cdot}363 \times 3{\cdot}142 = 7{\cdot}44 \text{ m. [24·41 ft.]},$$

and consequently the total delivery height will be

$$12 + 7{\cdot}44 = 19{\cdot}44 \text{ m. [63·78 ft.]}$$

§ 49. **Höll's Compressed Air Pump.**—As has already been mentioned, the air which is to raise the water is in this machine compressed by the action of falling water, as in the traditional *fountain of Hero.* Briefly stated, the function of the apparatus is to transmit the motive power stored in a water column, by means of compressed air, to another body of water in such a manner that when the former column descends the latter will be caused to rise. The essential arrangement of this water raising device will be seen from Fig. 190. The air enclosed in the reservoir M and the pipe CD is compressed by the water column in the supply pipe AB, and, by virtue of its expansive action, drives the water from the reservoir N up into the delivery pipe E, which at its lower end is provided with a valve V opening inward, and discharges the water at the top, at G. When the

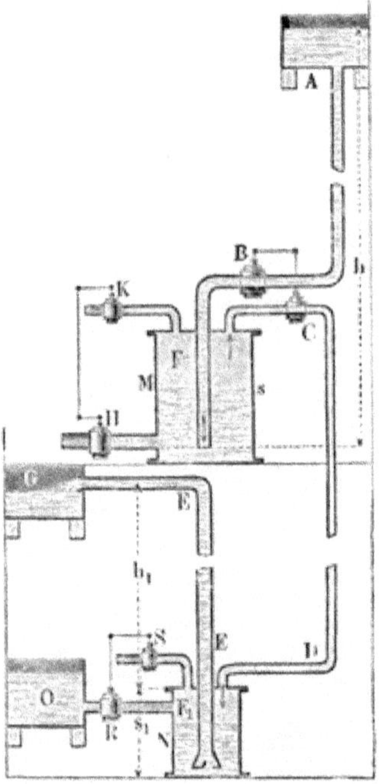

Fig. 190.

reservoir M is full of water the cocks B and C are closed, and H and K belonging to the upper reservoir, as well as R and S belonging to the lower one, are opened. The upper reservoir M then empties through H into the common receptacle G, while N is filled from the reservoir O, the foot valve V at the same time closing automatically, and the air which has collected in N escaping through S. When the upper tank M has been emptied and the lower N filled, the cocks H, K, R, and S are closed and B and C opened, and a new operation begins. *Höll* (in 1753) built a machine of this kind in the mine "Amalia" at Chemnitz (see N. Poda, *Kurzgefaszte Beschreibung der beim Bergbau zu Schemnitz errichteten Maschinen*, Prague, 1771), which required two attendants for opening the cocks and otherwise watching the machine. The first automatic apparatus of the kind is described by *Boswell* in 1796 (see Hachette's *Traité élémentaire des machines*).

Darwin's water-raising machine consists of a series of air machines forcing water one to the other; further, in *Detrouville's* pump the delivery pipe is replaced by a suction pipe, the water being lifted by suction. The hydraulic air compressors which were employed in modern times for driving the rock drills at Mount Cenis, but, owing to their insufficiency, were rapidly replaced by blowing engines, were, in the main, nothing but *Höll's* machines utilised to compress air instead of raising water (see article by F. Reuleaux in *Schweitzerischen polytechn. Zeitschrift*, Bd. II., on "Die Durchbohrung des Mont-Cenis"; also Rühlmann, *Allgem. Maschinenlehre*, Bd. IV.)

The effect of *Höll's* or the so-called *Hungarian* machine may be determined as follows. Let h be the fall, measured from the upper water-level at A to the lowest level in M, further, let h_1 be the height of delivery, counted from the discharge orifice G of the delivery pipe to the high water-level in N; let the water column corresponding to the atmospheric pressure be denoted by b, the sectional area of the tank M by F, and that of N by F_1, and finally, the height through which the water in M rises or falls during each operation by s, and the corresponding height in N by s_1. Then we have

$$b + h - s = b + h_1 + s_1, \text{ or } h - s = h_1 + s_1,$$

and, according to *Mariotte's* law,

$$\frac{b + h_1 + s_1}{b} = \frac{Fs}{F_1 s_1};$$

consequently the efficiency of the machine is

$$\eta = \frac{F_1 s_1 h_1}{Fsh} = \frac{bh_1}{(b + h_1 + s_1)h} = \frac{bh_1}{(b + h_1 + s_1)(h_1 + s + s_1)}.$$

To obtain the maximum efficiency it is necessary to make the variations of the tank levels as small as possible (nearly zero). Then we should have

$$h = h_1 \text{ and } \eta = \frac{b}{b + h_1},$$

and consequently a greater value of η, the smaller the fall h or the height of delivery h_1, though not equal to unity until $h = h_1 =$ zero. For greater lifts η sinks considerably below unity, as, for instance, for $h = h_1 = b$ we have $\eta = \frac{1}{2}$, for $h_1 = h = 3b$ we have $\eta = \frac{1}{4}$, etc., and therefore the machine is not suitable for raising water to great elevations. (Compare discussion of "Pneumatic Hoists" in the volume on *Hoisting Machinery*.)

§ 50. **The Pulsometer.**—The direct pressure of steam has long been used to force liquids to an elevation, the arrangements devised by *Savery* for lifting water being thus operated. Steam has also for a long time been employed to produce a vacuum in a vessel by replacing the air contained in the latter, and, by being subsequently condensed, allowing the atmospheric pressure to force the liquid up into the vessel. Such is the arrangement, for instance, made use of in sugar refineries for elevating the sugar juice, and the same plan has more recently been applied to emptying sewers and excavations whose thick contents will not allow of the use of valved pumps. In the *pneumatic dredging machines* steam is likewise employed to produce a vacuum in a chamber, into which the pasty dredging material is then forced by the atmospheric pressure.

A modern application of the direct pressure of steam for raising water has been made in the apparatus invented by *H. Hall* of New York (in 1872), and known by the name of the *pulsometer*. This device is very serviceable for certain purposes, because of its simplicity and the ease with which it may be set up, although it has the defect common to all devices for raising liquids *directly* by steam pressure, namely, of excessive steam consumption. The pulsometer works like a pump, exerting a suction as well as a forcing action, the greatest suction height

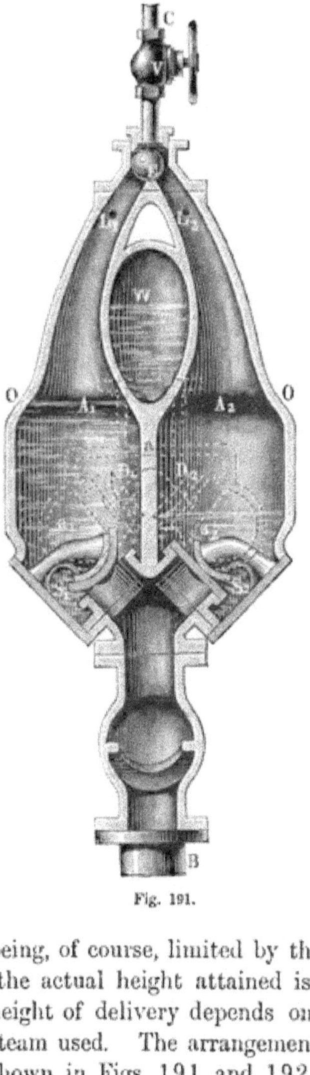

Fig. 191.

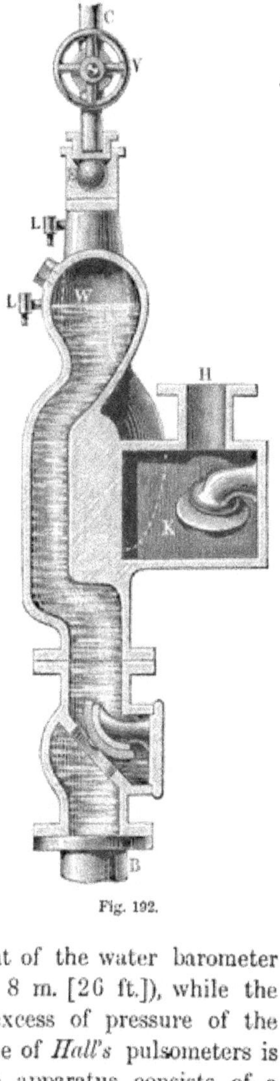

Fig. 192.

being, of course, limited by the height of the water barometer (the actual height attained is about 8 m. [26 ft.]), while the height of delivery depends on the excess of pressure of the steam used. The arrangement of one of *Hall's* pulsometers is shown in Figs. 191 and 192. The apparatus consists of a cast-iron casing divided into two pear-shaped spaces A_1 and A_2, separated by a partition a; the lower part of the casing is connected with the suction pipe B and the upper part with the steam pipe C, the latter containing a throttle valve V of ordinary construction, by means of which the supply of steam can be regulated. The chambers A are separated from the suction pipe by two rubber valves S_1 and S_2, which alternately open and close like the suction valves of a double-acting pump, so that either the water which is driven up the pipe B by the atmospheric pressure enters the chamber A_1 through the valve S_1 when open, while it is shut off from the other chamber A_2 by the valve S_2, or the reverse takes place. The space W between A_1 and A_2 is connected with the suction pipe B by a side passage, and performs for the latter the functions of an air chamber.

The chambers A_1 and A_2 are provided at one side with discharge passages a_1 and a_2 which lead to a valve chest K, each passage being shut off from this chest by a delivery valve D. These valves also open alternately upward, the water being forced through the valve chest K and up the delivery pipe H. It is evident from this arrangement that to drive the apparatus it is only necessary to alternately establish a vacuum and an excess of pressure in each chamber, in such a manner that while there is a vacuum in one chamber there will be an excess of pressure in the other. Thus, water can be drawn up to one chamber from B and forced out of the other up the delivery pipe H; the action will be similar to that of a double-acting pump, since the two chambers are comparable to the two sides of the piston in a reciprocating pump. In order to effect automatically this alternate action, a bronze ball E is placed at the point where the two chambers unite at the top; this ball is pressed first to one side and then to the other, now shutting off the chamber A_1 from the steam pipe and placing A_2 in communication with it, and then the reverse. This motion to and fro is effected automatically by the action of the steam, without the help of any special moving pieces, in the following manner. Let us suppose the apparatus filled with water, say to the height OO, which can be done through an opening in the air chamber W, the suction pipe B being provided with a foot valve to prevent the water from flowing out.

Now, if the throttle valve is opened steam will enter one of the chambers, say A_2, and will force the water through the passage a_2 and the delivery valve D_2 to the delivery pipe H. This operation will last till the water level in the chamber A_2 has fallen below the upper edge of the discharge passage a_2; from this instant considerable steam will escape through the latter opening into the delivery pipe, where it will be quickly condensed by the return flow of water. A vacuum will thus be created in the chamber A_2, in consequence of which the ball E will be thrown over to the right by the greater pressure in the chamber A_1. This completely shuts off steam from the chamber A_2, and the vacuum in A_2 now causes water to flow in from the suction pipe B through the opening valve S_2; the delivery valve D_2 closes, and thus prevents further return of the water delivered to H. Moreover, the shifting of the ball E admits steam to the chamber A_1 and the operation described above is repeated till the water-level sinks below the upper edge of the discharge passage a_1. At this instant there will likewise be a swift condensation of steam, and the ball valve will be pushed over to the left by the excess of pressure in the chamber A_2. During the process just described care must be taken to allow a small quantity of air to be drawn into the chamber where suction is exerted. For this purpose each chamber is provided at the upper end with a small air valve L_1, L_2 opening inward. The suction is thereby somewhat injured, but a means is at the same time provided of regulating the apparatus. For, by adjusting the air valve the quantity of air drawn in can be regulated, and thus the duration of suction controlled. On the other hand, the throttle valve enables us to control the duration of the forcing action, within certain limits, and it is in our power to so regulate the apparatus that the suction in one chamber and the forcing action in the other will be completed in the same space of time. The air entering the chamber during suction also performs another function, for in the following forcing period it acts as a non-conducting layer between the water below and the steam above, the assumption being made that when the latter enters it pushes the air before it. In consequence of the slight conducting power of air, this layer greatly diminishes the premature condensation of the steam during the forcing period, and when the water-level sinks below the upper edge of the discharge passage, it is always the greater portion of the air drawn in during the preceding suction that first escapes.

This air also plays an important part in moving the ball valve; its action may be described as follows. When a vacuum has been produced in a chamber by sudden condensation of the steam at the end of the forcing period, water rushes in from the suction pipe with an acceleration which is greater the more the pressure in the air chamber exceeds that which exists in the chamber A. The accelerated motion of the water only lasts till the pressures in the two spaces are equalised. The water-level in the chamber will not come to rest at this instant, however, for the water will pass beyond the position of equilibrium by virtue of its velocity, and, as it rises, will compress the air above it. It is the pressure thus brought to bear on the ball, combined with the vacuum in the other chamber, that causes the ball to shift its position.

Tests of the pulsometer, particularly those made by *Schaltenbrand*,[1] confirm the remark made above, that considerable loss of heat is connected with the direct action of steam on the water. In the tests referred to the lifts varied between 6 and 10·18 m. [19·68 and 33·39 ft.], and the consumption of steam per hour, for every horse-power actually expended in raising water, varied between 122 and 235 kg. [268 and 517 lbs.], which consumption greatly exceeds that of good pumps driven by steam engines.

The results obtained by *Eichler*[2] point to a more favourable performance due to certain improvements made in the pulsometer, the consumption of steam per minute being 1·43 kg. [3·15 lbs.] for every effective horse-power. Pulsometers may therefore be recommended for use in mines where a moderate cost, ease of erection, and occupation of the smallest possible space are prime requirements of the water-raising apparatus, and where the lift does not exceed 40 m. [130 ft.] They are also to be recommended for temporary service, for instance the draining of excavations connected with building operations. So far as simplicity of arrangement and ease of

[1] Schaltenbrand, *Der Pulsometer*, Berlin, 1877.

[2] C. Eichler, *Die Anwendung der Pulsometer auf Adolph-Schacht bei Reichenwalde.*

setting up are concerned, the pulsometer is all that can be desired, since it not even requires a fixed foundation, but may be set in action while kept suspended from a rope over the excavation. In *Schaltenbrand's* experiments the number of pulsations or single throws of the ball per minute varied between 43 and 128, while in the larger pulsometers tested by *Eichler*, which operated with a lift of 25 m. [82 ft.], each pulsation required from 4 to $4\frac{1}{2}$ seconds. For further data reference must be made to the sources mentioned.

§ 51. **Siphons.**—Strictly speaking, a siphon is not a water-raising machine, since by its use water is not raised to a higher point, but only transferred across an elevation; for this reason it is rather to be considered as a water conduit with an upward bend. The small siphons used in every laboratory are too well known to require explanation, and we will therefore only discuss the apparatus of this class employed on a large scale in hydraulic constructions. In the volume on Hydraulics a short siphon of this kind is described in connection with the means employed for regulating the water-level in canals. As another example may be mentioned the so-called periodical springs, as shown in ABCDE (Fig. 193). When the water contained in the cavity W has risen above the highest point D of the siphon-shaped, subterranean canal CDE, it will drive the air out of the latter and discharge through E until the water level in W has fallen below the inlet at C. The interruption which then takes place will last as long as the water-level remains below the point D; when the cavity again becomes full of water, which flows in through the cleft AB, up to the level of D, the discharge through CDE will begin anew.

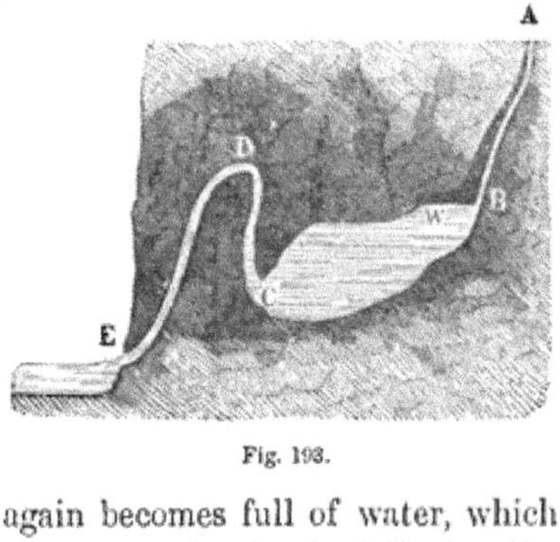

Fig. 193.

In Fig. 194 is shown the arrangement of a simple siphon employed for draining a marsh or ditch. It consists essentially of three cast-iron pipes, AB for the upward, CD for the downward flow, and the horizontal connection BC. A gate valve S is placed at the inlet A, a clack valve E at the outlet D, and in the middle pipe BC a short branch pipe K, which may be closed by a plug, is inserted. When the siphon is to be set in operation the valves S and E are closed and water is poured in through K until all the pipes are filled. If now K is closed up tight, S opened and E released, the action of the siphon begins, a continuous stream of water flowing through it in the direction ABCD.

This flow is, however, dependent on the fulfilment of two conditions. In the first place, it is necessary that the water-

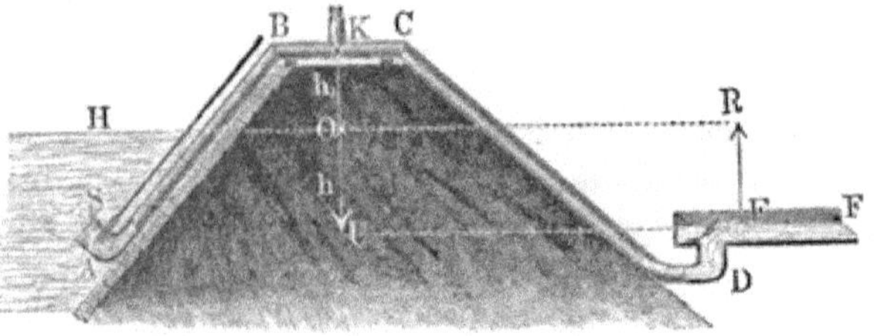

Fig. 194.

level H in the feed reservoir AH should be above the discharge orifice E, for the velocity of efflux v at E is

$$v = \mu \sqrt{2gh},$$

if μ is the coefficient of discharge, and h the head or vertical distance RE of the level H above the orifice E, and consequently for $h = 0$ we have $v = 0$.

In the second place, the height $KO = h_1$ of the middle pipe BC or summit K of the siphon above the level H of the feed reservoir must not exceed the height of the water barometer $b = 10{\cdot}34$ m. [33·9 ft.], for the pressure of the water at the highest point is $b - h_1$, and thus becomes zero for $h_1 = b$. In case h_1 should be greater than b, a vacuum would be created at K which would interrupt the continuity of the current ABCD.

As the pressure $b - h_1$ at the vertex always falls short of the atmospheric pressure, the air contained in the water supplied gradually separates from the latter at this point, and soon accumulates to such an extent as to impede the discharge of the water. It is therefore necessary to remove the air at intervals; in the simple apparatus shown in Fig. 194 this can be done only by closing the apertures S and E and refilling with water at the vertex, while in more perfect constructions an air pump is used for this purpose.

A larger and more complete siphon is illustrated in Fig. 195. It was constructed by *F. Ablay*,[1] of the Belgian Engineer Corps, for the purpose of furnishing water to the trenches of the fort St. Marie near Antwerp from the river Schelde, which runs close by. The conduit ABCD is made of cast-iron pipes 0·2 m. [7·87 ins.] in diameter, and of a total length of 36 m. [118 ft.] At the extremities there are two iron wells AH and DR of rectangular section, located in such a manner that their tops are on the same level HOR, and provided with iron

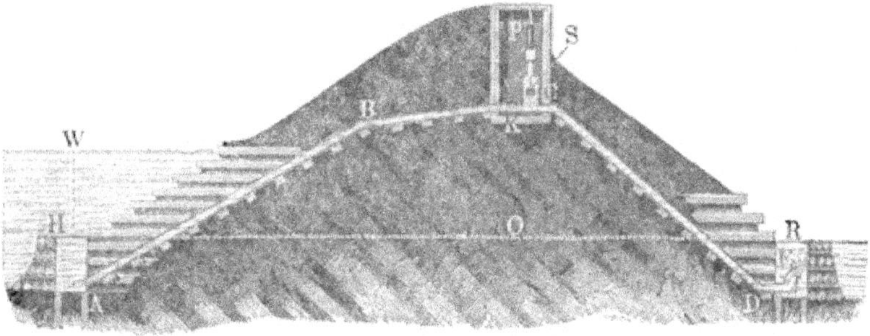

Fig. 195.

gratings. By means of this arrangement the two apertures A and D are always kept under water, and no air can therefore enter the siphon even when, at low tide, the surface W of the Schelde falls below the water-level R in the trench. To prevent the water in the siphon from running backwards when this occurs, the discharge orifice at D is provided with a clack valve E opening outward. At low tide the level of the river stands 1·4 m. [4·59 ft.] below that of the trench, while at high tide it rises from 2·7 to 3·2 m. [8·86 to 10·5 ft.] above the latter. To remove the air which accumulates at the vertex C of the siphon, an air chamber S and a suction pump P are made use of, the whole being enclosed in a shelter built of masonry. For military reasons the whole siphon, together with the shelter containing the air pump, is covered with

[1] See *Annales des Travaux publics de Belgique*, Tome ix., 1850 and 1851.

earth. By the cock K, which can be turned by means of gears operated by a crank, the flow of water in the siphon may be either started or discontinued, as may be required.

The construction of the suction pump used may be seen from the vertical section in Fig. 196. ACA is a receiver attached to the vertex of the siphon and provided with a float S, whose pointer (not visible in the figure) plays up and down in a glass tube and may be observed from without; further, B is the pump cylinder communicating with the receiver by means of a narrow pipe, K is the plunger, and EDF the hand lever having its fulcrum at D. At V the suction and at W the discharge valve are seen, and at H a small cock by means of which the pump may be shut off from the receiver with a view to preventing leakage of air into the latter.

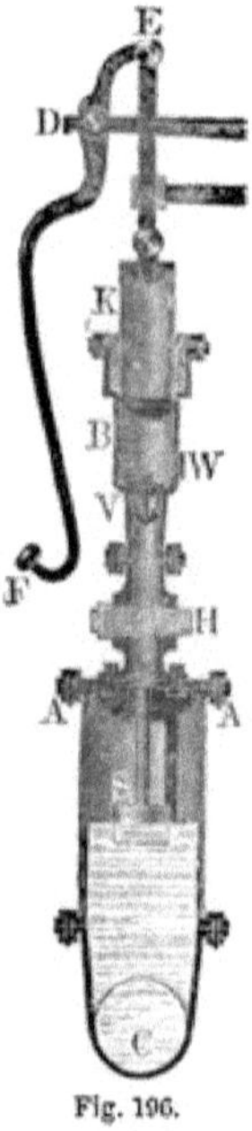

Fig. 196.

When the siphon is to be set in operation, the two reservoirs AH and RD (Fig. 195) must be filled with water and the air pump started; after a while the siphon becomes filled with water and its action begins. This action will be discontinued when the water-level W sinks below R, but commences again as soon as W rises above R. The air accumulating in ACA (Fig. 196) must evidently now and then be removed by the air pump.

It is especially difficult to maintain a steady flow of water in a siphon when the pipes are not perfectly air-tight. For this reason it is impossible to keep water flowing for any length of time in a siphon made of wooden pipes unless an air pump be permanently employed to remove the air. When the middle pipe is long there is a good chance for the air to accumulate, and in such cases it is necessary to give a slight pitch to the pipe and locate the air pump at the highest point directly above the branch in which the flow is downward, so as to carry the air bubbles to the receiver and away with the descending water. In regard to the results obtained in Saxon mines and in the Hartz with siphons having very long middle pipes, see *Jahrbuch für den Sächs. Berg- und Hüttenmann*, 1843, and *Die berg- und hüttenmännische Zeitung*, 1858.

The *theory of the siphon* is the same as that applying to the motion of water in ordinary pipe conduits (see vols. i. and ii. of Weisbach's *Mechanics*. Let h be the head or height RE, (Fig. 197) of the upper water-level H above the lower level, l the total length of pipes in the siphon, d their diameter, v the velocity of discharge, ζ_0 the coefficient of resistance for the

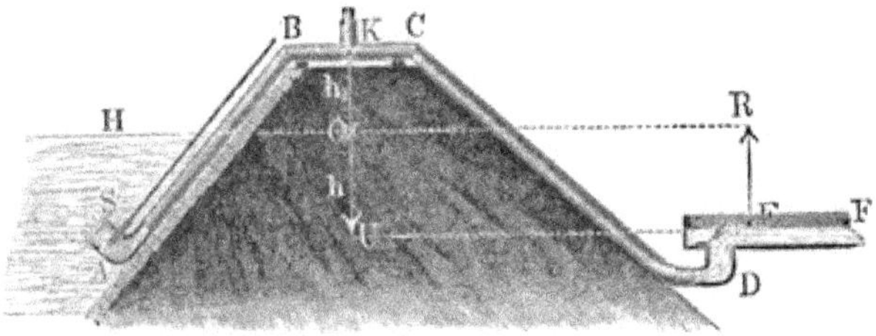

Fig. 197.

inlet at A, ζ the coefficient of friction for the flow of water in the pipes, and ζ_1 and ζ_2 the coefficients of resistance for the bends B and C. We then have

$$h = \left(1 + \zeta\frac{l}{d} + \zeta_0 + \zeta_1 + \zeta_2\right)\frac{v^2}{2g} = \phi\frac{v^2}{2g} \quad . \quad . \quad (1),$$

if we place $1 + \zeta\frac{l}{d} + \zeta_0 + \zeta_1 + \zeta_2 = \phi$. Hence follows the velocity of the discharging water

$$v = \sqrt{\frac{2gh}{\phi}},$$

and the quantity discharged per second

$$Q = Fv = \frac{\pi d^2}{4}\sqrt{\frac{2gh}{\phi}}.$$

Further, letting h_1 denote the height KO of the vertex above the high water-level, l_1 the length of the pipe leading from the aperture A to the summit, and z the pressure of the water at the summit, we obtain the equation

$$b - h_1 - z = \left(1 + \zeta\frac{l_1}{d} + \zeta_0 + \zeta_1\right)\frac{v^2}{2g} = \phi_1\frac{v^2}{2g} \quad . \quad (2),$$

if for the sake of brevity we place $1+\zeta\frac{l_1}{d}+\zeta_0+\zeta_1=\phi_1$. From (1) and (2) follows

$$\frac{b-h_1-z}{h}=\frac{\phi_1}{\phi},$$

whence we obtain the pressure z at the summit of the siphon

$$z=b-h_1-\frac{\phi_1}{\phi}h \qquad (3).$$

When the middle or summit pipe is very long, and consequently l_1 nearly equal to l, we can with sufficient accuracy place $\phi=\phi_1$, and then obtain

$$z=b-(h+h_1).$$

To obtain a steady flow, both h and z must be positive, and consequently we must have, according to (3),

$$\frac{\phi_1}{\phi}h+h_1<b,$$

or, when the middle pipe is very long,

$$h+h_1<b;$$

that is to say, the height $KU=h+h_1$ of the summit above the *lowest* water-level must be less than that of the water barometer b. For, if only the height $KO=h_1$ falls short of b, but $KU=h+h_1$ exceeds it, then, although water will rise to the summit pipe, it will not be supplied as rapidly as it is carried away by the other branch of the siphon. As a consequence, the flow of water will no longer be continuous, unless the discharge orifice D obtains a correspondingly smaller diameter d_1 than the pipes. For the latter event let v_1 be the velocity of discharge and ζ_m the coefficient of resistance for the nozzle, then we have,

$$h=\left(\zeta\frac{l}{d}+\zeta_0+\zeta_1+\zeta_2\right)\frac{v^2}{2g}+(1+\zeta_m)\frac{v_1^2}{2g}$$
$$=\left[\phi-1+(1+\zeta_m)\left(\frac{d}{d_1}\right)^4\right]\frac{v^2}{2g} \quad . \quad . \quad . \quad (4).$$

Combined with (2) this equation gives

$$\frac{b-h_1-z}{h}=\frac{\phi_1}{\phi-1+(1+\zeta_m)\left(\frac{d}{d_1}\right)^4},$$

whence we obtain the pressure z of the water in the summit pipe

$$z = b - h_1 - \frac{\phi_1 h}{\phi - 1 + (1 + \zeta_m)\left(\frac{d}{d_1}\right)^4}.$$

In order that a steady flow shall be maintained, z must be positive, *i.e.* we must have

$$h_1 + \frac{\phi_1 h}{\phi - 1 + (1 + \zeta_m)\left(\frac{d}{d_1}\right)^4} < b.$$

Now, if d_1 is much smaller than d, we can place

$$\frac{\phi_1}{\phi - 1 + (1 + \zeta_m)\left(\frac{d}{d_1}\right)^4} = 0,$$

and then obtain as the condition of *continuous* flow merely that $h_1 < b$, which relation must always be maintained in order that *any flow at all* may take place.

EXAMPLE.—A siphon is 100 m. [328 ft.] long and is made of pipes 0·1 m. [3·94 ins.] in diameter at every point; the coefficient of resistance $\zeta_0 = 0·100$, the coefficients for the two bends are $\zeta_1 = \zeta_2 = 0·3$, and the fall is $h = 3$ m. [9·84 ft.] The corresponding velocity of discharge is

$$v = \sqrt{\frac{2gh}{1 + \zeta\frac{l}{d} + \zeta_0 + \zeta_1 + \zeta_2}} = \sqrt{\frac{2 \times 9·81 \times 3}{1·7 + 1000\zeta}},$$

or, if we assume $\zeta = 0·02$,

$$v = \sqrt{\frac{58·86}{21·7}} = 1·65 \text{ m. } [5·41 \text{ ft.}]$$

According to Weisbach's *Mechanics*, vol. i., we should have for $v = 1·6$ m. $\zeta = 0·0219$, and therefore a more accurate value would be

$$v = \sqrt{\frac{58·86}{1·7 + 21·9}} = 1·58 \text{ m. } [5·18 \text{ ft.}]$$

Hence we obtain the quantity of water discharged per second

$$Q = 3·14 \times 0·05^2 \times 1·58 = 0·0124 \text{ cub. m.} = 12·4 \text{ litres } [0·438 \text{ cub. ft.}]$$

To produce a steady flow in the siphon, the vertex pipe of which is, say, $l_1 = 80$ m. [262·5 ft.] from the inlet, we must have

$$\frac{\phi_1}{\phi} h + h_1 < b$$

that is,

$$\frac{1+0\cdot0219\frac{80}{0\cdot1}+0\cdot1+0\cdot3}{1+0\cdot0219\frac{100}{0\cdot1}+0\cdot1+0\cdot6}\,3+h_1<10\cdot34.$$

or, in other words, the height h_1 must not exceed $10\cdot34-2\cdot41=7\cdot93$ m. [26 ft.] Otherwise, the discharge orifice must be correspondingly contracted.

NOTE.—A large number of special articles on water-raising machinery may be found in the various technical journals (compare pp. 232 and 233). In addition to the references made in the text to literary sources, we will here mention the following works on the subject: Eytelwein's *Handbuch der Mechanik und Hydraulik*, Berlin, 1842; Gerstner's *Handbuch der Mechanik*, vols. ii. and iii.; Kaiser's *Handbuch der Mechanik*, Karlsruhe, 1842; Langsdorf's *Vollständiges System der Maschinenkunde*, Leipzig, 1828; Jeep, *Der Bau der Pumpen und Spritzen*, Leipzig, 1891; Hagen's *Handbuch der Wasserbaukunde*, vol. i.; Hachette, *Traité élémentaire des Machines*, Paris, 1819; Borgnis, *Traité complet de Mécanique apliquée;* vol. iv. of *Machines hydrauliques*, Paris, 1819; Navier, *Résumé des Leçons sur l'application de la Méchanique*, part ii., Paris, 1838; D'Aubuisson, *Traité d'Hydraulique*, Paris, 1840; Morin, *Machines et appareils destinés à l'élévation des eaux*, Paris, 1863.

Mining pumps are extensively treated of in the large work *Die Wasserhaltungsmaschinen*, von J. v. Hauer, where a complete list of literature on the subject may also be found. We may further mention Serlo's *Bergbaukunde*, Rittinger's *Erfahrungen im berg- und hüttenmännischen Maschinenbau- und Aufbereitungswesen*, Riedler's *Excursionsbericht*, and Combes, *Traité de l'exploration des Mines*, T. iii., Paris, 1845.

With reference to city waterworks information may be obtained from Salbach, *Die Dresdener Wasserwerke, Halle*, and *Wasserwerk der Stadt Halle*, 1871, as well as several articles by the same author in Schilling's *Journal für Gasbeleuchtung; Hydraulica*, an historical and descriptive account of the waterworks of London, 1835. In regard to drainage see Treuding, *Ueber Ent- und Bewässerung von Ländereien, Zeitschr. des Hannov. Arch- und Ing.-Ver.*, 1864 and 1865; Gevers van Endegeest, *Over de droogmaking van het Haarlemer Meer*, Amsterdam, 1857; in German in Förster's *Bauzeitung*, 1865. In addition we may mention *Elementi di Meccanica e D'idraulica di* G. Venturoli, Naples, 1833; John Robison, *A System of Mechanical Philosophy*, with notes by Brewster, vol. ii., 1822. An extensive list of references pertaining to the subject is contained in the often cited work, Rühlmann's *Allgemeine Maschinenlehre*, vol. iv.

INDEX

THE END

Printed by R. & R. CLARK, LIMITED, *Edinburgh*

www.ingramcontent.com/pod-product-compliance
Lightning Source LLC
Chambersburg PA
CBHW021506210326
41599CB00012B/1152
* 9 7 8 3 7 4 3 6 8 6 9 2 2 *